Studies on *Eurytemora*: Proceedings of the *Eurytemora* conference, St. Petersburg, 2019

*Published in this series:*

| | |
|---|---|
| CRM 001 - *Stephan G. Bullard* | Larvae of anomuran and brachyuran crabs of North Carolina |
| CRM 002 - *Spyros Sfenthourakis et al. (eds.)* | The biology of terrestrial isopods, V |
| CRM 003 - *Tomislav Karanovic* | Subterranean Copepoda from arid Western Australia |
| CRM 004 - *Katsushi Sakai* | Callianassoidea of the world (Decapoda, Thalassinidea) |
| CRM 005 - *Kim Larsen* | Deep-sea Tanaidacea from the Gulf of Mexico |
| CRM 006 - *Katsushi Sakai* | Upogebiidae of the world (Decapoda, Thalassinidea) |
| CRM 007 - *Ivana Karanovic* | Candoninae (Ostracoda) from the Pilbara region in Western Australia |
| CRM 008 - *Frank D. Ferrari & Hans-Uwe Dahms* | Post-embryonic development of the Copepoda |
| CRM 009 - *Tomislav Karanovic* | Marine interstitial Poecilostomatoida and Cyclopoida (Copepoda) of Australia |
| CRM 010 - *Carrie E. Schweitzer et al.* | Systematic list of fossil decapod crustacean species |
| CRM 011 - *Peter Castro et al. (eds.)* | Studies on Brachyura: a homage to Danièle Guinot |
| CRM 012 - *Patricio R. De los Ríos-Escalante* | Crustacean zooplankton communities in Chilean inland waters |
| CRM 013 - *Katsushi Sakai* | Axioidea of the world and a reconsideration of the Callianassoidea (Decapoda, Thalassinidea, Callianassida) |
| CRM 014 - *Charles H.J.M. Fransen et al. (eds.)* | Studies on Malacostraca: Lipke Bijdeley Holthuis Memorial Volume |
| CRM 015 - *Akira Asakura et al. (eds.)* | New Frontiers in Crustacean Biology: Proceedings of the TCS Summer Meeting, Tokyo, 20-24 September 2009 |
| CRM 016 - *Danielle Defaye et al. (eds.)* | Studies on Freshwater Copepoda: a Volume in Honour of Bernard Dussart |
| CRM 017 - *Hironori Komatsu et al. (eds.)* | Studies on Eumalacostraca: a homage to Masatsune Takeda |
| CRM 018 - *Masahiro Dojiri & Ju-Shey Ho* | Systematics of the Caligidae, copepods parasitic on fishes |
| CRM 019 - *Darren Yeo et al. (eds.)* | Advances in Freshwater Decapod systematics and biology |
| CRM 020 - *Finn Viehberg et al. (eds.)* | The Recent and Fossil meet Kempf Database Ostracoda Festschrift Eugen Karl Kempf — Proceedings of the 15th International German Ostracodologists' Meeting |
| CRM 021 - *L.A.M. Neethling & A. Avenant-Oldewage* | Branchiura — A compendium of the geographical distribution and a summary of their biology |
| CRM 022 - *Célio Magalhães et al. (eds.)* | Studies on Decapoda and Copepoda in Memory of Michael Türkay |

# Studies on *Eurytemora*

# Proceedings of the *Eurytemora* conference, St. Petersburg, 2019

*By*

Natalia M. Sukhikh, Sami Souissi, Gesche Winkler
and Victor R. Alekseev (Editors)

CRUSTACEANA MONOGRAPHS, 23

BRILL

LEIDEN | BOSTON

This book was originally published as *Crustaceana* 93, 3-5 (2020), 241-547.

*Cover*: Logo of the *Eurytemora* Conference, St. Petersburg, 2019. Created by Vadim V. Semenovykh. The logo contains *Eurytemora affinis* image from the first species description by Poppe in 1880.

*Editors' addresses:*
Natalie M. Sukhikh, Zoological Institute of the Russian Academy of Science, Universitetskaya nab. 1, 199034 St. Petersburg, Russia; e-mail: susikh1@mail.ru
Sami Souissi, Laboratoire d'Océanologie et de Géosciences, Université de Lille, CRNS, Université Littoral Côte d'Opale, UMR 8187, 62930 Wimereux, France; e-mail: sami.souissi@univ-lille1.fr
Gesche Winkler, Institut des Sciences de la Mer de Rimouski, Université du Québec à Rimouski, Rimouski, QC, Canada G51 3A1; e-mail: gesche_winkler@uqar.ca
Victor R. Alekseev, Zoological Institute of the Russian Academy of Science, Universitetskaya nab. 1, 199034 St. Petersburg, Russia; e-mail: valekseev2@yahoo.com

The Library of Congress Cataloging-in-Publication Data is available online at http://catalog.loc.gov
LC record available at http://lccn.loc.gov/2021022011

ISBN 978 90 04 46555 8 (hardback)
ISBN 978 90 04 46556 5 (e-book)


Koninklijke Brill NV incorporates the imprints Brill, Brill Nijhoff, Brill Hotei, Brill Schöningh, Brill Fink, Brill mentis, Vandenhoeck & Ruprecht, Böhlau Verlag and V&R Unipress.



This book is printed on acid-free paper and produced in a sustainable manner.

CONTENTS

## ECOLOGY



## BIOLOGY IN TAXA RELATED TO *EURYTEMORA*

First published as a Special Issue of *Crustaceana*. The bracketed page numbers in this book refer to the page numbers in *Crustaceana* 93, 3-5 (2020). When citing from this book, refer to these bracketed page numbers.

## NOTE ON ORIGINAL PUBLICATION

The chapters in this book were originally published in the journal *Crustaceana*, Vol. 93, Issues 3-5 (2020), pp. 241-547.

Natalia M. Sukhikh, Sami Souissi, Gesche Winkler and Victor Alekseev, "Preface," *Crustaceana* 93(3-5): 241-244. Online Publication Date: 8 Jun 2020. DOI: https://doi.org/10.1163/15685403-00003966.

Victor R. Alekseev and Natalia M. Sukhikh, "On time and place of origin of continental calanoid families: a hypothesis," *Crustaceana* 93(3-5): 245-260. Online Publication Date: 8 Jun 2020. DOI: https://doi.org/10.1163/15685403-00004014.

V. I. Lazareva, "Distribution of *Eurytemora capsica* Sukhikh & Alekseev, 2013 (Copepoda, Calanoida) in the water reservoirs of the Volga and Don river basins," *Crustaceana* 93(3-5): 261-273. Online Publication Date: 8 Jun 2020. DOI: https://doi.org/10.1163/15685403-00003982.

Larysa Samchyshyna, Yulia Gromova and Kateryna Zorina-Sakharova, "Recent distribution of *Eurytemora velox* (Lilljeborg, 1853) (Copepoda, Calanoida) in brackish and fresh waters of Ukraine," *Crustaceana* 93(3-5): 275-281. Online Publication Date: 8 Jun 2020. DOI: https://doi.org/10.1163/15685403-00003992.

Vasil V. Vezhnavets and Anastasiya G. Litvinova, "Representatives of the genus *Eurytemora* Giesbrecht, 1881 (Calanoida, Temoridae) in aquatic ecosystems of Belarus," *Crustaceana* 93(3-5): 283-298. Online Publication Date: 8 Jun 2020. DOI: https://doi.org/10.1163/15685403-00004008.

Elena Fefilova, Natalia Sukhikh, Ekaterina Abramova and Ilya Velegzhaninov, "About the systematics of Palaearctic *Eurytemora* (Copepoda, Calanoida) based on morphological analysis, with focus on *Eurytemora gracilicauda* Akatova, 1949,"

*Crustaceana* 93(3-5): 299-315. Online Publication Date: 8 Jun 2020. DOI: https://doi.org/10.1163/15685403-00003976.

Dmitry Lajus, Natalia Sukhikh and Victor Alekseev, "Stochastic phenotypic variation: empirical results and potential use in *Eurytemora* research (Copepoda, Calanoida)," *Crustaceana* 93(3-5): 317-336. Online Publication Date: 8 Jun 2020. DOI: https://doi.org/10.1163/15685403-00003983.

Łukasz Sługocki, "Variability of mandible shape in the freshwater glacial relict *Eurytemora lacustris* (Poppe, 1887) (Copepoda, Calanoida, Temoridae)," *Crustaceana* 93(3-5): 337-353. Online Publication Date: 8 Jun 2020. DOI: https://doi.org/10.1163/15685403-00003985.

Anissa Souissi and Sami Souissi, "Abnormalities in shape and size of *Eurytemora affinis* (Copepoda, Calanoida) and its eggs under different environmental conditions," *Crustaceana* 93(3-5): 355-378. Online Publication Date: 8 Jun 2020. DOI: https://doi.org/10.1163/15685403-00003989.

Jory Cabrol, Réjean Tremblay and Gesche Winkler, "Differential eco-physiological performances of two pseudocryptic species of the *Eurytemora affinis* complex (Copepoda, Calanoida) in the St. Lawrence estuarine transition zone: a reciprocal transplant experiment," *Crustaceana* 93(3-5): 379-404. Online Publication Date: 8 Jun 2020. DOI: https://doi.org/10.1163/15685403-00003973.

Elena G. Krupa, "The ecological preferences of *Eurytemora affinis* (Poppe, 1880) in the water bodies of Kazakhstan (Central Asia) and some notes about *Eurytemora caspica* Sukhikh & Alekseev, 2013," *Crustaceana* 93(3-5): 405-428. Online Publication Date: 8 Jun 2020. DOI: https://doi.org/10.1163/15685403-00003967.

Lauri Kuismanen, Louise Forsblom, Jonna Engström-Öst, Ulf Båmstedt and Olivier Glippa, "Salinity effects on egg production, hatching, and survival of *Eurytemora affinis* (Copepoda, Calanoida)," *Crustaceana* 93(3-5): 429-445. Online Publication Date: 8 Jun 2020. DOI: https://doi.org/10.1163/15685403-00003988.

Astra Labuce, Solvita Strake and Inta Dimante-Deimantovica, "*Eurytemora affinis* (Poppe, 1880) (Copepoda, Calanoida) in the Gulf of Riga, Baltic Sea — elemental composition and diurnal vertical migration," *Crustaceana* 93(3-5): 447-466.

Online Publication Date: 8 Jun 2020. DOI: https://doi.org/10.1163/15685403-00003977.

Alexander W. Timpe and Bart T. De Stasio, "Comparison of respiration rate and electron transport system (ETS) enzyme-mediated reduction assay of the invasive copepod *Eurytemora carolleeae* Alekseev & Souissi, 2011 (Calanoida, Temoridae) in Green Bay, WI, U.S.A.," *Crustaceana* 93(3-5): 467-485. Online Publication Date: 8 Jun 2020. DOI: https://doi.org/10.1163/15685403-00003990.

Yan-Guo Wang, Li-Chun Tseng, Rou-Xin Sun, Zhi-Yong Liu, Mao Lin and Jiang-Shiou Hwang, "Effects of the China Coastal Current on the community structure of planktonic copepods in early spring, with notes on *Eurytemora pacifica* Sato, 1913 in the western Taiwan Strait," *Crustaceana* 93(3-5): 487-506. Online Publication Date: 8 Jun 2020. DOI: https://doi.org/10.1163/15685403-00004010.

Elena Yu. Naumova and Victor R. Alekseev, "How many generations does *Epischura baikalensis* (Copepoda, Calanoida) have in Lake Baikal?," *Crustaceana* 93(3-5): 507-518. Online Publication Date: 8 Jun 2020. DOI: https://doi.org/10.1163/15685403-00003981.

Li-Chun Tseng, Yan-Guo Wang, Guang-Shan Lian and Jiang-Shiou Hwang, "A multi-year investigation of the Temoridae (Copepoda, Calanoida) assemblage succession within the interplay waters of the northern South China Sea," *Crustaceana* 93(3-5): 519-540. Online Publication Date: 8 Jun 2020. DOI: https://doi.org/10.1163/15685403-00004009.

Denis S. Zavarzin, "*Pseudodiaptomus japonicus* Kikuchi, 1928 (Copepoda, Pseudodiaptomidae), a brackish-water copepod formerly known as *P. inopinus* Burckhardt, 1913, on Sakhalin Island (Russian Far East)," *Crustaceana* 93(3-5): 541-547. Online Publication Date: 8 Jun 2020. DOI: https://doi.org/10.1163/15685403-00004007.

[When citing this volume, please refer to *Crustaceana* 93 (2020), 241-547]

*Proceedings of the* Eurytemora *conference, St. Petersburg, 2019*

# PREFACE

The calanoid copepod genus *Eurytemora* to date contains 24-25 taxa at species level, and represents a challenging group of forms for constructing an accurate description of their phylogeny, explaining their Holarctic distribution, interpreting their possible Tethyan origin, and disclosing their ecology, with such a perfect adaptation to variable salinity values. During the last decades *Eurytemora*, in particular the former group of cryptic species provisionally denoted as *E. affinis* sensu lato, became a popular model group for molecular genetic investigations, especially in studies on the human mediated distribution of invasive aquatic organisms.

An international conference was held in St. Petersburg, Russia, 13-17 May 2019, of which this volume includes some of the papers. A wide spectrum of studies on *Eurytemora* and several other related calanoids was presented by scientists from 16 countries, mainly from Europe, but also from Canada, Kazakhstan, Taiwan, China, Bangladesh, Puerto Rico and the U.S.A. In this issue, most of the papers deal with *Eurytemora* studies and were organized in the conference programme in four sessions: Phylogeny (chairs: Carol Eunmi Lee and Victor R. Alekseev); Biogeography (chairs: Jiang-Shiou Hwang and Elena Kochanova); Taxonomy (chairs: Natalia Sukhikh and Łukasz Sługocki); and Ecology (chairs: Gesche Winkler and Dmitry Lajus). The present volume includes 17 selected papers from all these sessions. During the conference, two masterclasses on molecular-genetic and morphological tools in *Eurytemora* taxonomy and systematics were organized, for which we acknowledge in appreciation Dr. Natalia Abramson and Dr. Evgeniy Genelt-Yanovsky ("Taxon" Research Resource Center of ZIN RAS) for delivering nice lectures, and Dr. Vladimir Loskot and Konstantin Vesyolkin (Laboratory of Ornithology) for providing proper facilities.

We hereby extend our sincere thanks to all those who contributed to the success of the conference: our colleagues at the Department of Systematics, of the Laboratory of Freshwater and Experimental Hydrobiology, the Laboratory of Marine Biology, and the Research Resource Center «Taxon» of the Zoological Institute RAS, the organizers in St. Petersburg, the Russian Academy of Sciences, which made facilities available, the Russian Foundation for Basic Research, which sponsored the conference via grant N 19-04-20048, and following the conference those who, in addition to several of the participants, acted as referees of the

 | DOI:10.1163/9789004465565_002

contributions: Dr. Cornelius Hammer, Germany; Dr. Hans-Uwe Dahms, Taiwan; Dr. Hiroshi Ueda, Japan; Dr. R. Arthur James, India; and Dr. Syuhei Ban, Japan.

Finally, we hope the present compilation of papers may stimulate further research on these interesting fresh- and brackish water calanoids.

NATALIA SUKHIKH[1]), SAMI SOUISSI, GESCHE WINKLER and
VICTOR ALEKSEEV
Volume Editors

[1]) e-mail: susikh1@mail.ru

## LIST OF PARTICIPANTS

ABRAMOVA, EKATERINA N. — Lena Delta Nature Reserve, Tiksi, Russia

ABRAMSON, NATALIA I. — Zoological Institute of the Russian Academy of Sciences, Saint Petersburg, Russia

ALEKSEEV, VICTOR R. — Zoological Institute of the Russian Academy of Sciences, Saint Petersburg, Russia

BÅMSTEDT, ULF — Umeå Marine Sciences Centre, University of Umeå, Norrbyn and Hörnefors, Umeå, Sweden

BARABANSHCHIKOV, EVGENY I. — Pacific Branch Federal State Budget Scientific Institution "VNIRO", Vladivostok, Russia

CABROL, JORY — Institut des Sciences de la Mer de Rimouski, Rimouski, Québec, Canada

CHABAN, OLGA — Zoological Institute of the Russian Academy of Sciences, Saint Petersburg, Russia

DEMCHUK, ANNA — Zoological Institute of the Russian Academy of Sciences, Saint Petersburg, Russia

DE STASIO, BART T. — Lawrence University, Appleton, WI, U.S.A.

DIMANTE-DEIMANTOVICA, INTA — Latvian Institute of Aquatic Ecology, Agency of Daugavpils University, Riga, Latvia

ENGSTRÖM-ÖST, JONNA — Novia University of Applied Sciences, Ekenäs, Finland

FEFILOVA, ELENA — Institute of Biology of the Russian Academy of Sciences, Komi Science Centre, Syktyvkar, Komi Republic, Russia

FORSBLOM, LOUISE — Åbo Akademi University, Environmental and Marine Biology, Turku, Finland

GENELT-YANOVSKY, EVGENIY — Zoological Institute of the Russian Academy of Sciences, Saint Petersburg, Russia

GLIPPA, OLIVIER — Novia University of Applied Sciences, Ekenäs, Finland

GOLUBKOV, SERGEI M. — Zoological Institute of the Russian Academy of Sciences, Saint Petersburg, Russia

GROMOVA, YULIA — Institute of Hydrobiology, National Academy of Sciences of Ukraine, Kiev, Ukraine

HOLL, ANNE-CATHERINE — University of Lille, Lille, France

HWANG, JIANG-SHIOU — National Taiwan Ocean University, Keelung, Taiwan

KASKOVA, KSENIA — Saint Petersburg State University, Saint Petersburg, Russia

KOCHANOVA, ELENA S. — Zoological Institute of the Russian Academy of Sciences, Saint Petersburg, Russia

KRUPA, ELENA G. — Institute of Zoology, Ministry of Education and Science, Almaty, Kazakhstan

KUISMANEN, LAURI — Åbo Akademi University, Environmental and Marine Biology, Turku, Finland

LABUCE, ASTRA — Latvian Institute of Aquatic Ecology, Agency of Daugavpils University, Riga, Latvia

LAJUS, DMITRY L. — Saint Petersburg State University, Saint Petersburg, Russia

LAZAREVA, VALENTINA I. — Papanin Institute for Biology of Inland Waters, Russian Academy of Sciences, Borok, Russia

LEE, CAROL E. — University of Wisconsin, Madison, WI, U.S.A.

LIAN, GUANG-SHAN — Third Institute of Oceanography, Ministry of Natural Resources, Xiamen, P.R. China

LIN, MAO — Third Institute of Oceanography, Ministry of Natural Resources, Xiamen, P.R. China

LITVINOVA, ANASTASIYA G. — State Scientific Establishment "The Polesie Agrarian Ecological Institute of the National Academy of Sciences of Belarus", Minsk, Republic of Belarus

LIU, ZHI-YONG — Xiamen Marine Environment Monitoring Center Station of the SOA, Xiamen, P.R. China
NAUMOVA, ELENA YU. — Limnological Institute of the Siberian Branch of the Russian Academy of Sciences, Irkutsk, Russia
NOVIKOV, ALEKSANDR — Kazan Federal University, Kazan, Tatarstan, Russia
PARVEZ, IMRAN — Hajee Mohammad Danesh Science and Technology University, Dinajpur, Bangladesh
POLYAKOVA, NATALIYA V. — Saint Petersburg State University, Saint Petersburg, Russia
SAMCHYSHYNA, LARYSA — State Institution National Antarctic Scientific Center, Ministry of Education and Science of Ukraine, Kiev, Ukraine
SCHIZAS, NIKOLAOS — University of Puerto Rico, Mayagüez, Puerto Rico, U.S.A.
SEMENOVA, ANNA — Atlantic Branch of "VNIRO", Kaliningrad, Russia
SINEV, SERGEY YU. — Zoological Institute of the Russian Academy of Sciences, Saint Petersburg, Russia
SŁUGOCKI, ŁUKASZ — University of Szczecin, Szczecin, Poland
SOUISSI, ANISSA — University of Lille, Lille, France
SOUISSI, SAMI — University of Lille, Lille, France
STARKOV, ALEKSEY — Saint Petersburg State University, Saint Petersburg, Russia
STRAKE, SOLVITA — Latvian Institute of Aquatic Ecology, Agency of Daugavpils University, Riga, Latvia
STRELKOV, PETR P. — Saint Petersburg State University, Saint Petersburg, Russia
SUKHIKH, NATALIA M. — Zoological Institute of the Russian Academy of Sciences, Saint Petersburg, Russia
SUN, ROU-XIN — Third Institute of Oceanography, Ministry of Natural Resources, Xiamen, P.R. China
TIMPE, ALEXANDER W. — Lawrence University, Appleton, WI, U.S.A.
TREMBLAY, RÉJEAN — Institut des Sciences de la Mer de Rimouski, Rimouski, Québec, Canada
TSENG, LI-CHUN — Institute of Marine Biology, National Taiwan Ocean University, Keelung, Taiwan
VELEGZHANINOV, ILYA — Polytechnical Institute of Vyatka State University, Kirov, Russia
VEZHNAVETS, VASIL V. — Scientific and Practical Centre of the National Academy of Sciences of Belarus on Bioresources, Minsk, Republic of Belarus
WANG,YAN-GUO — Institute of Marine Biology, National Taiwan Ocean University, Keelung, Taiwan
WINKLER, GESCHE — Institut des Sciences de la Mer de Rimouski, Rimouski, Québec, Canada
ZAVARZIN, DENIS S. — Sakhalin Branch of the Federal State Budget Scientific Institution «Russian Federal Research Institute of Fisheries and Oceanography» ("SakhNIRO"), Yuzhno-Sahalinsk, Russia
ZORINA-SAKHAROVA, KATERYNA — Institute of Hydrobiology, National Academy of Sciences of Ukraine, Kiev, Ukraine
ZWERVER, SATU — Company Zwerver Plankton Analyses, Kemiö, Finland

[When citing this volume, please refer to *Crustaceana* 93 (2020), 241-547]

*Proceedings of the* Eurytemora *conference, St. Petersburg, 2019*

# ON TIME AND PLACE OF ORIGIN OF CONTINENTAL CALANOID FAMILIES: A HYPOTHESIS

BY

VICTOR R. ALEKSEEV[1]) and NATALIA M. SUKHIKH
Zoological Institute of the Russian Academy of Science, Universitetskaya nab. 1, 199034
Saint Petersburg, Russia

## ABSTRACT

Several independent methods: molecular-genetic, biogeographical, and morphological analyses — were applied to explain the origin of the continental calanoid fauna and the distribution of their recent genera. The theory of Continental Drift and the evolution of the Tethys Sea were also used for that purpose. The molecular-genetic-based phylogenetic tree that we constructed, as well as the largest number of genera and species in Diaptomidae, allow us to support the idea that this family of fresh- and brackishwater Copepoda arose earlier than the Temoridae and Centropagidae. The ancestors of the Diaptomidae likely invaded, and were distributed across, the supercontinent Pangaea before its division into two continental plates in the Mesozoic Era, i.e., not later than 180-200 Ma. Therefore, various genera of this family can be found almost everywhere on all continents, except Antarctica. The family Temoridae is known only from Europe, Asia and North America. These three continental plates stayed together long after separation of Pangaea into two parts: Laurasia and Gondwana (until circa 50 Ma). At approximately the same time (50 Ma), the genus *Eurytemora* should have been created, as its representatives are known from North America and Eurasia. The family Centropagidae seems to have invaded inland waters somewhere between Temoridae and Diaptomidae, as its representatives can be found on all continents except Africa. Also, as a possibly alternative option, this centropagid invasion could have happened independently in the northern and southern Pangaea blocks, by different marine ancestors, at the same time as Temoridae, as is shown herein in the molecular-genetic-based phylogenetic tree. Using the evolution model of the Tethys Sea proved to be very productive for explaining the modern ranges of continental calanoids, both within families and in individual genera, including the genus *Eurytemora*.

## RÉSUMÉ

Plusieurs méthodes indépendantes : analyses de génétique moléculaire, biogéographiques et morphologiques — ont été appliquées pour expliquer l'origine de la faune continentale de calanoïdes et la répartition de leurs genres actuels. La théorie de la dérive des continents et l'évolution de la mer de Téthys ont aussi été utilisées dans ce but. L'arbre phylogénétique fondé sur les études de génétique moléculaire que nous avons construit, ainsi que le plus grand nombre de genres et d'espèces chez

[1]) Corresponding author; e-mail: valekseev2@yahoo.com

 | DOI:10.1163/9789004465565_003

les Diaptomidae, nous permettent de soutenir l'idée que cette famille de copépodes d'eaux douces et saumâtres est apparue plus tôt que les Temoridae et les Centropagidae. Les ancêtres des Diaptomidae ont probablement envahi, puis se sont répartis dans le supercontinent de la Pangée avant sa division en deux plaques continentales au Mésozoïque au plus tard vers 180-200 Ma. Ainsi, divers genres de cette famille peuvent être trouvés presque partout sur tous les continents, excepté l'Antarctique. La famille des Temoridae est seulement connue d'Europe, d'Asie et d'Amérique du Nord. Ces trois plaques continentales sont restées ensemble longtemps après la séparation de la Pangée en deux parties : Laurasie et Gondwana (jusqu'à environ 50 Ma). Approximativement au même moment (50 Ma), le genre *Eurytemora* serait apparu, puisque ses représentants sont connus de l'Amérique du Nord et de l'Eurasie. La famille des Centropagidae semble avoir envahi les eaux intérieures, entre les Temoridae et les Diaptomidae, puisque ses représentants peuvent être trouvés sur tous les continents excepté l'Afrique. Comme autre option possible, cette invasion des Centropagidae aurait pu se produire indépendamment dans les blocs nord ou sud de la Pangée, par différents ancêtres marins, en même temps que les Temoridae, comme il est montré ici sur l'arbre phylogénétique fondé sur la génétique moléculaire. L'utilisation du modèle d'évolution de la mer de Téthys s'est avérée très utile pour expliquer la répartition actuelle des Calanoïdes continentaux, à la fois au sein des familles et des genres, y compris le genre *Eurytemora*.

## INTRODUCTION

The components in the recent fauna of continental Copepoda have different ages and origins. One of the fascinating faunistic elements is the relict fauna formed in the inland waterbody called the Tethys Sea, which penetrated the ancient supercontinent Pangaea and was dispersed along with the various continental plates as these became separated.

The idea and name of the Tethys Sea was suggested by the Austrian palaeontologist Eduard Suess (1893). He discovered similarities in fossils on different continents and hypothesized that Pangaea was divided into two subcontinents, Gondwana [also known as "Gondwanaland"] and Laurasia. A sea belt appeared between those continental masses that Suess named the "Tethys Sea". His productive ideas soon turned into several significant studies of the palaeo-fauna of the Eurasian region (Rögl, 1998; Dercourt et al., 2000). Based on this hypothesis, geologists have specified several continental movement models, one of which we used in this work (Besse, 1995).

The extensive area of water belonging to the Tethys Sea finally turned into an inland waterbody about 20 Ma (Hou & Li, 2018) due to the strong compression of the central part of the Euro-Asian continent on account of the African and Indian subcontinents approaching the Asian continent from the south (fig. 1). This model can be used to date the divergence of ancient faunal complexes, especially in some genera like *Eurytemora*. The fauna's of waterbodies in North America, Europe and Asia exhibit features that are somewhat similar, primarily among the most ancient elements: some species of copepods, ostracods, sponges, sturgeon fish, etc. (Berg, 1922; Vereshchagin, 1940). Existing large-scale molecular-genetic

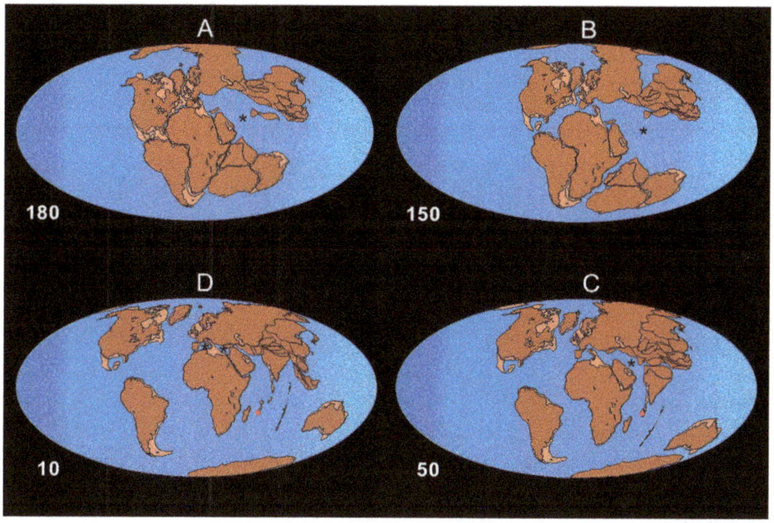


Fig. 1. Evolution of the Pangaea supercontinent and the Tethys Sea in accordance with the model of Besse (1995). A, 180 Ma; B, 150 Ma; C, 50 Ma; D, 10 Ma; asterisk indicates Tethys Sea.

studies of copepods are mainly devoted to the confirmation of monophyletic origin in large taxonomic groups (Khodami et al., 2017). In this paper we try to apply the methods of modelling the movement of continental plates, along with molecular-genetic, biogeographic and morphological methods, to estimate periods or times of penetration of different calanoid copepods into continental waters. Similar approaches (combinations of morphology and molecular-genetic tools) have been used to study historical roots of faunistic complexes in other Eurasian aquatic organisms: amphipods (Hou & Sket, 2016), ariid catfishes (Betancur-R, 2010). But their reconstructions were limited by the late Mesozoic period (60 Ma). Our hypothesis is based on palaeo-evolution of much more ancient crustaceans: copepods. So, we apply the Continental Drift hypothesis to an earlier time, i.e., from 200 Ma.

## MATERIAL AND METHODS

Numerous methods have been applied to test, or validate, our hypothesis:

1. Biogeographical data on the distribution of calanoid genera across continents at the present time. As a database for this analysis, the well-known "World directory of Crustacea Copepoda of inland waters, I — Calaniformes" by Dussart & Defaye (2002) was used. A table was created with information on one pure freshwater family (Diaptomidae), and two brackish water families (Centropagidae and Temoridae) as groups comparative to the Diaptomidae (table I).

TABLE I

Present-day continental spreading in selected calanoid families (data from Dussart & Defaye, 2002)

| | | Eur | Asia | NA | SA | Afr | Aus | Ind | Ant |
|---|---|---|---|---|---|---|---|---|---|
| | TEMORIDAE | | | | | | | | |
| 1 | *Eurytemora* | o | o | o | | | | | |
| 2 | *Epischura* | | o | o | | | | | |
| 3 | *Heterocope* | o | o | o | | | | | |
| | CENTROPAGIDAE | | | | | | | | |
| 4 | *Parabroteas* | | | | o | | | | o |
| 5 | *Limnocalanus* | o | o | o | | | | | |
| 6 | *Isias* | o | o | | | | o | | |
| 7 | *Osphranticum* | | | o | | | | | |
| 8 | *Boeckella* | o | o | | o | | o | | o |
| 9 | *Hemiboeckella* | | | | | | o | | |
| 10 | *Neoboeckella* | | | | o | | | | |
| 11 | *Calamoecia* | | | | | | o | | |
| 12 | *Sinocalanus* | | o | o | | | | | |
| 13 | *Gladioferens* | | | | | | o | | o |
| 14 | *Gippslandia* | | | | | | o | | |
| 15 | *Parathalassius* | | | | | | o | | |
| | DIAPTOMIDAE (57 genera) | | | | | | | | |
| | PARADIAPTOMINAE | | | | | | | | |
| 16 | *Paradiaptomus* | | o | | | o | | o | |
| 17 | *Lovenula* | | | | | o | | | |
| 18 | *Neolovenula* | o | | | | o | | | |
| 19 | *Metadiaptomus* | | o | | | o | o | | |
| | DIAPTOMINAE | | | | | | | | |
| 20 | *Diaptomus* | o | o | o | | o | | | |
| 21 | *Hemidiaptomus* | o | o | | | o | | | |
| 22 | *Eudiaptomus* | o | o | o | | o | | | |
| 23 | *Copidodiaptomus* | o | | | | o | | | |
| 24 | *Dussartius* | o | | | | | | | |
| 25 | *Acanthodiaptomus* | o | o | o | | | | | |
| 26 | *Arctodiaptomus* | o | o | o | | o | | o | |
| 27 | *Mixodiaptomus* | o | o | o | | o | | | |
| 28 | *Eodiaptomus* | | o | | | | o | o | |
| 29 | *Keraladiaptomus* | | | | | | | o | |
| 30 | *Phyllodiaptomus* | | o | | | | | o | |
| 31 | *Spicodiaptomus* | | | | | | | o | |
| 32 | *Megadiaptomus* | | | | | | | o | |
| 33 | *Neodiaptomus* | | o | | | | | o | |
| 34 | *Vietodiaptomus* | | o | | | | | | |
| 35 | *Heliodiaptomus* | | o | | | | | o | |
| 36 | *Allodiaptomus* | | o | | | | | o | |
| 37 | *Sinodiaptomus* | o | o | o | | | | o | |
| 38 | *Neutrodiaptomus* | | o | | | | | | |

TABLE I
(Continued)

| | | Eur | Asia | NA | SA | Afr | Aus | Ind | Ant |
|---|---|---|---|---|---|---|---|---|---|
| 39 | *Mongolodiaptomus* | | o | | | | | | |
| 40 | *Nordodiaptomus* | | o | o | | | | | |
| 41 | *Ligulodiaptomus* | | o | | | | | o | |
| 42 | *Dentodiaptomus* | | o | | | | | | |
| 43 | *Dolodiaptomus* | | o | | | | | | |
| 44 | *Thermodiaptomus* | | | | | o | | | |
| 45 | *Tropodiaptomus* | | o | | | o | o | o | |
| 46 | *Filipinodiaptomus* | | o | | | | | | |
| 47 | *Leptodiaptomus* | | o | o | | | | | |
| 48 | *Aglaodiaptomus* | | | o | | | | | |
| 49 | *Hesperodiaptomus* | | o | o | | | | | |
| 50 | *Onychodiaptomus* | | | o | | | | | |
| 51 | *Skistodiaptomus* | | | o | | | | | |
| 52 | *Mastigodiaptomus* | | | o | | | | | |
| 53 | *Prionodiaptomus* | | | o | o | | | | |
| 54 | *Argyrodiaptomus* | | | | o | | | | |
| 55 | *Calodiaptomus* | | | | o | | | | |
| 56 | *Notodiaptomus* | | | | o | | | | |
| 57 | *Pectenodiaptomus* | | | | o | | | | |
| 58 | *Rhacodiaptomus* | | | | o | | | | |
| 59 | *Austrinodiaptomus* | | | | o | | | | |
| 60 | *Dactylodiaptomus* | | | | o | | | | |
| 61 | *Odontodiaptomus* | | | | o | | | | |
| 62 | *Idiodiaptomus* | | | | o | | | | |
| 63 | *Aspinus* | | | | o | | | | |
| 64 | *Tumeodiaptomus* | | | | o | | | | |
| 65 | *Dasydiaptomus* | | | | o | | | | |
| 66 | *Scolodiaptomus* | | | | o | | | | |
| 67 | *Colombodiaptomus* | | | | o | | | | |
| 68 | *Spelaeodiaptomus* | o | | | | | | | |
| 69 | *Stygodiaptomus* | o | | | | | | | |
| | MICRODIAPTOMINAE | | | | | | | | |
| 70 | *Troglodiaptomus* | o | | | | | | | |
| 71 | *Microdiaptomus* | | | o | | | | | |
| | SPEODIAPTOMINAE | | | | | | | | |
| 72 | *Speodiaptomus* | o | | | | | | | |
| | PSEUDODIAPTOMIDAE | | | | | | | | |
| 73 | *Calanipeda* | o | o | | | | | | |

Eur, Europe; NA, North and Central America; SA, South America; Afr, Africa and Madagascar; Aus, Australia and New Zealand; Ind, India; An, Antarctica.

2. To estimate the comparative time of origin and separation of these three selected families from (a) possible marine ancestor(s) (family Calanidae), we applied a molecular-genetic analysis. A maximum likelihood tree for several families of Calaniformes was constructed using the nuclear gene 18SrRNA. We used ten 18SrRNA sequences obtained from GenBank. For the family Diaptomidae, we used five sequences (genera *Hesperodiaptomus*, *Eudiaptomus*, *Aglaodiaptomus*, *Leptodiaptomus*, and *Skistodiaptomus*); for Temoridae, Centropagidae, and Pseudodiaptomidae sequences of three genera (*Calanipeda*, *Eurytemora*, and *Limnocalanus*) were used. As an outgroup, the sequences of two typically marine species, *Calanus helgolandicus* (Claus, 1863) and *Clausocalanus arcuicornis* (Dana, 1849) of, respectively, the families Calanidae and Clausocalanidae were used.
3. The evolution of continental copepods, and indications thus far of their separation from marine ancestors, can be observed first of all in the transformation of morphological features related to sexual reproduction (Andronov, 2015). The fifth pair of legs in Calanidae, which in most continental families is highly specialized for mating, did, in the ancestral form, not differ in structure or function from the other swimming legs according to Andronov. The same applies to the structure and function of the right (geniculate) antennule of males and the genital double-somite of females. From the extent to which the structure of these taxonomically important parts of the body deviates from the supposedly ancestral condition, we can approximately estimate the time of separation of this family from the original ancestral forms, fide Andronov. We compared important male and female structures related to mating, copulation, and breeding in three continental families and one marine family. In males it was the transformation in P5 and the geniculate antennule; while in females the transformation was studied for P5 and the genital somite. As an outgroup, morphological structures of *Calanus finmarchicus* (Gunnerus, 1770) were used. For morphological comparison among marine copepods (*Calanus finmarchicus*, Calanidae), brackish-water species (*Eurytemora affinis* (Poppe, 1880), Temoridae; and *Limnocalanus macrurus* Sars G.O., 1863, Centropagidae) and freshwater forms (*Mixodiaptomus laciniatus* (Lilljeborg in Guerne & Richard, 1889), Diaptomidae), we used drawings modified from Brodsky (1983) and Gurney (1931).
4. In accordance with our hypothesis about the different time of introduction of the selected families of calanoids into continental waters, it was necessary to determine the time of separation of Pangaea into two supercontinents, the further division of Laurasia and Gondwana, and the formation of the modern position of the continents, as well as the evolution of the Tethys Sea over the past 200 Ma. Jean Besse's model (1995) turned out to be the most suitable for these purposes.

## RESULTS AND DISCUSSION

### Molecular-genetic analysis and phylogeny

For the molecular-genetic analysis, we selected gene sequences of several genera of Diaptomidae, Centropagidae and Temoridae that inhabit inland waters.

According to the resulting tree (fig. 2) the species associated with brackish water of the Centropagidae (*Limnocalanus macrurus*) and Temoridae (*Eurytemora affinis*), are grouped together, and they separated from their marine ancestor (*Calanus*) together with the Diaptomidae, but they split-up much later. These families (Centropagidae and Temoridae) include much fewer genera than Diaptomidae (10 and 6, respectively). In our view, this can be an indirect confirmation of the longer time of evolution and exploration of the continental waters of Diaptomidae in comparison with the two other families. Among the species of Centropagidae and Temoridae, there are truly marine forms: some genera are even purely marine, for instance, *Temora* and *Centropages*. At the same time, among diaptomids only purely freshwater genera and species are known, except *Arctodiaptomus salinus* (Daday, 1885)

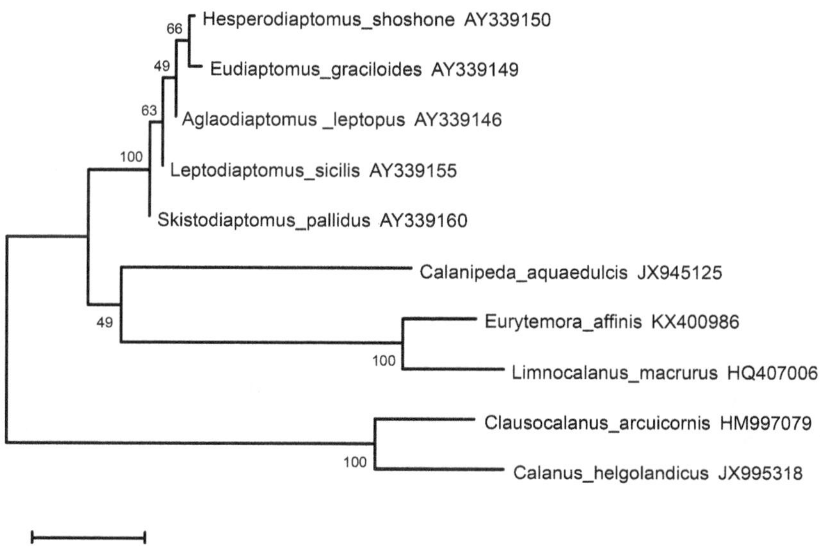


Fig. 2. The phylogenetic tree constructed by maximum likelihood analysis using the GTR + G model on the 10 species based on nuclear 18S rRNA (1210 base pairs). Bootstrap values for maximum likelihood (ML) (⩾45%) are given for the nodes. The species mentioned and their (sub)familial relationships are (from top to bottom) as follows. Diaptomidae: Diaptominae: *Hesperodiaptomus shoshone* (Forbes S.A., 1893); *Eudiaptomus graciloides* (Lilljeborg, 1888); *Aglaodiaptomus leptopus* (Forbes S.A., 1882); *Leptodiaptomus sicilis* (Forbes S.A., 1882); *Skistodiaptomus pallidus* (Herrick, 1879); Pseudodiaptomidae: *Calanipeda aquaedulcis* Krichagin, 1873; Temoridae: *Eurytemora affinis* (Poppe, 1880); Centropagidae: *Limnocalanus macrurus* Sars G.O., 1863; Clausocalanidae: *Clausocalanus arcuicornis* (Dana, 1849); Calanidae: *Calanus helgolandicus* (Claus, 1863).

that is adapted to the salinity of continental salt lakes. This difference in structure of the families also can be related to the process of adaptation to continental water conditions. Specialization in any field of adaptation is usually evidence of a longer evolution (Simpson, 1953; Futuyma & Moreno, 1988).

The phylogenetic tree shows that the mostly marine family Pseudodiaptomidae separated from its ancestor approximately at the same time as Centropagidae and Temoridae (see fig. 2). At the same time, from a biogeographical point of view, we can see that the numbers of species that adapted to continental conditions are very different among these three families. Only *Calanipeda aquadulcis* is known as a mass form of fresh- and brackish water bodies in the Mediterranean and Ponto-Caspian regions, which likely seem to be the principal remains of the Tethys Sea in Eurasia. Among Temoridae and Centropagidae, the number of such species is much higher (see table I).

Earlier, it was proposed that Alaska, where the highest number of known *Eurytemora* species is found, should be recognized as the place of origin of this genus (Dodson et al., 2010). Our analysis partly confirms this point of view, in that the ancestors of the *Eurytemora* settled in the northern part of the mainland plate, then comprising present-day Europe, Asia and North America (see fig. 1). The question is, when this could have happened. In our opinion, given the current distribution of representatives of the *Eurytemora affinis* group (Holarctic), this happened before the separation of North America from Asia, not later than 50 Ma. On the other hand, the molecular-genetic tree of the *Eurytemora affinis* group indicates that Caspian *E.* cf. *affinis* (= *E. caspica* Sukhikh & Alekseev, 2013) is genetically closer to *Eurytemora* species group from the Pacific area than to *Eurytemora* from the Baltic Sea or from the North Sea (the type locality of *E. affinis*) (Lee, 1999; Sukhikh et al., 2016). This possibly means that speciation in this group had been finished before the separation of the Caspian Sea from the remains of the Tethys Sea in the east. That separation happened when the Indian subcontinent came into contact with Asian plates (10-15 Ma). All these dates, that have come from tectonic geology, likely seem very important for the dating of molecular-genetic evidence of species divergence (Ho, 2008).

## Morphological analysis

In *Calanus finmarchicus*, the morphological structures selected (in males the geniculate antennule and P5, and in females the genital double-somite and P5) are practically not specialized (fig. 3). In Centropagidae and Temoridae, the geniculate antennule and P5 of the males are specialized for capturing females, whereas in female the genital somite and P5 are hardly specialized and still close to the structures in the outgroup. In diaptomids, we find the highest level of specialization

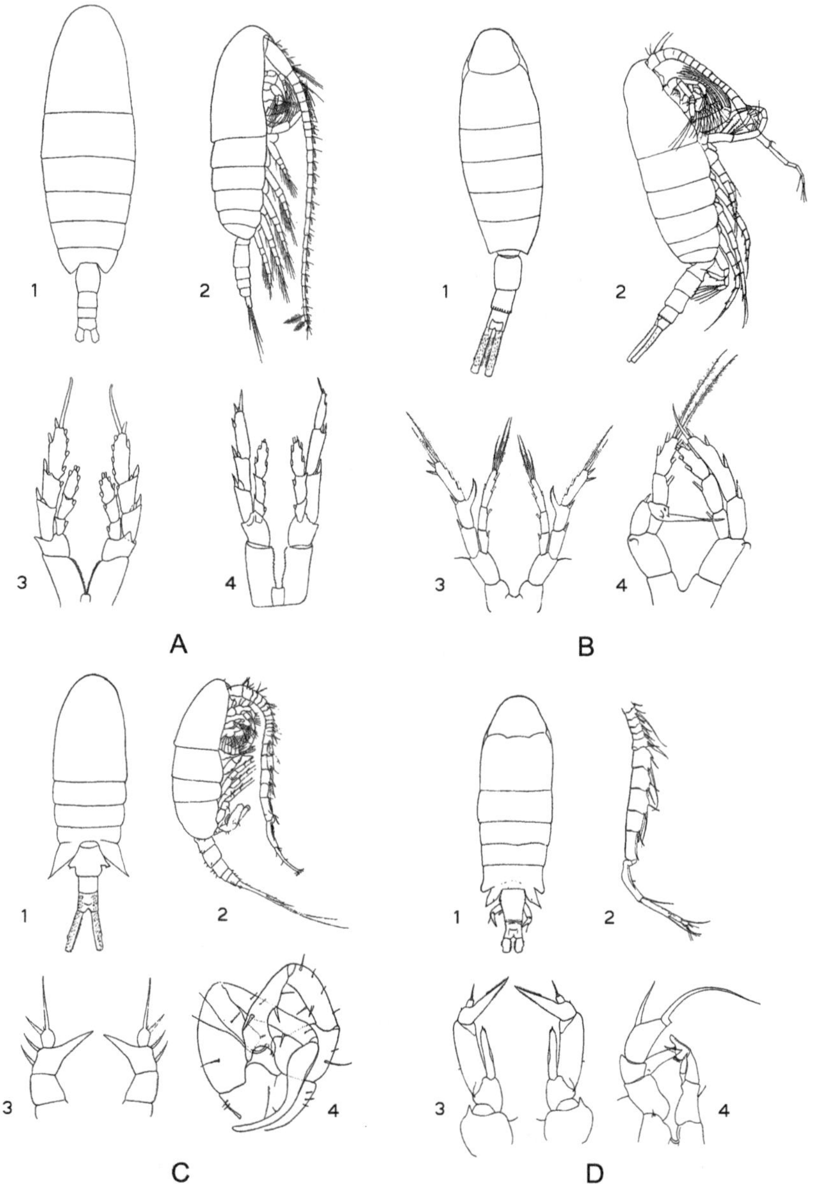


Fig. 3. Selected elements of calanoid morphology in males and females of species from four families: A, Calanidae (*Calanus finmarchicus* (Gunnerus, 1770)): 1, habitus of female, dorsal view; 2, habitus of male, lateral view; 3, P5 of female; 4, P5 of male [after Brodsky et al., 1983]. B, Centropagidae (*Limnocalanus macrurus* (Sars, G.O., 1863)): 1, habitus of female, dorsal view; 2, habitus of male, lateral view; 3, P5 of female; 4, P5 of male [after Gurney, 1931]. C, Temoridae (*Eurytemora affinis* (Poppe, 1880)): 1, habitus of female, dorsal view; 2, habitus of male, lateral view; 3, P5 of female; 4, P5 of male [after Gurney, 1931]. D, Diaptomidae (*Mixodiaptomus laciniatus* (Lilljeborg in Guerne & Richard, 1889)): 1, habitus of female, dorsal view; 2, geniculate antennule of male; 3, P5 of female; 4, P5 of male [after Gurney, 1931].

among the selected structures, which should be indicative of a longer evolutionary path. The geniculate antennule of male is morphologically transformed and is used not only for capturing females, but also for species recognition, by virtue of a special group of strong spines of different lengths on segments 10-16. The P5 of males is also highly specialized, and not only different for the various genera, but also specific for each species. In females, one can find significant transformations of the genital somites due to the fusion of 2-3 segments: the urosome no longer consists of 4-5 free somites, but of 2-3. The P5 of the females is quite small and very different from the other swimming legs, being specialized in differently in every species as well.

## Biogeographical distribution of continental calanoids

In the continental waters of the world, eight families of calanoids can be found (Dussart & Defaye, 2002).

The family Aetideidae, mainly marine but also found in continental waters, is represented with the single genus *Senecella*. In estuaries, coastal marine waterbodies and the large, cold lakes of North America and eastern Siberia, two morphologically close species have been discovered: *S. calanoides* Juday, 1923 and *S. siberica* Vyshkvartzeva, 1994. This family is represented only in the Northern Hemisphere, so it should most probably have originated after the division of Pangaea into the northern and southern blocks (not before 150 Ma).

The family Clausocalanidae also includes one genus, *Drepanopus*, with three species inhabiting Antarctic lakes and one circumpolar Arctic species (*D. bungei* Sars G.O., 1898), inhabiting Siberia, Canada and the U.S.A., with possibly the same time and place of origin as the previous family.

The family Centropagidae is represented in the world with twelve genera. The most species-abundant of these is the genus *Boeckella*, which mainly inhabits the Southern Hemisphere: Australia and South America (Bayly, 1991). In accordance with Bayly (1991), this genus originated in the southern block of continents before the separation of Australia and Southern America. After the model of Besse (1995), this should have happened earlier than about 100 Ma. Bayly (1991) so far was one of the first who applied the theory of Continental Drift to copepod origin and distribution. The genus *Limnocalanus* includes only two species: *L. macrurus* and *L. johanseni* Marsh, 1920. But these species' distribution area covers the whole Holarctic region. This genus seems to have originated after the separation of Pangaea into two blocks (150 Ma), but the time of its real introduction into continental waters may have been much later, as this form until now lives mainly in cold waters that are tightly connected. Confirmation of a later time of origin of this genus can be found by comparing it with other calanoids, both in morphology and in the small number of species.

Representatives of the family Temoridae are found in the Northern Hemisphere only: it includes three valid genera that inhabit continental waters:

The genus *Epischura* includes four Asian (subgenus *Epischurella*) and four North American species (subgenus *Epischura*). *Epischura* (*Epischurella*) *baikalensis* Sars G.O., 1900 is a mass plankton species (filter feeders) in Lake Baikal (Russia). Two other Asian species are known from the River Amur and its related waterbodies (Lake Khanka and Lake Udyl) and one from Kamchatka (all in Russia). The American species of *Epischura* are widespread in different waterbodies across the U.S.A. and Canada, from Alaska to Nevada.

The genus *Heterocope* includes six large-sized carnivorous Calanoida; two of them, *H. appendiculata* Sars G.O., 1863 and *H. borealis* (Fischer, 1851), are widely distributed throughout Eurasia. *H. soldatovi* Rylov, 1922 is known only from the Russian Far East, while *H. septentrionalis* Juday & Muttkowski, 1915 is found in the Arctic and Subarctic regions of North America.

The genus *Eurytemora* includes 24-25 valid species, depending on the source (Kos, 2016; Razouls et al., 2005-2020; Walter & Boxshall, 2020). Most of them live in brackish waters, but two species widely distributed in the Ponto-Caspian area, *E. velox* (Lilljeborg, 1853) and *E. lacustris* (Poppe, 1887), inhabit purely fresh water, whereas several other species are known from marine coastal areas with oceanic salinity. Previously, the Holarctic-distributed *E. affinis* was split into a flock of sibling species (Lee, 2000). Some of them, however, were described later on as valid species by using a set of new micro-characters (Alekseev & Souissi, 2011; Sukhikh & Alekseev, 2013).

This genus is also the most studied in terms of molecular genetics, primarily due to the pioneering activity of the American researcher Carol Lee (Lee, 1999, 2000). The results obtained by Lee and her followers allow us to consider the history of settlement and speciation in this genus, associated with the last residuals of the Tethys Sea. First of all, it should be pointed out that the ancestor of this genus moved into the inland waters of Pangaea after its separation into the northern and southern parts. This happened about 150 Ma, according to the model of Continental Drift (see fig. 1). Most species described in this genus are somehow related to living in cold waters. Apparently, also judging by the distribution of the species of the *affinis* group, the ancestor inhabited the northern part of the continent. This is largely consistent with the opinion of Stanley Dodson and co-authors, who considered the centre of origin, and hence of speciation, to be the waterbodies in Alaska (Dodson et al., 2010). However, this clearly happened before the American block system was separated from the Eurasian one, and it is not entirely correct to say that Alaska became the centre of origin, since representatives of this genus are also present in other, more southern inland reservoirs of Eurasia (five species in the Caspian Sea region, one species in Issyk-Kul Lake). In our opinion, it is more correct to indicate the northern part of Laurasia as the centre of origin of this genus. If we correlate the points of the modern locations of species of the genus *Eurytemora* with the projection of the Tethys Sea as it was before the connection of the African and Indian subcontinents to Eurasia, it becomes quite probable that the last phase in the existence of the Tethys Sea and its compression due to the convergence of the aforementioned subcontinents with Eurasia, were one of reasons for the recent distribution of species in this part of the world. It is noteworthy that the representative of the *affinis* group from the Caspian Sea, now described as *E. caspica*, is genetically closest to the representatives of this group found in the Far East, on the islands of Sakhalin and Hokkaido (Lee, 2000). The Caspian species, in contrast to what we anticipated, turned out to be genetically farther from the territorially closer representatives of this group, which currently inhabit the Baltic, North, and White seas, including *E. affinis*, with respect to a typical habitat (Elba River delta) (Sukhikh et al., 2016). Another example is the distribution of local populations of *E. composita* Keiser, 1929, described from Lake Issyk-Kul and later found in the brackish coastal waters of Alaska (Wilson, 1953), the island of St. Lawrence (Bering Sea) and the island of Paramushir (Sea of Okhotsk) (Kiefer, 1938). The places of occurrence of isolated populations of *Eurytemora* relict species in Eurasia, allows to recognize a belt, possibly showing a representation of their extension in the old time. This belt (40-50°N), covering the Mediterranean, Black and Caspian seas in the west and the Pacific islands of Sakhalin, Paramushir and Hokkaido in the east, roughly reflects the extent of the last phase of the existence of the in**tra**continental or, better put, the in**ter**continental Tethys Sea. This conclusion is, to a certain extent in accordance with the model of Continental Drift for that period (see fig. 1).

We may say that a similar pattern of isolated occurrences of a number of African and Ponto-Caspian species in Eurasia falls into approximately the same belt. Examples include the *Afrocyclops* species outside Africa, *Ochridacyclops* outside Europe, and the southern species of *Eucyclops* and *Mesocyclops* (described from Africa, in the vicicity of the Mediterranean) in Central Asia and the Far East (Mirabdullayev, 1996; Karaytug, 1999; Alekseev & Sanoamuang, 2006). In our opinion, the distribution of the genus *Eurytemora* in the southern part of the area can act as a marker of the residual waterbodies of the Tethys Sea, which should be taken into account when conducting relevant studies of the relict fauna of this ancient waterbody.

The largest family Diaptomidae is comprised of two-thirds of all continental species of Calaniformes (about 400); it is subdivided into four subfamilies.

The subfamily Paradiaptominae includes four genera. *Paradiaptomus*, with one exception (*P. greeni* that was found only in Asia), inhabits Africa (Rayner, 2000) and possibly originated on this continent. Four species of the genus *Lovenula* were also described as originating from Africa. The monotypic genus *Neolovenula* with its only species *N. alluaudi* (Guerne & Richard, 1890) inhabits North Africa and the south of Europe. The genus *Metadiaptomus* includes 11 species, only one of which, *M. asiaticus* (Uljanin, 1875), can be found outside of Africa, from the Caspian Sea region to China, while the other species are endemic to Africa. So far, this subfamily seems tightly linked to the African continent. The highest species diversity thus possibly indicates Africa as a place of origin for this subfamily, and this could have happened after the division of Pangaea in two continents, and also after the separation of South America and Australia from Africa (circa 100 Ma).

The subfamily Diaptominae is the largest family-group taxon among all continental Copepoda, and includes 58 genera widely distributed throughout the world. Two vicariant species, *Eudiaptomus gracilis* (Sars G.O., 1863) and *E. graciloides* (Lilljeborg, 1888), are widely distributed in the Palaearctic, and the former species is also known from North America (Dussart & Defaye, 2002). The two largest genera, *Arctodiaptomus* and *Tropodiaptomus*, include more than 50 species each.

Several hundred other species of the subfamily cover all continents of the world, except Antarctica, where they possibly became extinct due to the harsh climatic conditions.

## CONCLUSIONS

Three independent methods — molecular-genetic, biogeographical and morphological analyses — brought us to the same conclusions in regard to the world's continental Calanoida:

1. The most divergent with respect to numbers of species and the most widely distributed across all continents, is the family Diaptomidae. With the highest level of morphological specialization and the probably earliest time of separation from marine ancestors, it seems likely that they were the first continental-water colonizing calanoids. In accordance with the model of Continental Drift, this happened in the Mesozoic era (before 150 Ma), prior to the division of Pangaea into two blocks by the Tethys Sea.
2. Ancestors of the two other most important continental families, the Centropagidae and Temoridae, should have invaded continental waters significantly later than the Diaptomids. For the genus *Eurytemora*, a possible time of origin was estimated as prior to 50 Ma when North America separated from Eurasia. At the same time, it is possible that an invasion could have occurred independently in the northern and southern Pangaea blocks by different marine ancestors, as was shown in the phylogenetic tree.

3. Usage of the Continental Drift model can be useful for determination of the time of origin, not only for calanoids but also for other copepods, especially those with a limited distribution range. This approach provides a convenient tool for the molecular-genetic dating technique.
4. Our hypothesis, if confirmed by direct observation of fossil copepod remains 15-25 Ma in amber and underground salt lakes (Selden et al., 2010; Huys et al., 2016), could significantly increase the postulated time of colonization of inland waters by Calaniformes.

## ACKNOWLEDGEMENTS

This work was conducted in accordance with the national research initiative AAAA-A19-119020690091-0 of the Russian Academy of Science and was supported by the Russian Foundation for Basic Research grants 20-04-00035, 17-04-00027 and 20-34-70020. The Federal Collection of the Zoological Institute RAS was also used. We are thankful to Olga Chaban and to Dr. Dmitry Lajus for significant help with the preparation of the manuscript for publication.

## REFERENCES

ALEKSEEV, V. R. & L.-O. SANOAMUANG, 2006. Biodiversity of cyclopoid copepods in Thailand — with a description of *Afrocyclops henrii* sp. n. Arthropoda Selecta, **15**(4): 277-290.

ALEKSEEV, V. R. & A. SOUISSI, 2011. A new species within the *Eurytemora affinis* complex (Copepoda: Calanoida) from the Atlantic coast of USA, with observations on eight morphologically different European populations. Zootaxa, **2767**: 41-56.

ANDRONOV, V. N., 2014. Phylogeny and revision of [the] calanoid copepods system: 1-204. (Smartbooks, Kaliningrad). [In Russian.]

BAYLY, I. A. E., 1991. The fauna of athalassic saline waters in Australia and the Altiplano of South America: comparisons and historical perspectives. In: H. HURLBERT (ed.), Saline lakes V: proceedings of the V$^{th}$ international symposium on inland saline lakes: 225-231. (Springer Science & Business Media, Bolivia).

BERG, L. S. (ed.), 1922. The fauna of Baikal and its origin, climate and life: 28-53. (Moscow). [In Russian.]

BESSE, J., 1995. Continental Drift. [Retrieved 5 March 2020, from: http://www.tectonics.caltech.edu/outreach/animations/drift.html.

BETANCUR, R. R., 2010. Molecular phylogenetics supports multiple evolutionary transitions from marine to freshwater habitats in ariid catfishes. Molecular Phylogenetics and Evolution, **55**(1): 249-258.

BRODSKY, K. A., N. V. VYSHKVARTSEVA, M. S. KOS & E. L. MARKHASEVA, 1983. Copepod crustaceans (Copepoda: Calanoida) of the USSR seas and adjacent waters. Opredeliteli po Faune SSSR, **135**: 1-357. (Nauka, Leningrad). [In Russian.]

DERCOURT, J., M. GAETANI & B. VRIELYNCK (eds.), 2000. Atlas peri-Tethys. (Palaeogeographical Maps, Commission de la Carte Géologique du Monde (CCGM), Paris).

DODSON, S. I., D. A. SKELLY & C. E. LEE, 2010. Out of Alaska: morphological diversity within the genus *Eurytemora* from its ancestral Alaskan range (Crustacea, Copepoda). In: L. NASELLI-FLORES & G. ROSSETTI (eds.), Fifty years after the "Homage to Santa Rosalia": old and new paradigms on biodiversity in aquatic ecosystems. Developments in Hydrobiology, **213**: 131-148. (Springer, Dordrecht).

DUSSART, B. & D. DEFAYE, 2002. World directory of Crustacea Copepoda of inland waters, **1** — Calaniformes: i-viii, 1-276. (Backhuys, Leiden).

FUTUYMA, D. J. & G. MORENO, 1988. The evolution of ecological specialization. Annual Review of Ecology and Systematics, **19**(1): 207-233.

GURNEY, R., 1931. British fresh-water Copepoda. I. General; Calanoida: i-lii, 1-238. (The Ray Society, London).

HO, S., 2008. The molecular clock and estimating species divergence. Nature Education, **1**(1): 168.

HOU, Z. & B. SKET, 2016. A review of Gammaridae (Crustacea: Amphipoda): the family extent, its evolutionary history, and taxonomic redefinition of genera. Zoological Journal of the Linnean Society, **176**(2): 323-348.

HOU, ZH. & S. LI, 2018. Tethyan changes shaped aquatic diversification: aquatic diversification in the Tethyan region. Biol. Rev. Camb. Philos. Soc., **93**(2): 874-896. DOI:10.1111/brv.12376.

HUYS, R., E. SUÁREZ-MORALES, M. SERRANO-SÁNCHEZ, E. CENTENO-GARCÍA & F. J. VEGA, 2016. Early Miocene amber inclusions from Mexico reveal antiquity of mangrove-associated copepods. Sci. Rep., **6**: 34872. DOI:10.1038/srep34872.

KARAYTUG, S., 1999. Copepoda: Cyclopoida. Genera *Paracyclops*, *Ochridacyclops* and key to the Eucyclopinae. In: H. J. F. DUMONT (ed.), Guides to the identification of the microinvertebrates of the continental waters of the world, **14**: 1-217. (SPB Academic Publishing, The Hague).

KHODAMI, S., J. MCARTHUR, L. BLANCO-BERCIAL & P. MARTINEZ ARBIZU, 2017. Molecular phylogeny and revision of copepod orders (Crustacea: Copepoda). Sci. Rep., **7**: 9164. DOI:10.1038/s41598-017-06656-4.

KIEFER, F., 1938. Freilebende Süsswassercopepoden von den Nordkurilen. Bulletin of the Biogeographical Society of Japan, **8**: 75-94.

KOS, M. S., 2016. Calanoid copepods of the families Stephidae and Temoridae of the seas of Russia and adjacent waters: 1-108. (Zoological Institute of the Russian Academy of Science, St.-Petersburg). [In Russian.]

LEE, C. E., 1999. Rapid and repeated invasions of fresh water by the copepod *Eurytemora affinis*. Evolution, **53**(5): 1423-1434.

LEE, C. E., 2000. Global phylogeography of a cryptic copepod species complex and reproductive isolation between genetically proximate "populations". Evolution, **54**(6): 2014-2027.

MIRABDULLAYEV, I. M., 1996. The genus *Mesocyclops* (Crustacea: Copepoda) in Uzbekistan (Central Asia). Internationale Revue der Gesamten Hydrobiologie, **81**(1): 93-100.

RAYNER, N. A., 2000. Distribution and biogeography of the Paradiaptominae (Copepoda: Calanoida: Diaptomidae). Southern African Journal of Aquatic Sciences, **25**(1): 93-97.

RAZOULS, C., F. DE BOVÉE, J. KOUWENBERG & N. DESREUMAUX, 2005-2020. Diversity and geographic distribution of marine planktonic copepods. (CNRS, Sorbonne University, Paris). Available at http://copepodes.obs-banyuls.fr/en. [Accessed 14 January 2020.]

RÖGL, F., 1999. Mediterranean and Paratethys. Facts and hypotheses of an Oligocene to Miocene paleogeography (short overview). Geologica Carpathica, **50**(4): 339-349.

SELDEN, P. A., R. HUYS, M. H. STEPHENSON, A. P. HEWARD & P. N. TAYLOR, 2010. Crustaceans from bitumen clast in Carboniferous glacial diamictite extend fossil record of copepods. Nat. Commun., **1**: 50. DOI:10.1038/ncomms1049.

SIMPSON, G. G., 1953. The major features of evolution: 1-434. (Columbia University Press, New York, NY).

SUESS, E., 1893. Are great ocean depths permanent? Nat. Sci., **2**: 180-187.

SUKHIKH, N. M. & V. R. ALEKSEEV, 2013. *Eurytemora caspica* sp. nov. from the Caspian Sea — one more new species within the *E. affinis* complex (Copepoda: Calanoida: Temoridae). Proceedings of the Zoological Institute Russian Academy of Sciences, **317**(1): 85-100.

SUKHIKH, N. M., V. CASTRIC, N. V. POLYAKOVA, S. SOUISSI & V. R. ALEKSEEV, 2016. Isolated populations of *Eurytemora americana* Williams (Crustacea, Copepoda) in the White Sea rock pools — postglacial relicts or anthropogenic invasions? Rus. J. Biol. Invasions, **7**(4): 396-404.

VERESHCHAGIN, G. Y., 1940. The origin and history of Baikal, its fauna and flora. Proceedings of the Baikal Limnological Station, Academy of Sciences of the USSR, **10**: 73-239. [In Russian.]

WALTER, T. C. & G. BOXSHALL, 2020. World of Copepods database. *Eurytemora* Giesbrecht, 1881. Accessed at: http://www.marinespecies.org/copepoda/aphia.php?p=taxdetails&id=104240 on 2020-01-14.

WILSON, M. S., 1953. New Alaskan records of *Eurytemora* (Crustacea, Copepoda). Pacific Sci., **7**: 504-512.

First received 30 December 2019.
Final version accepted 26 March 2020.

[When citing this volume, please refer to *Crustaceana* 93 (2020), 241-547]

*Proceedings of the* Eurytemora *conference, St. Petersburg, 2019*

# DISTRIBUTION OF *EURYTEMORA CASPICA* SUKHIKH & ALEKSEEV, 2013 (COPEPODA, CALANOIDA) IN THE WATER RESERVOIRS OF THE VOLGA AND DON RIVER BASINS

BY

V. I. LAZAREVA[1])
Papanin Institute for Biology of Inland Waters, Russian Academy of Sciences, Borok, Nekouzskii raion, Yaroslavl oblast 152742, Russia

## ABSTRACT

The zooplankton composition of the Volga, Kama and Don rivers was studied in the summers of 2015-2018. The copepod *Eurytemora caspica* Sukhikh & Alekseev, 2013, recently isolated from the group of species contained in *Eurytemora affinis*, represents a widespread (50-100% of samples) and relatively abundant (maximum abundance up to $18 \times 10^3$ ind./m$^3$) species. It is found in the Lower and Middle Volga from the city of Astrakhan to the mouth of the Kama River, in the Kama River to the upper reaches of the Kama reservoir, in the lower and middle sections of the Don River within the Tsimlyansk reservoir, as well as in the reservoirs of the Volga-Don shipping canal. In the Kama River, *E. caspica* was discovered by the author for the first time in 2016. In other habitats, it was previously identified as *Eurytemora affinis* (Poppe, 1880), although this particular species was not found in any of the analysed samples. The distribution features and habitat conditions of *Eurytemora caspica* in the rivers of the basins of the Caspian and Azov seas are discussed.

## RÉSUMÉ

La composition du zooplancton de la Volga, la Kama et le Don a été étudiée au cours de l'été de 2015-2018. Le copépode *Eurytemora caspica* Sukhikh & Alekseev, 2013, récemment isolé du groupe d'espèces d'*Eurytemora affinis*, constitue une espèce largement représentée (50-100% des échantillons) et relativement abondante (abondance maximale jusqu'à $18*10^3$ ind./m$^3$). Il a été trouvé dans la Volga inférieure et moyenne, de la ville d'Astrakhan jusqu'à l'embouchure de la Kama, dans la rivière Kama jusqu'aux parties supérieures du réservoir de la Kama, dans les parties inférieure et moyenne du Don dans le réservoir de Tsimlyansk, ainsi que dans les réservoirs du chenal de navigation Volga-Don. Dans la Kama, *E. caspica* a été découvert par l'auteur pour la première fois en 2016. Dans les autres habitats, il était auparavant identifié comme *Eurytemora affinis* (Poppe, 1880), bien que cette espèce particulière n'ait pas été trouvée dans aucun des échantillons analysés. Les caractères de répartition et les conditions de l'habitat d'*Eurytemora caspica* dans les fleuves des bassins des mers d'Azov et Caspienne sont discutés.

[1]) e-mail: lazareva_v57@mail.ru

 | DOI:10.1163/9789004465565_004

## INTRODUCTION

The copepod *Eurytemora affinis* (Poppe, 1880), was earlier suspected as a complex of cryptic species (Lee & Frost, 2002), inhabiting fresh- and brackish waters of the Holarctic (Sukhikh & Alekseev, 2013). In the early 1980s, *Eurytemora* cf. *affinis* appeared in the Volga (Kuibyshev reservoir), and the species dominated among Calanoida in the first half of the summer since 1984 (Timokhina, 2000). In the 2000s, the species was common (occurrence of >40%) in the Saratov reservoir (Popov, 2007; Mukhortova, 2011). By the beginning of the 2010s, *E.* cf. *affinis* was recorded in the Volgograd, Saratov and Kuibyshev reservoirs (Popov, 2011). There were no reports on the appearance of the species in the reservoirs of the Kama River (Presnova & Khulapova, 2015; Seletkova, 2015).

A new species, *E. caspica* Sukhikh & Alekseev, 2013, was isolated from the group of "*affinis*" species in 2013 based on material from the Volga delta and the northern part of the Caspian Sea (Sukhikh & Alekseev, 2013). In 2015, it was first established (Lazareva et al., 2018) that the reservoirs of the Middle and Lower Volga are inhabited by *E. caspica*, not *Eurytemora* cf. *affinis*. *E. caspica* was first registered in all three Kama reservoirs in 2016 (Lazareva et al., 2018). The morphological analysis of copepod populations by the first descriptors of *E. caspica*, as well as the analysis of the mitochondrial cytochrome oxidase 1 (CO1) gene region, confirmed the findings of this particular species in the Volga and Kama rivers, while *E.* cf. *affinis* is not found at any of the studied stations (Sukhikh et al., 2020).

Back in the 1960s, it was shown (Mordukhai-Boltovskoi, 1964) that the Caspian autochthonous fauna originates from the relics of the Pontic Sea, which combined the waters of today's Caspian, Azov and Black seas in the Pliocene. Closely related, or even the same, species of crustaceans are distributed both in the Caspian Sea and along the coasts of the Black and Azov seas, as well as in the lower reaches of their tributaries. Similar to other Caspian crustaceans (for example, *Heterocope caspia* Sars, 1897, *Cercopagis pengoi* (Ostroumov, 1891)), *E. caspica* probably inhabits the reservoirs of the Ponto-Azov basin (including the Don River). However, a number of studies indicate that only *E.* cf. *affinis* is found in the Sea of Azov and Tsimlyansk reservoir on the Don River (Povazhnyi, 2009; Svistunova & Sayapin, 2010; Selifonova, 2013; Vekhov et al., 2014). This is possibly caused by the fact that the identification of *Eurytemora* species in Russia and neighbouring countries relies on outdated keys (Borutzky et al., 1991; Alekseev, 2010).

The aim of this work was to clarify the distribution of *E. caspica* in the reservoirs of the Volga, Kama and Don rivers, to analyse the distribution of the species' abundance, and to compare its population characteristics in different habitats.

## MATERIAL AND METHODS

### Region of studies

There are 11 reservoirs in the Volga-Kama cascade, eight of them (Ivan'kovo, Uglich, Rybinsk, Gor'ky, Cheboksary, Kyibyshev, Saratov and Volgograd) are situated on the Volga, three (Kama, Votkinsk and Nizhnekamsk), on the Kama (fig. 1). Drainage from the Kama reservoirs enters the Kama reach of the Kyibyshev reservoir. The creation of the Volga-Kama cascade was finished in the early 1980s after impoundment of the Nizhnekamsk (1979) and Cheboksary (1981) reservoirs (Edelstein, 1998). The following sections of the Volga River are distinguished: Upper Volga, the lower boundary of which is the dam of the Rybinsk Hydroelectric Power Plant (HPP); the Middle Volga with the southern border along the Zhiguli Hydroelectric Power Plant and the Lower Volga, which includes two reservoirs (Saratov and Volgograd) and an unregulated river section below the Volga Hydroelectric Power Plant (Mordukhai-Boltovskoy, 1979; Edelstein, 1998).

The only reservoir on the Don River, Tsimlyansk (fig. 1), impounded in 1952 (Edelstein, 1998), is situated in the middle course of the Don River (about 200 km upstream from the river's mouth). The entrance to the Volga-Don shipping canal is located in its upper section below the city of Kalach. From the Volga's side, the canal begins below the city of Volgograd, along the valley of the river Sarpa (tributary of the Volga) passes to the watershed of the Volga and the Don, then goes into the valley of the Chervlenaya River (a tributary of the Don River), in which two small reservoirs, Varvarovskoye and Breslavskoye, are formed by a system of dams and sluices; then the canal goes into the valley of the Karpovka River (a tributary of the Don River) and the Karpovskoye reservoir (Morukov, 2006). The canal is fed from the Don waters by three pumping stations at the watershed in the Varvarovskoye reservoir, from where it flows into the Volga.

The results of zooplankton survey of all eight Volga reservoirs in August 2015-2017, of the unregulated section of the Lower Volga between the cities of Volgograd and Astrakhan in August 2017, of the three Kama reservoirs in August 2016, as well as of the Volga-Don Canal and the Tsimlyansk reservoir in August 2018, are analysed in this article. Samples were taken mainly in the pelagial, 8-20 stations per reservoir (more than 130 stations in total). The mouth areas of large tributaries were examined additionally in 2017-2018. *E. caspica* was found in the three lower reservoirs of the Volga (Kuibyshev, Saratov and Volgograd) and a section of the river below the city of Volgograd, as well as in all Kama River reservoirs, the Volga-Don canal and the Tsimlyansk reservoir.

The Volga and Kama reservoirs are large (over 1000 km$^2$), deep (average depth more than 6 m, maximum up to 41 m) with a high (more than four times per year) and very high (up to 19 times per year) water exchange rate (table I). In

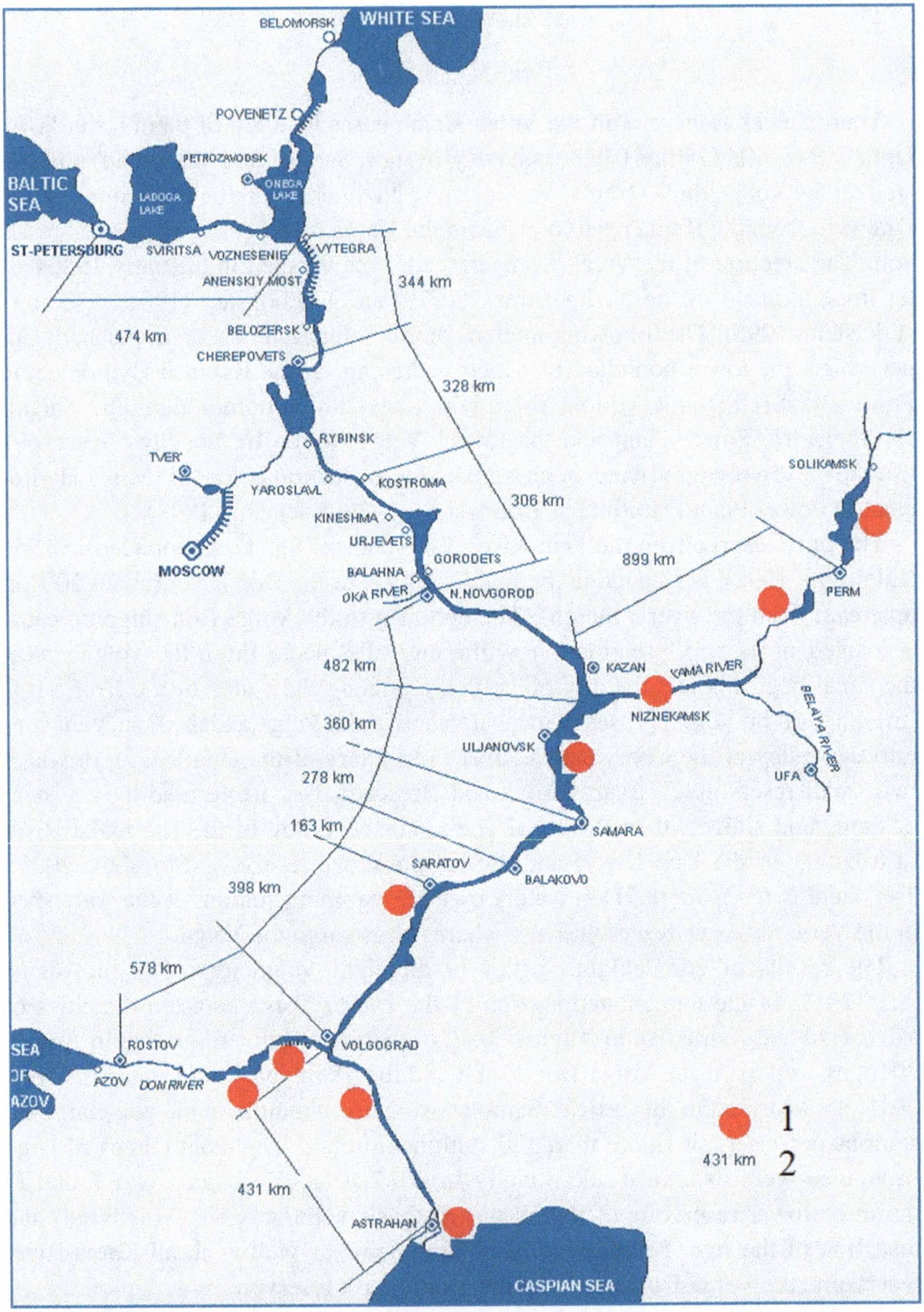


Fig. 1. Spread of *Eurytemora caspica* Sukhikh & Alekseev, 2013 in the basins of the Volga, Kama and Don rivers. Legends: 1, Locations of occurrence of *Eurytemora caspica*; 2, distance, in km. [Figure adapted from: http://www.w-fish.ru/info/reka-volga-verhnyaya-srednyaya-nizhnyaya.php].

TABLE I

Characteristics of the reservoirs of the Volga, Kama and Don rivers that *Eurytemora caspica* Sukhikh & Alekseev, 2013 was found in (i.e., found by: Avakyan et al., 1987; Edelshtein, 1998; Vekhov et al., 2014)

| Reservoir | River | Years of creation of reservoir | Distance from the Caspian Sea (km) | Area ($km^2$) | Depth (m) | | $C_{we}$ ($year^{-1}$) |
|---|---|---|---|---|---|---|---|
| | | | | | Mean | Max. | |
| Kama | Kama | 1954-1956 | 2430-2600 | 1915 | 6.4 | 30 | 4.4 |
| Votkinsk | Kama | 1961-1964 | 2165-2420 | 1120 | 8.4 | 28 | 5.7 |
| Lower Kama | Kama | 1979-1981 | 1905-2160 | 1000 | 8.0 | 14 | 6.6 |
| Kuibyshev | Volga | 1955-1957 | 1400-1900 | 6150 | 9.3 | 41 | 4.2 |
| Saratov | Volga | 1967-1968 | 1100-1850 | 1830 | 7.0 | 31 | 19.1 |
| Volgograd | Volga | 1958-1960 | 530-1050 | 3117 | 10.0 | 41 | 8.0 |
| Tsimlyansk | Don | 1952-1953 | 1140-1310 | 2620 | 8.7 | 31 | 1.1 |
| Varvarovskoye | Chervlenaya* | 1948-1952 | 1080-1100 | 16 | 7.8 | 12 | – |
| Breslavskoye | Chervlenaya* | 1948-1952 | 1100-1110 | 14 | 3.4 | 11 | – |
| Karpovskoye | Karpovka* | 1948-1952 | 1115-1140 | 42 | 3.8 | 5 | – |

$C_{we}$, water exchange coefficient.
*Volga-Don shipping canal (Don River basin).

terms of area and depths, the Tsimlyansk reservoir is of the same order as the Volga's reservoirs, but it has a lower water exchange rate (1.1 times a year). The total area of three reservoirs of the Volga-Don canal is less than 75 $km^2$ and they are relatively shallow; only in Varvarovskoye the average depth is comparable to that of the large reservoirs surveyed (table I). At present, according to the phytoplankton content of chlorophyll *a*, the Middle Volga and Kama reservoirs are characterized as eutrophic or close to eutrophic, and the Lower Volga as mesotrophic (Belyaeva et al., 2018; Mineeva & Makarova, 2018). The Tsimlyansk reservoir is characterized as hypertrophic according to the level of phytoplankton development (Vekhov et al., 2014). The Volga-Don canal reservoirs also belong to highly productive ecosystems, but information about their current trophic status is not available in the literature.

## Research methods

The crustaceans were taken from the total zooplankton samples, which were taken with a Juday net (12 cm opening diameter, 105 $\mu$m mesh). The entire water column from the bottom to the surface of the reservoir was sampled. The samples were fixated with 4% formaldehyde and viewed in the laboratory under a StereoDiscovery-12 stereomicroscope (Carl Zeiss, Jena). Taxonomic identification was performed using keys (Borutzky et al., 1991; Alekseev, 2010; Sukhikh &

Alekseev, 2013). Water temperature and electrical conductivity were measured with YSI ProODO and YSI-85 hand-held probes (YSI, Inc., Yellow Springs, OH).

## RESULTS

### Hydrophysical characteristics

The depth at the sampling sites in the main course of the Kama River varied from 8 to 23 m, and the average values in the three Kama reservoirs were in that same range (11-16 m). The depth along the Volga below its confluence with the Kama was 6-28 m, the average values for individual reservoirs were 14-18 m, and for the unregulated part of the Volga, the depth was 12 m. The depth along the channel of the Don within the Tsimlyansk reservoir increased from 6-8 m in its upper reaches to 11-18 m near the hydroelectric dam close to the city of Tsimlyansk, and was 10 m on average over the reservoir.

The electrical conductivity of the water reservoirs (reduced to 25°C) reflects their total salts content. In the summer low water period, it varied in the Kama in the range of 270-490 $\mu$S/cm, in the Volga River — 315-395 $\mu$S/cm, and its maximum values were noted in the Kama (860 $\mu$S/cm) and Kuibyshev (400 $\mu$S/cm) reservoirs. In the Don River, conductivity (610-810 $\mu$S/cm) was twice as high as that in the Volga, and it was even higher in the Volga-Don channel (750-1130 $\mu$S/cm), especially in the Varvarovskoye reservoir.

In summer, the warming of the reservoir's waters increases from north to south; in August, the mean multiannual values of water surface temperature are maximal in the Lower Volga and the Tsimlyansk reservoir (22.5°C), and minimal in the Kama (18-19°C) (Gotlib et al., 1976; Hydro-Meteorological Service, 1988; Litvinov & Roshchupko, 1993). At the end of August 2015, the water temperature in the Middle (18.9 ± 0.3°C) and the Lower Volga (19.1 ± 0.2°C) was 1.5°C below the norm for this month. In the same period of 2016-2017 it significantly exceeded (by 1.4-4.4°C in the Volga and by 5-6°C in the Kama) the norm, and reached 24.8 ± 0.1°C in the Kuibyshev reservoir, 24.0 ± 0.2°C in the Lower Volga, and 24.5 ± 0.3°C in the Kama reservoirs. At the end of August 2018, water temperature was close to normal in the Tsimlyansk reservoir (22.0 ± 0.2°C) and the Volga-Don canal (22.5 ± 0.2°C). The bottom water temperature in all reservoirs was only 0.1-2.3°C lower than that at the surface.

### Distribution and abundance

In the month of August of 2015-2017, *E. caspica* was an ordinary (50-90% of samples) inhabitant of the pelagic zone of the Kuibyshev and Volgograd reservoirs, as well as of the section of the unregulated Volga below the city of Volgograd (table II). In the Saratov reservoir, the species was not found in the years indicated. Up the Volga, *E. caspica* settled only up to the mouth of the Kama River (55°12′N). In the summer of 2016, *E. caspica* was first recorded in all three Kama reservoirs (occurrence of >70%). The northern limit of the species' distribution is located in the Kama reservoir upstream of the city of Perm' (58°52′N). In August 2018, *E. caspica* was first registered along the whole Volga-Don canal. The species was also found in the Tsimlyansk reservoir (70% of samples), mainly inhabiting the middle and lower parts of this reservoir. *E. caspica* was not found in the Don River north of the Tsimlyansk reservoir. The southern boundary of species' distribution in the Don River is located near the dam of the Tsimlyansk HPP (47°43′N), and in the Volga, in this river's delta below the city of Astrakhan (46°18′N). Up the Volga, *E. caspica* was distributed to a distance of 1400 km from the Caspian Sea, and along the Kama to a distance of 1200 km from the Volga (more than 2500 km from the Caspian Sea). In the basin of the Don River this species has spread over a distance of 30-250 km from the Volga (over 700 km from the Caspian Sea).

The abundance of *E. caspica* (adults, copepodids and nauplii) varied greatly from reservoir to reservoir, as well as within those. The maximum values (over $18 \times 10^3$ ind./m$^3$) were seen in the Karpovskoye reservoir (table II), and in other parts of the Volga-Don channel the abundance was 3100-3800 ind./m$^3$. A comparable number (up to $5 \times 10^3$ ind./m$^3$) was noted in the southern part of the Tsimlyansk reservoir. In the Volga, a high abundance of *E. caspica* was recorded in the Volga-Kama reach of the Kuibyshev Reservoir (more than $5 \times 10^3$ ind./m$^3$) and in the middle part of the Volgograd Reservoir (more than 1500 ind./m$^3$). In the Kama, the highest abundance of *E. caspica* (more than $3 \times 10^3$ ind./m$^3$) was observed in the middle part of the Votkinsk reservoir. The abundance of the species in other reservoirs of this river was 20-800 ind./m$^3$.

In August, adult breeding individuals formed 5-70% of the population of *E. caspica* (table II). The maximum individual fecundity of 10-30 eggs in an egg sac ($23 \pm 5$ on average) was recorded for the relatively large copepods from the Kama (table II). In the Volga, fecundity was the largest (10-25 eggs, on average $15 \pm 1$) in the Kuibyshev reservoir, while in Volgograd it dropped to 7-15 eggs (on average $11 \pm 2$), and in the unregulated part of the river below the city of Volgograd *E. caspica* did not breed, mainly only adult males and juvenile females were observed. In the Tsimlyansk reservoir and the Volga-Don canal, *E. caspica* females carried 10-35 eggs each. The average fecundity in the canal ($22 \pm 3$ eggs/♀) is comparable to that noted in the Kama River, and values observed in the Tsimlyansk reservoir ($16 \pm 1$ eggs/♀) are close to those in the Kuibyshev reservoir.

One congener of *E. caspica*, a larger species, *E. velox* (Lilljeborg, 1853) (female body length reaching 1.3 mm) was found in the Kuibyshev, Kama and Votkinsk reservoirs in small numbers (<500 ind./m$^3$).

TABLE II

Abundance and characteristics of the *Eurytemora caspica* Sukhikh & Alekseev, 2013 population in the Volga, Kama and Don River reservoirs in the summers of 2015-2018

| Parameter | Reservoir | | | | | | | |
|---|---|---|---|---|---|---|---|---|
| | Km | Vt | LK | K | V | LV | VD | Ts |
| Occurrence (%) | 75 | 100 | 70 | 70 | 90 | 50 | 100 | 70 |
| Average density (ind./m$^3$) | 560 ± 130 | 780 ± 470 | 80 ± 30 | 330 ± 110 | 240 ± 150 | 90 ± 40 | 7000 ± 3000 | 2530 ± 1000 |
| Maximum density (ind./m$^3$) | 820 | 3050 | 180 | 5710 | 1580 | 330 | 18200 | 5200 |
| Adults (%) | 30 | 10 | 60 | 30 | 20 | 70 | 5 | 15 |
| Body length ♀ (mm) | 1.1-1.3 | 1.1-1.2 | 1.0-1.2 | 1.0-1.1 | 0.9-1.2 | 1.0 | 0.9-1.1 | 0.9-1.1 |
| Body length ♂ (mm) | 1.0-1.1 | 1.0-1.1 | 1.0 | 0.8-0.9 | 0.8-1.1 | 0.8-1.0 | 0.8-1.0 | 0.9-1.1 |
| Eggs/♀ | 8-20 | 15-30 | 10-20 | 8-25 | 7-15 | 0 | 14-35 | 10-23 |

Reservoirs: Km, Kama; Vt, Votkinsk; LK, Lower Kama; K, Kuibyshev; V, Volgograd; Ts, Tsimlyansk; LV, unregulated Lower Volga below the dam of the Volga hydroelectric station; VD, Volga-Don shipping canal.

## DISCUSSION

It has long been shown that the Sea of Azov and Tsimlyansk reservoir on the Don River are inhabited by *Eurytemora* cf. *affinis*, considering it as a common species (Povazhnyi, 2009; Svistunova & Sayapin, 2010; Selifonova, 2013; Vekhov et al., 2014). A representative of this genus, which settled in the Volga reservoirs in the 1980s, was also identified as *E.* cf. *affinis* (cf. Thimokhina, 2000; Popov, 2011). Our research indicates that *E. caspica* was a mass species in 2015-2018, in the reservoirs of the Volga and Don rivers, while *E.* cf. *affinis* was not found in any plankton samples. We did not find *E. caspica* in the zooplankton of the Saratov reservoir, but *E.* cf. *affinis* was recorded in numbers of up to 2000 ind./m$^3$ in the pelagic zone and the coastal area of this reservoir in the second half of the 2000s (Mukhortova, 2011; Popov, 2011).

Apparently, *E. caspica* has invaded the reservoirs of the Kama River from the Kuibyshev reservoir in the early 2010s. The proximity of the Volga and Kama populations of this species was confirmed genetically (Sukhikh et al., 2020). However, until 2016, there were no reports of finds of copepods of the *E. affinis* complex and other Ponto-Caspian species in the Kama River (Presnova & Khulapova, 2015; Seletkova, 2015). *E. caspica* was found for the first time in the Kama River in 2016, the northernmost find of the species is located in the Kama reservoir (58°52′N) (Lazareva et al., 2018). By that time, the invader had already naturalized and became numerous in virtually the entire Nizhnekamsk, Votkinsk and Kama reservoirs within 1000 km from the Volga. Another Caspian copepod, *Heterocope caspia* has penetrated into the Kama River from the Volga simultaneously with *E. caspica* (cf. Lazareva et al., 2018). After impoundment of the Kuibyshev and Volgograd reservoirs in 1955-1960, *Heterocope caspia* spread up the Volga to the confluence zone of the Volga and Kama (Mordukhai-Boltovskoy, 1979). From the mid-1960s, the species became numerous in the Kuibyshev, Saratov and Volgograd reservoirs in the warm season (Thimokhina, 2000; Popov, 2011).

*E. caspica* was recognized as a new species, isolated from the group of cryptic species of the *E. affinis* complex, only in 2013, based on material from the Volga delta and the northern Caspian Sea (Sukhikh & Alekseev, 2013). According to the established concepts (Mordukhai-Boltovskoi, 1964), the Caspian autochthonous fauna and forms derived from it inhabit many parts of the Ponto-Caspian basin, including the Sea of Azov, the mouth areas of its tributaries, and tributaries of the Black Sea. If we assume that *E. caspica* is an ancient Caspian species, although recently discovered, then the finds of this species in the Don River outside of the current Caspian basin fully correspond to the distribution area of the autochthonous Caspian fauna. Two results follow from this consideration. Firstly, *E. caspica* is a native species for the whole Ponto-Caspian region and possibly it inhabited the reservoirs of the Volga-Don Canal, filled with Don River water, but was not found before this work. Secondly, the Sea of Azov and the Don River downstream of the Tsimlyansk reservoir (possibly the tributaries of the Black Sea as well) are also inhabited by *E. caspica*, although only *E.* cf. *affinis* has been recorded in these habitats at present (Povazhnyi, 2009; Selifonova, 2013; Svistunova et al., 2014).

*E. caspica* is the smallest among the Ponto-Caspian copepods inhabiting the Volga, Kama and Don (fig. 2). The length of adult copepods varied from reservoir to reservoir (table II), the largest females (body length 1.1-1.3 mm) and males (1.0-1.1 mm) were recorded in the Kama and Votkinsk reservoirs, the smallest (1.0-1.1 mm, 0.8-1.0 mm) in the Kuibyshev reservoir and unregulated Volga. The body length of adults of *E. caspica* from the Volga-Don canal and the Tsimlyansk reservoir, was comparable to that in the Volga. Individuals from the northern Caspian Sea also have a shorter body length (0.90-0.98 mm, 0.88-0.96 mm) (Sukhikh & Alekseev, 2013). For comparison, the body length of adults of

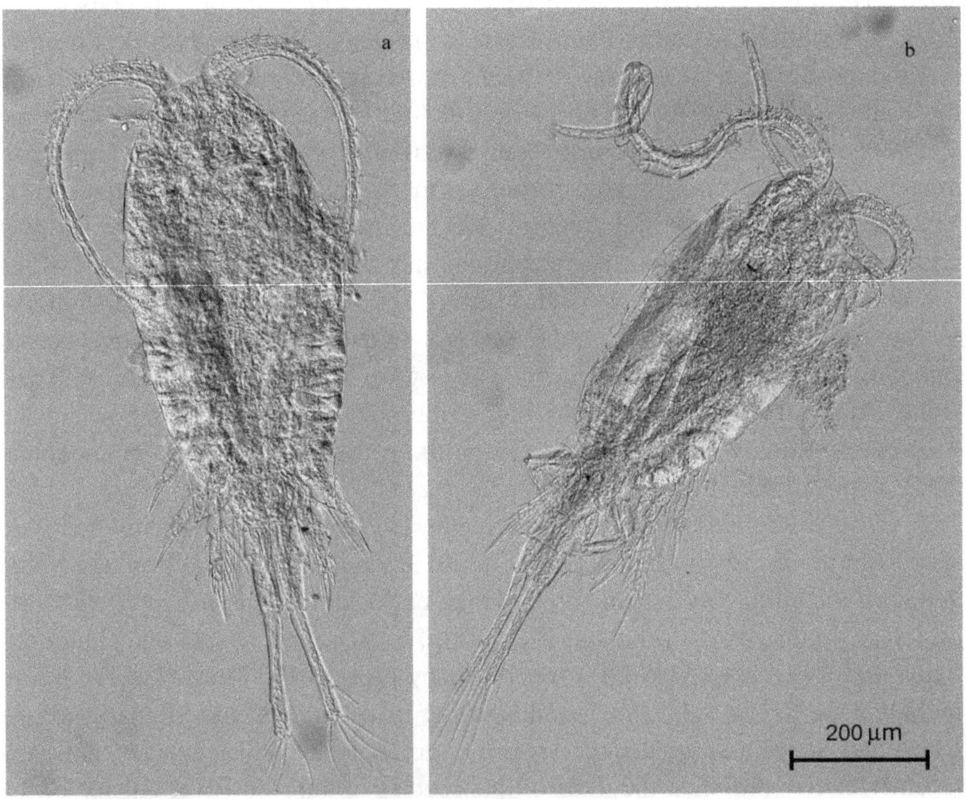


Fig. 2. a, Female; and, b, male, of *Eurytemora caspica* Sukhikh & Alekseev, 2013 from the Kama River reservoirs.

*Heterocope caspia* is 1.4-1.8 mm, and of *Calanipeda aquaedulcis* Kritschagin, 1873, 1.0-1.4 mm. One more species is indicated for the Kuibyshev and Saratov reservoirs: *E. lacustris* (Poppe, 1887) (cf. Timokhina, 2000; Mukhortova, 2011; Popov, 2011). However, *E. lacustris* was not found in the Volga, Kama and Don rivers in the summers of 2015-2018. *Eurytemora affinis* (Poppe, 1880) sensu stricto has not been found anywhere, either.

In the Volga, Kama and Don rivers, *E. caspica* lives in areas with a depth of 8-28 m, in fresh waters with medium and high salt content (electrical conductivity within 270-1130 $\mu$S/cm). Many Ponto-Caspian brackish-water crustaceans are capable of living in fresh water (Mordukhai-Boltovskoi, 1964). When the water temperature at the time of observation (August) was 18-25°C, this species multiplied intensively in almost all studied reservoirs. Females carried up to 35 eggs in the egg sac, the proportion of juveniles (copepodids and nauplii) reached 90% of the population. For the Kuibyshev reservoir (Volga River), it is noted that the greatest reproduction intensity and abundance of copepods of the genus *Eurytemora* is observed in the first half of the summer (Thimokhina, 2000). In the Sea of Azov and the Don delta, *E.* cf. *affinis* is abundant from late May to early October with water temperatures above 15°C, and copepodids are also encountered in the winter plankton (Povazhnyi, 2009; Svistunova et al., 2014).

## CONCLUSION

Thus, the recently described copepod *E. caspica* is a widespread and abundant species in the Lower and Middle Volga from the city of Astrakhan to the mouth of the Kama River, in the Kama River to the upper reaches of the Kama reservoir, in the lower and middle sections of the Don River within the Tsimlyansk reservoir, as well as in the reservoirs of the Volga-Don shipping canal. *E.* cf. *affinis* that was previously presumed to occur in these habitats was not found in any of the samples analysed.

Apparently, *E. caspica* inhabits the middle and lower reaches of the Don River from a very long time ago and this must be considered its natural habitat. This copepod penetrated into the Middle Volga in 1980s, and it can be considered an invader here. In 2016, this species was first registered in the Kama River, and it settled in virtually the whole Nizhnekamsk, Votkinsk and Kama reservoirs up to 59°N. *E. caspica* has probably invaded the Kama reservoirs several years earlier, possibly in the early 2010s.

## ACKNOWLEDGEMENTS

The author is grateful to V. A. Gusakov for his help in photographing the copepods.

## FINANCIAL SUPPORT

The study was carried out within the state assignment (theme Nos AAAA-A18-118012690106-7 and AAAA-A18-118012690105-0).

## ADHERENCE TO ETHICAL STANDARDS

All international and national principles for the use of animals for scientific purposes were followed.

## REFERENCES

ALEKSEEV, V. R. (ed.), 2010. Guide to identifying zooplankton and zoobenthos of fresh water in European Russia, **1**, Zooplankton: 1-495. (KMK Scientific Press, Moscow). [In Russian.]

AVAKYAN, A. B., V. P. SALTANKIN & V. A. SHARAPOV, 1987. Reservoirs: 1-325. (Mysl', Moscow). [In Russian.]

BELYAEVA, P. G., N. M. MINEEVA, L. E. SIGAREVA, N. A. TIMOFEEVA & O. S. SHARAPOVA, 2018. Content of the plant pigments in water and bottom sediments of the Kama River reservoirs. Transactions of IBIW RAS, **81**(84): 97-104. [In Russian.]

BORUTZKY, E. S., L. A. STEPANOVA & M. S. KOS, 1991. Guide to identifying freshwater Calanoida of USSR: 1-504. (Nauka, Leningrad). [In Russian.]

EDELSTEIN, K. K., 1998. Reservoirs of Russia: ecological problems and ways of their solution: 1-277. (GEOS, Moscow). [In Russian.]

GOTLIB, YA. L., V. M. ZHIDKIKH & N. M. SOKOL'NIKOV, 1976. Thermal regime of hydroelectric reservoirs: 1-204. (Gidrometeoizdat, Leningrad). [In Russian.]

[HYDRO-METEOROLOGICAL SERVICE], 1988. Kama River basin. In: Long-term data on the regime and resources of surface land waters, **1**(25): 1-707. (Gidrometeoizdat, Leningrad). [In Russian.]

LAZAREVA, V. I., R. Z. SABITOVA, S. V. BYKOVA, S. M. ZHDANOVA & E. A. SOKOLOVA, 2018. Distribution of summer zooplankton in cascade of Volga and Kama reservoirs. Transactions of Papanin Institute for Biology of Inland Waters, Russian Academy of Sciences, **83**(86): 62-84. [In Russian.]

LEE, C. E. & B. W. FROST, 2002. Morphological stasis in the *Eurytemora affinis* species complex (Copepoda: Temoridae). Hydrobiologia, **480**: 111-128.

LITVINOV, A. S. & V. F. ROSHCHUPKO, 1993. Thermal characteristics of the Volga cascade reservoirs. Transactions of Papanin Institute for Biology of Inland Waters, Russian Academy of Sciences, **63**(66): 3-24. [In Russian.]

MINEEVA, N. M. & O. S. MAKAROVA, 2018. Chlorophyll content as an indicator of the modern (2015-2016) trophic state of Volga River reservoirs. Inland Water Biology, **11**(3): 107-110.

MORDUKHAI-BOLTOVSKOI, PH. D., 1964. Caspian fauna beyond the Caspian Sea. Int. Revue ges. Hydrobiol., **49**(1): 139-176.

MORDUKHAI-BOLTOVSKOY, PH. D. (ed.), 1979. The River Volga and its life: 1-473. (Junk Publishers, The Hague).

MORUKOV, YU. N., 2006. Volga-Don canal. In: Big Russian encyclopedy, **5**: 625-626. (Moscow). [In Russian.]

MUKHORTOVA, O. V., 2011. Some singularities of distribution of zooplankton species-colonizers in littoral biotopes (macrophyte thickets, pebbly foreshore) of Saratov reservoir. Izvestiya Samarskogo Nauchnogo Tsentra, Russian Academy of Sciences, **13**(1): 188-193. [In Russian.]

POPOV, A. I., 2007. Some data on fauna and structure of zooplankton of Saratovskoe reservoir. Izvestiya Samarskogo Nauchnogo Tsentra, Russian Academy of Sciences, **9**(4): 1013-1019. [In Russian.]

POPOV, A. I., 2011. Alien species of zooplankton in Saratov reservoir (Russia, Volga River). Russian Journal of Biological Invasions, **2**(2-3): 126-129.

POVAZHNYI, V. V., 2009. Percularities of Taganrog Bay zooplankton dynamics. Vestnik Yuzhnogo Nauchnogo Tsentra, **5**(2): 94-101. [In Russian.]

PRESNOVA, E. V. & A. V. KHULAPOVA, 2015. Structure and distribution of zooplankton in the central district of Votkinsk reservoir. Vestnik Permskogo Universiteta, (Biologiya) **4**: 366-370. [In Russian.]

SELETKOVA, E. B., 2015. Zooplankton of Kama reservoir. In: Biocenoses of rivers and reservoirs of the Western Urals: 136-151. (Gos. nauch.-issled. in-t ozern. rechn. rybnogo khoz-va, St.-Petersburg). [In Russian.]

SELIFONOVA, J. P., 2013. State pelagic copepods community in the Azov Sea. Vestnik Zoologii, **47**(5): 421-430. [In Russian.]

SUKHIKH, N. M. & V. R. ALEKSEEV, 2013. *Eurytemora caspica* sp. nov. from the Caspian Sea — one more new species within the *E. affinis* complex (Copepoda: Calanoida, Temoridae). Proceedings of the Zoological Institute RAS, **317**(1): 85-100.

SUKHIKH, N. M., V. I. LAZAREVA & V. R. ALEKSEEV, 2020. The copepod *Eurytemora caspica* Sukhikh et Alekseev, 2013 (Crustacea, Calanoida) in Volga and Kama reservoirs. Inland Water Biology, **13**(2): 132-165.

SVISTUNOVA, L. D., V. A. BRYN'KO & M. V. NABOZHENKO, 2014. The current conditions of summer zooplankton in the Don River delta. Vestnik Yuzhnogo Nauchnogo Tsentra, **10**(3): 75-82. [In Russian.]

SVISTUNOVA, L. D. & V. V. SAYAPIN, 2010. The eutrophication problem of the Tsimlyansk water storage reservoir (Rostov region) and new data on its zooplankton. Vestnik Yuzhnogo Nauchnogo Tsentra, **6**(4): 61-67. [In Russian.]

TIMOKHINA, A. F., 2000. Zooplankton as a component of the ecosystem of the Kuibyshev reservoir: 1-193. (Institute of Ecology of the Volga Basin, Russian Academy of Sciences, Togliatti). [In Russian.]

VEKHOV, D. A., A. N. NAUMENKO, V. P. GORELOV, T. B. GOLOKOLENOVA & T. P. SHEVLYAKOVA, 2014. Modern condition and use of water bioresources of Tsimlansk reservoir (2009-2013). In: Fish research on water facilities of the European part of Russia: 116-145. (Gos. nauch.-issled. in-t ozern. rechn. rybnogo khoz-va, St.-Petersburg). [In Russian.]

WORLD FISH, 2019. Great Russian Volga River. URL: http://www.w-fish.ru/info/reka-volga-verhnyaya-srednyaya-nizhnyaya.php (accessed 30 September 2019).

First received 29 May 2019.
Final version accepted 20 September 2019.

[When citing this volume, please refer to *Crustaceana* 93 (2020), 241-547]

*Proceedings of the* Eurytemora *conference, St. Petersburg, 2019*

# RECENT DISTRIBUTION OF *EURYTEMORA VELOX* (LILLJEBORG, 1853) (COPEPODA, CALANOIDA) IN BRACKISH AND FRESH WATERS OF UKRAINE

BY

LARYSA SAMCHYSHYNA[1,3,4]), YULIA GROMOVA[2]) and KATERYNA ZORINA-SAKHAROVA[2])

[1]) State Institution National Antarctic Scientific Center, Ministry of Education and Science of Ukraine, Tarasa Shevchenka blvd. 16, Kyiv-01601, Ukraine[5])

[2]) Institute of Hydrobiology, National Academy of Sciences of Ukraine, Heroiv Stalingradu 12, Kyiv-04210, Ukraine

[3]) Institute of Fisheries and Marine Ecology, Konsulska str., 8, Berdiansk-71118, Ukraine

## ABSTRACT

The brackish water calanoid copepod *Eurytemora velox* has become a widespread species in inland fresh waters of Ukraine. Nowadays it is one of the dominant crustacean species in the littoral zooplankton of the main Ukrainian rivers: Dnieper, Danube and Pivdenny Buh. We present a survey on *E. velox* localities in, mainly, fresh waters of Ukraine, data on some quite isolated populations of *E. velox* in the Carpathian region, as well as new records of the species in the upper parts of some rivers. These data will provide a more complete view of the inland distribution of *E. velox* in Ukraine.

## RÉSUMÉ

Le copépode calanoïde d'eau saumâtre *Eurytemora velox* est devenue une espèce largement répandue dans les eaux douces d'Ukraine. Aujourd'hui, c'est l'une des espèces de crustacés dominante dans le zooplancton littoral des principaux fleuves d'Ukraine : Dniepr, Danube et Pivdenny Buh. Nous présentons ici une étude de la présence de *E. velox*, dans principalement les eaux douces d'Ukraine, des données sur des populations isolées de *E. velox* dans la région des Carpathes, ainsi que de nouvelles mentions de l'espèce dans les cours supérieurs de quelques rivières. Ces données fourniront une vue plus complète de la répartition intérieure de *E. velox* en Ukraine.

## INTRODUCTION

The genus *Eurytemora* Giesbrecht, 1881 constitutes a highly evolvable genus with respect to salinity (Lee & Gelembiuk, 2008). It is represented in Ukraine by

[4]) Corresponding author; e-mail: larysa.samchyshyna@gmail.com

[5]) "Kyiv" was formerly also known as "Kiev".

 | DOI:10.1163/9789004465565_005

four species and one subspecies (Samchyshyna, 2011). One of these, *Eurytemora velox* (Lilljeborg, 1853) is widely distributed in the brackish waters from the Arctic Ocean and the Baltic Sea shore to south-eastern Europe, the Sea of Azov, and the Caspian Sea (Vranovský, 1994). *E. velox* is believed to be a generalist species, which constantly broadens its area of distribution, occupies new ecological niches, and colonizes not only the saline and brackish water bodies, but freshwater habitats away from the marine coasts as well (Pandourski & Evtimova, 2006).

In Ukraine, the geographic range of *E. velox* extends beyond the coastal waters of the Black Sea and Sea of Azov and the lower streams of the main Ukrainian rivers. The species inhabits a variety of water bodies from hypersaline salt marshes to fresh waters (Lyashenko et al., 2018). In the Middle Dnieper River and its tributaries, the species was found in the second half of the 20th century by many authors (Melnikov, 1948; Monchenko, 1974; Polishchuk et al., 1978; etc.). The species is widely distributed, especially in streamlets, channels, marshes, ponds, and lakes of the Danube Delta, in waters with a large range of salinities (Spandl, 1926; Markovsky, 1955; Polishchuk, 1974; Parchuk, 1985; Kharchenko et al., 2005; Zorina-Sakharova et al., 2014).

In this paper, we present a survey on the distribution of *E. velox* in, mainly, fresh waters of Ukraine, data on some quite isolated populations of *E. velox* in the Carpathian region, as well as new records of the species in the upper parts of rivers; then, where possible, data on the species' abundance are given.

## MATERIAL AND METHODS

Zooplankton samples were collected using a hand-thrown, small plankton tow net with a 100 $\mu$m mesh, during various sampling campaigns of both the Institute of Hydrobiology (IHB AN) and the Institute of Zoology NASU (IZAN), in 1997-2019. The samples were preserved in 4% formaldehyde solution. Part of the processed samples from earlier years (1956-1983) originate from the depository of the IZAN and were collected by Prof. Vladyslav Monchenko and Dr. V. Polishchuk. Samples from the Danube Delta were collected as a part of the IHB AN project "Climatogenic restructuring of hydrobiont communities and its impact on ecological situation and bioproductivity of EU cross-border rivers in Ukraine". Samples from Khmelnitsky nuclear power plant cooling pond were collected in 2014-2018 as a part of a multi-year monitoring programme led by Prof. O. Protasov (IHB AN).

## RESULTS

The abundance of *Eurytemora velox* in the Vita River varied from 7 up to 106 ind/m$^3$ in 1997-1998 and from 30 up to 300 ind/m$^3$ in 2011-2014, depending on the season and on the speed of the water current. The abundance of *E. velox* in this river reached 420 ind/m$^3$ in 2016.

At the Ukrainian part of the Danube Delta, *E. velox* was recorded in the Kiliya (= Chilia) Arm with 10-340 ind/m$^3$ (2011, 2018, 2019) and in the following Kiliya Delta branches: Starostambulskyi (= Stambulsky) with 10-1060 ind/m$^3$ (2009); Ochakovskyi, 10-160 ind/m$^3$ (2009, 2011, 2012); Bystryi, 10-400 ind/m$^3$ (2009, 2010, 2011); Tsyganka, 40 ind/m$^3$ (2009); Vostochnyi, 10-280 ind/m$^3$ (2009, 2011); and some channels and lakes of the Small Tataru and Ermakov Islands, 10-40 ind/m$^3$ (2018). *E. velox* is wide spread also in the saline lagoons of the Kiliya Delta, e.g., Solonyi Kut, 80-130 ind/m$^3$ (2011); and Badika Kut, 40-80 ind/m$^3$ (2011). Finally, we recorded species at the following freshwater bodies of the delta: Anankin Kut Lake, 10-120 ind/m$^3$ (2009, 2011); and the lagoons Potapiv Kut, 120 ind/m$^3$ (2009); Deliukiv Kut, 60-120 ind/m$^3$ (2009, 2011); and Bystryi Kut, 90-150 ind/m$^3$ (2011, 2014).

There are also some new records of *E. velox* in the middle and upper parts of several Ukrainian rivers (fig. 1). In 2000 the species was found in the middle part of the Salgir River (Crimean peninsula); in 2001 and 2014-2018 in some areas of the Horyn River basin (33-50 ind/m$^3$ in the cooling pond of Khmelnitsky nuclear power plant in 2014) and at the Dniester River near Kamenetc-Podolsky. In 2018-2019 the species was discovered in the Olexandriiske water reservoir and

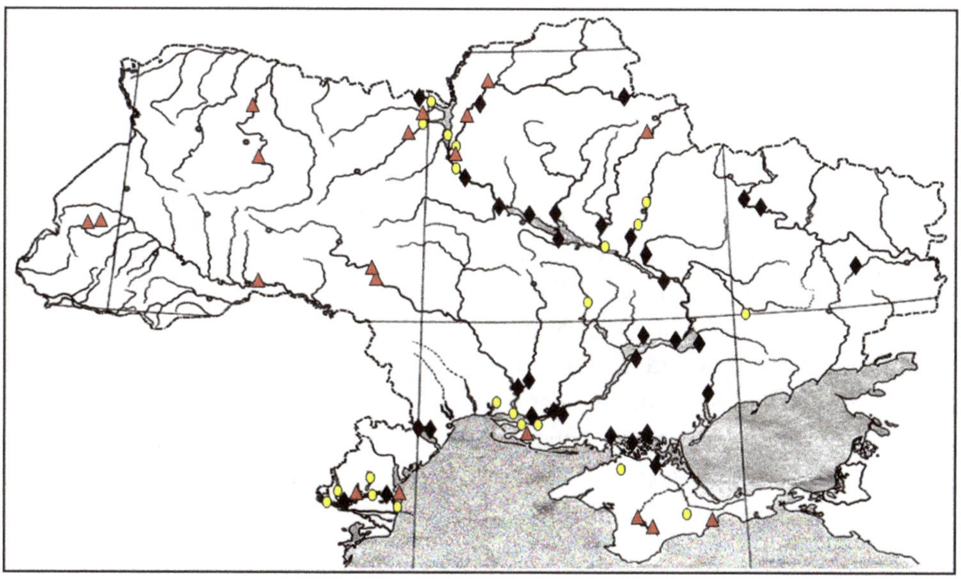

Fig. 1. Map depicting collection sites of *Eurytemora velox* (Lilljeborg, 1853) in fresh and brackish waters in Ukraine. Circles represent sites where samples from existing collections were processed (1953-1981). Triangles represent personal observations (1998-2019). Rhombi represent literature data.

in the Pivdenny Buh River. In the region of the Eastern Carpathian Foothills of Ukraine, *E. velox* was recorded in 2001 in the pond between Drogobych and Boryslav, while according to stored samples from 1976-1978, deposited at IZAN, the species had earlier been absent in that watershed.

Between other records from processed archived samples are: Dnieper River near Kyiv [= Kiev] and Kaniv (1953-2019), Prypiat River (in 1963, 5 ind/m$^3$, 1982, 2000-2003, in 2010-2014, 10 ind/m$^3$), Kahovsky water reservoir (1962), Dnieper-Bug estuarine area (1966), Black Sea Biosphere Reserve (1983), Staro-Krymske water reservoir (1974), Symferopilske water reservoir (2000), Inhuletc River (1973), Holtva River (1956), Vorona River (1956), Desna River (1956, 2017), Psel River (1997-1998, 10-600 ind/m$^3$), Vovcha River (1971), Teteriv River (2000), Stryzhen River (2000), Kuhurluy Lake (1966, 1981), Kytai Lake (1966), Katlabuh Lake (1963, 1981), Kagul Lake (1964), Berezansly Lake (1979), and lower part of Prut River (1965). Some additional old records in the Danube Delta are from the Zavodnitske branch (1966), Zhebriiansky Lagoon (1966), Shabash Kut Lagoon (1965), Taranov Kut Lagoon (1965), Lazorkin Kut Lake (1965) and Poludenoe branch (1965).

## DISCUSSION

The idea of a strong relationship between the level of penetration of invertebrates into continental waters and the adaptation to fresh water on the one hand, and the ancientness of the taxon, on the other, was expressed by Robert Gurney (Gurney, 1931). The family Temoridae includes four genera, of which only one is marine: *Temora* Baird, 1850. The other three are brackish and freshwater genera: *Eurytemora* Giesbrecht, 1881, *Heterocope* Sars, 1863 and *Epischura* Forbes, 1882, of which *Eurytemora* at all events and in all aspects appears to be a relatively recent immigrant into fresh water. This genus is considered as a neo-limnetic form, according to the classification of Starobogatov (1970).

Siewerth (1926) found *Eurytemora velox* in Ukraine for the first time at the mouth of the Seversky Donetc River. Later Markovsky (1928) revealed the presence of the species in the Inhuletc River and considered this species as a remnant of the Post-Pliocene marine basin. The Black Sea area has a complex evolution from subduction to collisions during the closure of the Neo-Tethys Ocean, and was exposed to multiple water-level fluctuations in successive glaciations (Hippolyte et al., 2018). Markovsky assumed that *E. velox*, together with *E. affinis* (Poppe, 1880), *Heterocope caspia* Sars, 1897 and *Calanipeda aquaedulcis* Krichagin, 1873, are adaptive relicts that adapted well to changing environmental conditions and will be extending their range over the relict area into fresh waters. In spite of the fact that subsequent records of *E. velox* in Ukraine were made in the range of marine fluctuations since the Last Glacial Maximum, Sabaneev (1930) also supported the suggestion about future migration of this species up along rivers.

Outside the ancient marine transgressions, *E. velox* were recorded many times in successive years (Taran, 1931; Melnikov 1948). In the Middle Dnieper basin near Kiev, the species was mentioned by many authors (Travyanko & Tseeb, 1967; Radzimovsky & Polishchuk, 1970; Samchishina, 2001; etc.). In the Dnieper tributaries, the Psel, Desna, Prypiat, Holtva and Vorskla Rivers, this species was mentioned by Monchenko (1974). That last-mentioned author concluded that the absence of the species in the upper parts of the rivers indicates relatively recent migration into fresh waters. In the Prypiat River basin, *E. velox* shows high abundance in summer in the cooling pond of the Khmelnitsky nuclear power plant on the Hnylyi Rih River, the inflow of the Horyn River (Protasov et al., 2011).

Elsewhere in the Eastern Europe, the species was also found, e.g., in Hungary (Forró & Gulyás, 1992) and Slovakia (Vranovský, 1994) in 1991 for the first time, and has recently largely expanded its range in those countries (Hudec & Illyová, 1998; Kiss et al., 2015). Simultaneously, in North America, another species of the genus, *Eurytemora carolleeae* Alekseev & Souissi, 2011 (formerly considered part of the *Eurytemora affinis* species complex) is, reportedly, recently invading the Laurentian Great Lakes. It has been in the Great Lakes since the 1950s (Lee, 1999), and is continuously being reported from new inland localities since that time. It was commonly accepted that the species has spread by shipping traffic through the discharge of ballast waters. Furthermore, the species recently is invading also the Gulf of Finland in Europe (Sukhikh et al., 2019). Similar biological invasions of Ponto-Caspian *Eurytemora* species are observed in the Volga River in Russia. Thus, *Eurytemora caspica* Sukhikh & Alekseev, 2013 (also recently separated from the *E. affinis* species complex) had invaded the Volga, Saratov and Kuibyshev Reservoirs on Volga River until the 2010s. The species was next revealed in the Kama Reservoir (2700 km removed from the Caspian Sea) for the first time in 2016, and has become a dominant species in the zooplankton in summertime in most of those reservoirs (Lazareva, 2019).

We have assumed that the recent success of the natural invasion into fresh water by the copepod *E. velox* in Ukraine can be interpreted as a result of a mainly passive route (e.g., transit of undigested eggs in fish's guts, shipping traffic, etc.), however, the active way of range extension, i.e., by actively moving through swimming might also be considered. Obviously, this oligohaline species is able to cope well with new environments and with freshwater conditions, through being equipped with effective adaptations for the required osmoregulation. We have noticed euryoxybiontic abilities of the species in wintertime. Finally, acquiring the ability to survive in fresh water, might be a beginning of speciation for populations entering the fresh waters, since contact between populations inhabiting upper reaches of rivers may become limited. Accordingly, a scenario of allopatric evolution might be starting to develop. Evidently, further research should be conducted to explain and monitor this phenomenon, specifically also through using molecular genetic tools.

## ACKNOWLEDGEMENTS

The Khmelnitsky nuclear power plant cooling pond zooplankton samples were investigated through courtesy of Prof. O. Protasov (IHB AN), and the abundance data were kindly provided by the group of technical hydrobiology. Sampling campaigns at the Danube Delta were funded by Program 6541230 (No. 0118U002287) of the IHB AN. Existing preserved samples were provided by the depository of IZAN, for processing.

## REFERENCES

FORRÓ, L. & P. GULYÁS, 1992. *Eurytemora velox* (Lilljeborg, 1853) (Copepoda, Calanoida). Miscellanea Zoologica Hungarica, **7**: 53-58.

GURNEY, R., 1931. British fresh-water Copepoda, I. Calanoida: i-lii, 1-238. (The Ray Society, London).

HIPPOLYTE, J.-C., A. MUROVSKAYA, YU. VOLFMAN, T. YEGOROVA, O. GINTOV, N. KAYMAKCI & E. SANGU, 2018. Age and geodynamic evolution of the Black Sea Basin: tectonic evidences of rifting in Crimea. Marine and Petroleum Geology, **93**: 298-314.

HUDEC, I. & M. ILLYOVÁ, 1998. *Pleuroxus denticulatus* (Crustacea: Anomopoda: Chydoridae): a new invader in the Danube Basin. Hydrobiologia, **368**: 65-73.

KHARCHENKO, T. A., A. V. LYASHENKO, E. E. ZORINA-SAKHAROVA, V. V. MAKOVSKY, A. V. SHALAEVA & YU. N. VOLIKOV, 2005. Structural and functional parameters of zoocenoses waterbodies and streams of different types (biocenotic diversity, water quality, selfrestoration potential). A. Ukrainian part of Lower Danube. (IHB NAS Ukraine, Kiev).

KISS, A., E. ÁGOSTON-SZABÓ, M. DINKA & A. BERCZIK, 2015. Microcrustacean diversity in the Gemenc-Béda-Karapancsa Floodplain (Danube-Drava National Park, Hungary): rare and alien species. Opuscula Zoologica Budapest, **46**: 183-197.

LAZAREVA, V. I., 2019. Spreading of alien Ponto-Caspian zooplankton species in reservoirs of the Volga and Kama rivers. Russian Journal of Biological Invasions, **3**: 29-52.

LEE, C., 1999. Rapid and repeated invasions of fresh water by the copepod *Eurytemora affinis*. Evolution, **53**: 1423-1434.

LEE, C. & G. GELEMBIUK, 2008. Evolutionary origins of invasive populations. Evolutionary Applications, **1**: 427-448.

LIASHENKO, A. V., S. O. AFANASIEV, K. SANDU, ET AL., 2018. Hydrobiocenoses of the trans-boundary sections of the Ukrainian and Romanian parts of the Danube delta. (Kafedra, Kiev).

MARKOVSKY, YU., 1928. On the zooplankton of inhuletc river. Zbirnyk Pratc Dniprovskoi Biologichnoi Stantcii, **3**: 113-156.

MARKOVSKY, YU., 1955. Fauna of invertebrates of Ukrainian rivers lower parts, conditions of its existence and ways of use. Water bodies of Kilia part of Danube Delta. (AN USSR, Kiev).

MELNIKOV, G., 1948. On the zooplankton of Samara-Dneprovska. Vestnik n.-i. Instituta Hydrobiologii, **8**: 117-121.

MONCHENKO, V., 1974. Gnathostome cyclopoids (Cyclopidae). Fauna of Ukraine, **27**(3): 1-452.

PANDOURSKI, I. & V. EVTIMOVA, 2006. First record of *Eurytemora velox* (Lilljeborg, 1853) (Crustacea: Copepoda: Calanoida) in Iceland with morphological notes. Historia Natur. Bulg., **17**: 39-42.

PARCHUK, G., 1985. Zooplankton of the Soviet part of Danube. Hydrob. Zhurn., **21**: 76-81.

POLISHCHUK, V. V., 1974. Hydrofauna of Lower Danube of Ukraine. (Naukova dumka, Kiev).

POLISHCHUK, V. V., V. S. TRVIANKO, G. D. KONENKO, ET AL., 1978. Hydrobiology and hydrochemistry of right-side of Prydniprovie rivers. (Naukova dumka, Kiev).

PROTASOV, A. A., V. P. SEMENCHENKO, A. A. SILAEVA, ET AL., 2011. Techno-ecosystem of NPP. Hydrobiology, abiotic factors, ecological evaluations. (Institut Hydrobiologii NAN Ukrainy, Kiev).

RADZIMOVSKY, D. O. & V. V. POLISHCHUK, 1970. Plankton of the Pripiat River. (Naukova Dumka, Kiev).

SABANEEV, P. P., 1930. *Eurytemora velox* at the Lower Dnieper. Zbirnyk Pratc Dniprovskoi Biologichnoi Stantcii, **5**: 331-335.

SAMCHISHINA, L. V., 2000. Assumed recent invasion of oligohaline Copepoda *Eurytemora velox* (Lill.) (Copepoda, Calanoida) in the Middle Dnieper River and its large tributaries. Ecologiya Moray, **52**: 52-55.

SAMCHYSHYNA, L. V., 2011. Faunistical overview of calanoid copepods (Crustacea) from continental waters of Ukraine. Vestnik Zoologii, **45**(4): 299-305.

SIEWERTH, M. W., 1926. Zur geographische Verbreitung und Biologie von *Eurytemora velox* (Lill.) und *Laophonte mohammed* Blanch. et Rich. Zool. Anz., **65**(7/8): 201-212.

SPANDL, H., 1926. Wissenschaftliche Forschungsergebnisse aus dem Gebiete der unteren Donau und des Schwarzen Meers. II. Die Susswasser-Microfauna. Arch. Hydrobiol., **26**: 528-604.

STAROBOGATOV, YA. I., 1970. Fauna of molluscs and zoogeographic division of inland water bodies. (Nauka, Leningrad).

SUKHIKH, N., A. SOUISSI, S. SOUISSI, A. C. HOLL, N. V. SCHIZAS & V. ALEKSEEV, 2019. Life in sympatry: coexistence of native *Eurytemora affinis* and invasive *Eurytemora carolleeae* in the Gulf of Finland (Baltic Sea). Oceanologia, **61**: 227-238.

TARAN, M. K., 1931. On spreading of *Eurytemora velox* (Lill.) in Dnieper River. Zbirnyk Pratc Dniprovskoi Biologichnoi Stantcii, **6**: 67-70.

TRAVYANKO, V. S. & YA. YA. TSEEB, 1967. Zooplankton of Upper Dnieper and its basin. (Naukova Dumka, Kiev).

VRANOVSKÝ, M., 1994. *Eurytemora velox* (Lilljeborg, 1853) (Crustacea, Copepoda), a new immigrant in the middle Danube. Biologia, Bratislava, **49**: 167-172.

ZORINA-SAKHAROVA, K., A. LIASHENKO & I. MARCHENKO, 2014. Effects of salinity on the zooplankton communities in the foredelta of Kyliya Branch of the Danube River. Acta Zool. Bulg., **7**: 129-133.

First received 29 October 2019.
Final version accepted 25 November 2019.

[When citing this volume, please refer to *Crustaceana* 93 (2020), 241-547]

*Proceedings of the* Eurytemora *conference, St. Petersburg, 2019*

# REPRESENTATIVES OF THE GENUS *EURYTEMORA* GIESBRECHT, 1881 (CALANOIDA, TEMORIDAE) IN AQUATIC ECOSYSTEMS OF BELARUS

BY

VASIL V. VEZHNAVETS[1,3]) and ANASTASIYA G. LITVINOVA[2,4])

[1]) Scientific and Practical Centre of the National Academy of Sciences of Belarus on Bioresources, Minsk, Republic of Belarus

[2]) State Scientific Establishment «The Polesie Agrarian Ecological Institute of the National Academy of Sciences of Belarus», Brest, Republic of Belarus

## ABSTRACT

The main biological features of two species of the genus *Eurytemora* have been studied: the relict *E. lacustris* (Poppe, 1887) and the alien *E. velox* (Lilljeborg, 1853), which can be found in natural environments on the territory of the Republic of Belarus.

The relict species *E. lacustris* has been found only in two mesotrophic dimictic lakes belonging to the Baltic Sea basin: Volchin (maximum depth 32.9 m) and Vecheliye (35.9 m). *E. velox* has now colonized the main rivers of the south of Belarus and their accessory reservoirs belonging to the basins of the Baltic and Black seas.

*E. lacustris* is characterized by an uneven distribution in the water column and location in deep water layers with temperatures below 13°C. Seasonal and daily vertical migrations have been established for *E. lacustris*. With age, the amplitude of migrations increases, reaching values of more than 10 m of depth for adults. *E. velox* is recorded in the coastal biotopes of watercourses and shallow water bodies.

In the recent two years, the average density of *E. lacustris* in Lake Vecheliye has been established as about 2000 ind./$m^3$, while their percentage in the zooplankton was 6%. In Lake Volchin, its density has been 2-5 times lower in the same period, and there this species' percentage in the plankton did not exceed 2%. The density of *E. velox* varies from 20 to 12 000 ind./$m^3$, their average density is 1631 ind./$m^3$, and their percentual share in the zooplankton does not exceed 1%.

For *E. lacustris*, the body length at all stages of development has been determined. It has been established that males have a greater body length in comparison with females, due to the relatively longer male abdomen, which is not typical for other species of freshwater planktonic copepods.

In Belarus, *E. lacustris* is a bicyclic species, whereas *E. velox* is polycyclic.

[3]) Corresponding author; e-mail: vezhn47@mail.ru

[4]) e-mail: nastya_litvinova_1986@mail.ru

 | DOI:10.1163/9789004465565_006

## RÉSUMÉ

Les principaux traits biologiques de deux espèces du genre *Eurytemora* ont été étudiés : l'espèce relique *E. lacustris* (Poppe, 1887) et l'espèce exotique *E. velox* (Lilljeborg, 1853), qui peuvent être trouvées dans les environnements naturels sur le territoire de la République de Belarus.

L'espèce relique *E. lacustris* a été trouvée seulement dans deux lacs mésotrophes dimictiques appartenant au bassin de la Mer Baltique : Volchin (profondeur maximale 32,9 m) et Vecheliye (35,9 m). *E. velox* a maintenant colonisé les principales rivières du sud de Belarus et leurs retenues d'eau associées appartenant aux bassins de la mer Baltique et de la mer Noire.

*E. lacustris* est caractérisée par une répartition hétérogène dans la colonne d'eau et par sa présence dans les couches d'eaux profondes avec des températures inférieures à 13°C. Des migrations verticales journalières et saisonnières ont été établies pour *E. lacustris*. Avec l'âge, l'amplitude des migrations augmente, atteignant des valeurs de plus de 10 m de profondeur pour les adultes. *E. velox* est rapportée des biotopes côtiers des cours d'eau et des plans d'eaux peu profondes.

Ces deux dernières années, la densité moyenne de *E. lacustris* dans le lac Vecheliye a été établie à environ 2000 ind./m$^3$, tandis que son pourcentage dans le zooplancton était de 6%. Dans le lac Volchin, sa densité était 2-5 fois plus faible dans la même période, et le pourcentage de cette espèce dans le zooplancton ne dépassait pas 2%. La densité de *E. velox* variait de 20 à 12 000 ind./m$^3$, sa densité moyenne étant de 1631 ind./m$^3$, et sa part en pourcentage dans le zooplancton ne dépassait pas 1%.

Chez *E. lacustris*, la longueur du corps à tous les stades de développement a été déterminée. Il a été établi que les mâles ont une longueur du corps en comparaison supérieure à celle des femelles, en raison d'un abdomen mâle relativement plus long, ce qui n'est pas le cas pour d'autres espèces de copépodes planctoniques d'eaux douces.

Dans le Belarus, *E. lacustris* est une espèce bicyclique, tandis que *E. velox* est polycyclique.


## INTRODUCTION

The genus *Eurytemora* Giesbrecht, 1881 includes between 20 and 30 species (Borutsky, 1960; Kos, 1977; Costello et al., 2001; Dodson, 2010). The considerable divergence in indications of that number could be explained by frequent changes in taxonomy of already known forms, and the description of some new ones, utilizing the results of genetic studies: for example, *E. carolleeae* Alekseev & Souissi, 2011. The wide range of tolerance towards various environmental parameters (salinity, temperature) is characteristic of representatives of this genus, which also influences their broad morphological variability. The genus *Eurytemora* can be found in waters of the widest range of salinity among all known copepod genera (Dodson et al., 2010). It is currently supposed that the forms of this genus have spread from coastal areas of the Northern Hemisphere and so are Holarctic (Borutsky et al., 1991). Nowadays their distribution areas have extended from subtropical to subarctic latitudes, with an increase in species diversity toward the north.

There are two species of this genus in Belarus: the relict *E. lacustris* (Poppe, 1887), that remained in the area since the times of glacial transgressions, and the alien *E. velox* (Lilljeborg, 1853), of Ponto-Caspian origin. Belarusian studies have repeatedly confirmed the presence of these two species in the water bodies of the country (Petrovich, 1956; Mitrakhovich & Lyakhnovich, 1979; Vezhnovets et al., 1996; Karatayev, 1999; Vezhnavets, 2004, 2009). The specifics of their distribution borders in basins of the main rivers of Belarus require further clarification, due to a general lack of data on the dynamics of their density, dimensional characteristics, and the main ecological and biological features of these species, in Belarus as well as in neighbouring regions.

There have been four *Eurytemora* species known from the freshwater bodies of Europe for a long time: *E. affinis* (Poppe, 1880), *E. lacustris*, *E. velox* and *E. grimmi* (Sars G. O., 1897) (Illies, 1978). Now it is necessary to add to the list of species for Europe the recently validated *E carolleeae* and another new species described from the Volgograd Caspian region, *E. caspica* Sukhikh & Alekseev, 2013. The four previously mentioned species are known to also inhabit the regulated Dnieper River in Ukraine. As the process of colonization of freshwater reservoirs continues, invasion of these species into Belarus can be expected. Due to Belarus' geographic location, invasive species are attracting additional interest for study. This region is both inhabited and transited by alien species, as it covers the central water corridor for the invasive penetration of alien species from the Black Sea-Caspian Basin into Central and Western Europe.

In Belarus, there currently are two known species of the genus *Eurytemora*, which differ in geographical origin: one is alien, the other is a relict. For Belarus, the only existing data are related to their general distribution, while the other aspects of their biology and ecology are not yet studied. Long-term investigations of the start of their populations in reservoirs of Belarus were not conducted either. Therefore, comparative biological study of these two species is essential, especially as they are representatives of two faunas very different in origin.

The purpose of this study was to identify features of the biology of the relict and the alien species of the genus *Eurytemora*, to assess the state of their populations in Belarus, and to determine trends in their numbers and distribution in todays conditions.

## MATERIAL AND METHODS

The main sampling of *Eurytemora lacustris* was conducted in the mezotrophic dimictic lakes Volchin (maximum lake depth 32.9 m) and Vechelye (35.9 m). Samples were collected in the last decade of July, at noon. Life cycles and seasonal migrations were determined for 2014 and 2015. A plankton net with a mouth diameter of 25 cm and a mesh size of the filtering cone of 100 $\mu$m was used. The whole water column was sampled, from a surface to the bottom, but fractionally, in steps of 5 m.

To quantify the population localization of *E. lacustris*, the average depth of the "core population" at a certain time of day was calculated as:

$$H = \frac{\sum(h \times n)}{\sum n},$$

where $H$ is the average depth of the "core population"; $h$ the average depth of the sampled layer; and $n$ the density of individuals in this layer.

Plankton samples of *E. velox* were collected in water bodies of the south of Belarus in the summers of 2007, 2012 and 2013. Sample collection for life cycle estimation was carried out monthly. In the rivers we surveyed the current zone and various sites along the banks, in the lakes, both the littoral and the pelagic zone. Quantitative samples were collected via filtration of 50 liters of water through a net with a mesh size of 45 $\mu$m, qualitative ones by dragging a net of mesh size 100 $\mu$m. Concurrently, we measured habitat parameters: temperature, pH, oxygen content, water conductivity, i.e., the main hydrochemical characteristics, using equipment of Hanna Instruments (see Appendix).

The quantitative processing of zooplankton samples was carried out in a count camera of Bogorov under a stereo research microscope MBS-10 with 8 $\times$ 4 magnification power. For morphological analyses, we used a compound light microscope with 600 $\times$ magnification power. We identified nauplii (developmental stages NI-NVI) based on the number of bristles on the first antenna of a nauplius and on the "arms" at the back end of the body (Sazhina, 1985). We identified the copepodite stages (CI-CVI) based on the number of pairs of appendages and the number of somites of the body. Body length was defined as the distance from the tip of the cephalothorax to the ends of the furcal branches, with an accuracy of $\pm$4 $\mu$m (fig. 1). MS Excel was used for the processing of statistical data.

## RESULTS AND DISCUSSION

### Distribution

*Eurytemora lacustris* was recorded in only two lakes, belonging to the basin of the western Dvina River: Lake Volchin (Myadelsky district, Minsk Region) and Lake Vechelye (Ushachsky district, Vitebsk Region) (fig. 2). *E. velox* appeared to inhabit the main rivers of the south of Belarus, belonging to the catchment basins of the Baltic and Black seas. The highest frequency of occurrence was noted in the River Pripyat, where the species was found throughout its Belarusian sites, from of Mikashevichi in the west to Narovlia in the southeast. *E. velox* was not found in the tributaries of the Pripyat (with one exception: the River Pina) or in the main course of the Dnieper (with as an exception: the Sozh River).

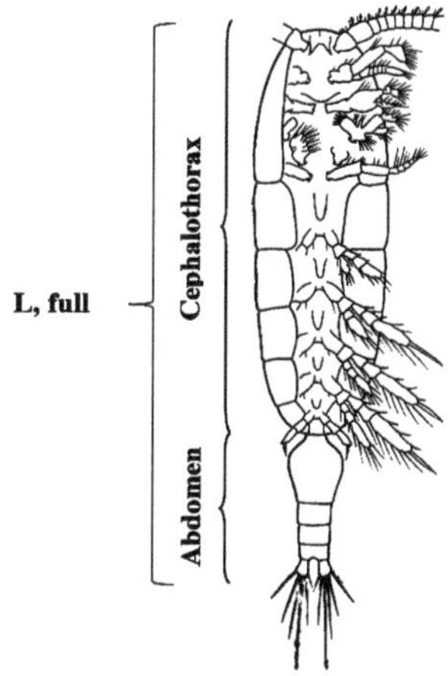


Fig. 1. Schematic for determining the length of the body of copepodites of species of the copepod order Calanoida.

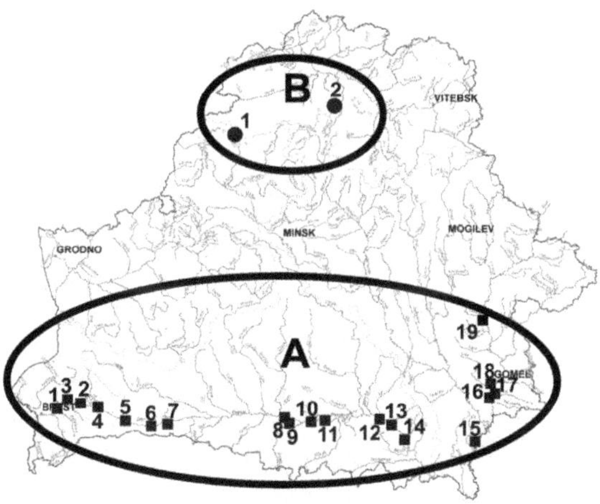


Fig. 2. Sites from which the *Eurytemora* species have been reported that are known from the water bodies of Belarus. A, *Eurytemora velox* (Lilljeborg, 1853): 1, Mukhavets River, in the city of Brest; 2, Mukhavets River, near village Bulkovo, main stream; 3, Mukhavets River, quarry near village of Bulkowo; 4, Dnieper-Bug Canal, near city of Kobrin, main stream; 5, Dnieper-Bug Canal, near village of Vygoda, main stream; 6, Dnieper-Bug Canal, near village of Duboy, main stream; 7, Pina River, in city of Pinsk, thermal power station, both the main stream and backwater in the port; 8, Pripyat River, near urban village of Mikashevichi, in the main river, at the entrance to Mikashevichskiy Channel; 9, Pripyat River, near city of Turov, old canal; 10, Pripyat River, near city of Petrikov, backwater; 11, Pripyat River, Lake Chernetskoe, near city of Petrikov; 12, Pripyat River, near village of Kastsyukovichy, main stream and backwater; 13, Pripyat River, in city of Mozyr, port, main stream and backwater; 14, Pripyat River, in city of Naroulya, main stream; 15, Sozh River, in city of Loew, backwaters; 16, Sozh River, Uzkoe Lake, in village of Tchonki; 17, Sozh River, in village of Tchonki, main stream; 18, Sozh River, in city of Gomel; 19, Sozh River, in village of Ippolitovka, the old bed. B, *E. lacustris* (Poppe, 1887): 1, Lake Volchin; 2, Lake Vechelye.

## Density of populations

The average annual density of *E. lacustris* in the watercolumn of Lake Vechelye comprised 1939 ind./m$^3$ (minimum: 654 ind./m$^3$; maximum: 3772 ind./m$^3$), which was 6.7% of the zooplankton and 12.1% of total Copepoda. The density of *E. lacustris* in Lake Volchin was 2-5 times as low, and there its proportion in the plankton density did not exceed 2%.

The density of *E. velox* in the studied water bodies was measured as 20-12 000 ind./m$^3$. The average density at all studied sites was: in 2007, 1224 ind./m$^3$; in 2012, 2028 ind./m$^3$; and in 2013, 1642 ind./m$^3$. However their percentual proportion in the zooplankton did not exceed 1%. The proportion of this species was higher in the rivers of the Baltic Sea basin, than in the rivers belonging to the Black Sea basin.

Data on the abundance of *E. velox* in water bodies of Belarus are comparable with data for other freshwater habitats: e.g., the water bodies of Ukraine: 50-300 ind./m$^3$ (Polishchuk, 1987); Moldova: from several individuals to 6000 ind./m$^3$ (Alykova, 2015); Russia (the Rybinsk reservoir): 114 ind./m$^3$ (Kuzin, 1972); whereas in salty and brackish waters, the density of *E. velox* may exceed 1 000 000 ind./m$^3$ (Thiéry et al., 1995; Quintana et al., 1998).

## Biotope preference and distribution in water bodies

*Eurytemora lacustris*.— Features of a seasonally varying spatial distribution in the water (fig. 3) are typical for *E. lacustris* (cf. Mitrakhovich & Lyakhnovich, 1979).

Temperature and dissolved oxygen content seem to be the limiting factors for *E. lacustris*' distribution (Vezhnavets & Litvinova, 2015b). The core of the population was located below 10 m at a temperature range of 2 to 13°C, which confirms that the species is cold-stenothermic. The average annual depth of immersion for a population was 14.6 m. Younger stages occupied shallower depths, the older ones deeper strata: the average annual depth of their stay was for nauplii 11.5 m, for the copepodites that was 19.9 m.

We observed daily vertical migrations of *Eurytemora lacustris*, i.e., after sunset, the density of individuals rises in the higher strata, whereas in the afternoon they descend to the deeper layers (Vezhnavets & Litvinova, 2015b). The main movements take place in the low-temperature layers of the water. With increasing age of the animals, their average daily depth of dwelling and the amplitude of their migrations also increased (table I).

The vertical daily distribution of the crustaceans was different in each year of our study, and appeared to be correlated with the oxygen concentration in the various sublayers of the water (fig. 4).

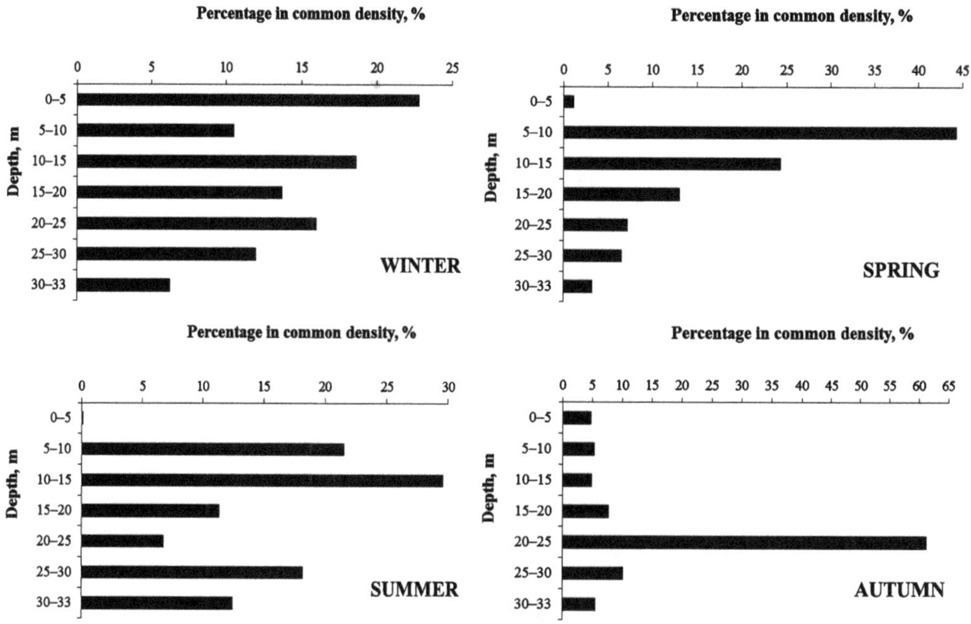


Fig. 3. Changes of daytime vertical distribution of *Eurytemora lacustris* (Poppe, 1887) in the different seasons of the year.

*Eurytemora velox.*— In river ecosystems, the density of *E. velox* was 4.3 times as high in verdant zones along the banks or shores, as it was in zones in the current (fig. 5).

In non-flowing water bodies, the density of *E. velox* was 8.5 times as high in littoral zones as in pelagic zones (fig. 6). The preference for near-shore biotopes probably allows *E. velox* to avoid the competition with pelagial representatives of family Diaptomidae occupying the same body of water. Age differences in the spatial distribution, depending on the biotope type, are characteristic of the species. Nauplii inhabit mainly sites in open water, and copepodites are found in zones with plants, as later stages and adult individuals probably use macrophytes as a refuge from predators.

TABLE I

The amplitude of migrations (A, m) and the average daily depth of occurrence (B, m) for the developmental stages of *Eurytemora lacustris* (Poppe, 1887)

| Index | Developmental stage | | | | | | | | | | | |
|---|---|---|---|---|---|---|---|---|---|---|---|---|
| | NI | NII | NIII | NIV | NV | NVI | CI | CII | CIII | CIV | CV | CVI |
| A | 1.3 | 2.7 | 1.7 | 1.7 | 7.2 | – | 12.6 | – | 5.4 | 11.2 | 11.5 | 13.3 |
| B | 10.9 | 9.7 | 10.7 | 11.4 | 13.1 | – | 12.6 | – | 14.1 | 19.2 | 21.3 | 20.9 |

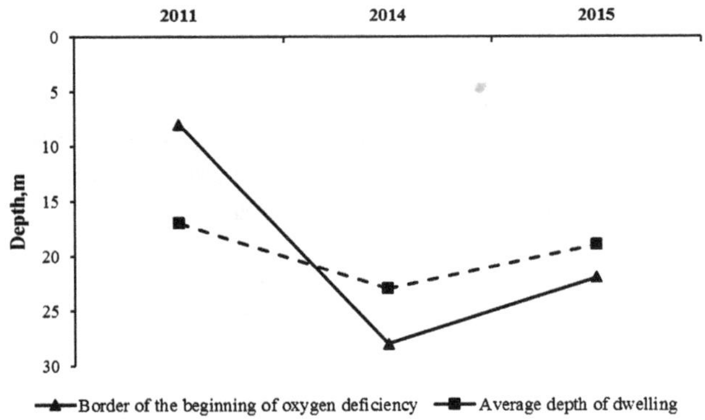


Fig. 4. Changes in *Eurytemora lacustris* (Poppe, 1887) average daily depth of dwelling, and the depth of the oxygen deficiency (oxygen content of less than 2 mg/l).

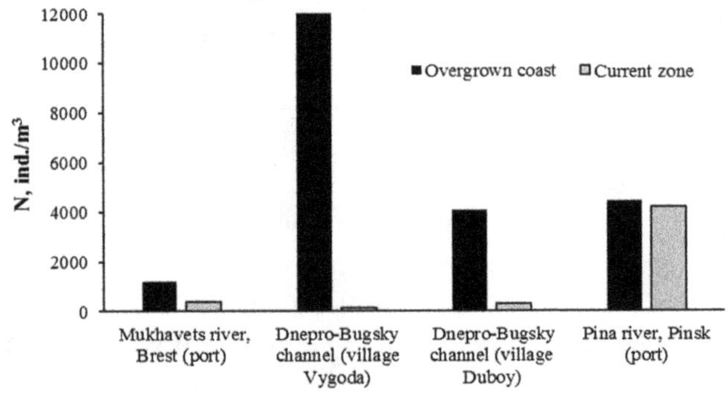


Fig. 5. Density ($N$, ind./m$^3$) of *Eurytemora velox* (Lilljeborg, 1853) in the various biotopes of lotic water systems.

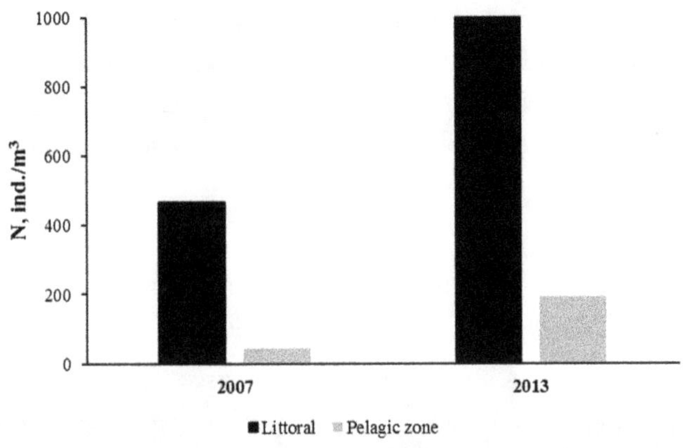


Fig. 6. Density ($N$, ind./m$^3$) of *Eurytemora velox* (Lilljeborg, 1853) in the various zones of lentic water systems.

Thus, in Belarusian water bodies, *E. velox* inhabits mainly littoral biotopes.

## Body dimensions and age characteristics

For the first time, we have, in this study, determined the dimensional characteristics of all developmental stages *E. lacustris* (table II). The body length of adult individuals of *E. lacustris* in the lakes of Belarus is comparable to the sizes found for the lakes of Russia (1.000-1.400 mm) and some other countries of Europe: Poland (♀♀: 0.923-1.425 mm; ♂♂: 1.245-1.290 mm) (Wolska & Piasecki, 2007) and Sweden (♀♀: 1.500-1.640 mm; ♂♂: 1.200 mm) (Northkote & Hammar, 2006). Larger sizes of females are typical for the majority of freshwater calanoid Copepoda (Borutsky et al., 1991). However, the average body length of males of *E. lacustris* was 0.021-0.056 mm longer, because of their greater abdomen length, which is typical for this species as we have shown previously (Vezhnovets & Litvinova, 2015a).

The body sizes of all developmental stages of *E. velox*, from nauplii to mature individuals, were determined for the populations in the freshwater reservoirs (table III). In general, the sizes were smaller than described in the literature for specimens from brackish waters (Gurney, 1931). Similarly to most other species of copepods, the body length of females was larger in *E. velox*.

TABLE II

Body length ($L$), range and average $\pm \sigma$, of *Eurytemora lacustris* (Poppe, 1887) specimens of all developmental stages from Lake Vechelye

| Developmental stage | Sample size, quantity of individuals | $L$, min.-max. (mm) | $L$ av. $\pm \sigma$ |
|---|---|---|---|
| NI | 50 | 0.120-0.136 | 0.130 ± 0.005 |
| NII | 178 | 0.136-0.184 | 0.162 ± 0.010 |
| NIII | 157 | 0.184-0.224 | 0.199 ± 0.013 |
| NIV | 130 | 0.232-0.280 | 0.250 ± 0.016 |
| NV | 42 | 0.280-0.344 | 0.311 ± 0.019 |
| NVI | 19 | 0.352-0.408 | 0.376 ± 0.019 |
| CI | 65 | 0.442-0.546 | 0.508 ± 0.026 |
| CII | 122 | 0.546-0.728 | 0.648 ± 0.038 |
| CIII | 219 | 0.702-0.910 | 0.812 ± 0.043 |
| CIV | 411 | 0.832-1.118 | 0.960 ± 0.050 |
| CV (♂) | 122 | 1.040-1.300 | 1.164 ± 0.052 |
| CV (♀) | 183 | 0.988-1.248 | 1.099 ± 0.082 |
| CVI (♂) | 271 | 1.092-1.534 | 1.340 ± 0.077 |
| CVI (♀) | 415 | 1.092-1.534 | 1.308 ± 0.069 |

TABLE III

Body length ($L$), average $\pm\ \sigma$ and range, of *Eurytemora velox* (Lilljeborg, 1853) specimens of all developmental stages

| Developmental stage | Belarus, original data | | | England (Gurney, 1931), $L$ av | |
|---|---|---|---|---|---|
| | Sample size, quantity of individuals | $L$ av. $\pm\ \sigma$ | $L$, min.-max. (mm) | ♀ | ♂ |
| NI | 253 | 0.109 ± 0.008 | 0.104-0.128 | – | |
| NII | 535 | 0.138 ± 0.008 | 0.128-0.152 | 0.162 | |
| NIII | 230 | 0.168 ± 0.010 | 0.152-0.184 | 0.208 | |
| NIV | 207 | 0.209 ± 0.012 | 0.184-0.240 | 0.243 | |
| NV | 129 | 0.258 ± 0.013 | 0.240-0.280 | 0.288 | |
| NVI | 166 | 0.317 ± 0.017 | 0.280-0.360 | 0.322-0.358 | |
| CI | 417 | 0.468 ± 0.031 | 0.450-0.525 | 0.560 | 0.490 |
| CII | 367 | 0.580 ± 0.041 | 0.550-0.650 | 0.746 | 0.667 |
| CIII | 274 | 0.718 ± 0.055 | 0.625-0.775 | 0.953 | 0.850 |
| CIV | 228 | 0.896 ± 0.070 | 0.750-1.000 | 1.240 | 1.070 |
| CV | ♀♀ 52 ♂♂ 75 | ♀♀ 1.116 ± 0.080 ♂♂ 0.997 ± 0.051 | ♀♀ 0.975-1.250 ♂♂ 0.925-1.125 | 1.570 | 1.330 |
| CVI | ♀♀ 86 ♂♂ 100 | ♀♀ 1.331 ± 0.108 ♂♂ 1.113 ± 0.064 | ♀♀ 1.125-1.525 ♂♂ 1.000-1.175 | 1.808 | 1.497 |

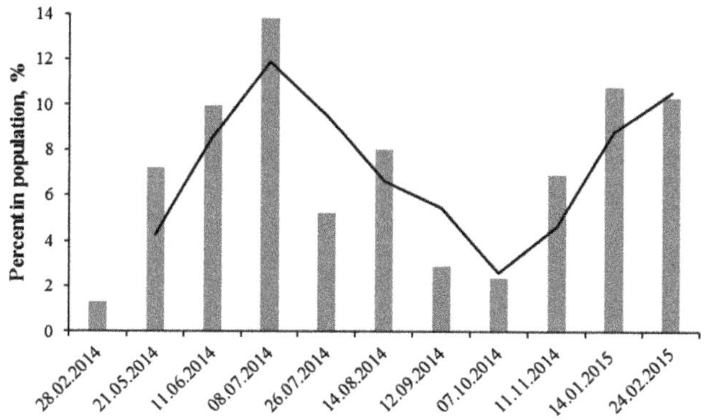


Fig. 7. Changes in the share of CVI copepodites, i.e., adults, in the population of *Eurytemora lacustris* (Poppe, 1887).

## Features of life cycle and fecundity

*E. lacustris* is a bicyclic species, due to two prominent maxima in numbers of mature individuals (fig. 7) and nauplii of the first stage (fig. 8), noticeable in the yearly population dynamics. The development of both generations lasts for about six months.

The individual fecundity of *E. lacustris* in the lakes of Belarus varies from 8 to 48 eggs, with an average of 13 eggs per female. These values were lower in comparison to the values from other sites, in which the number of eggs in an egg sac went as high as 32-56, with an average of 44 eggs (Maier et al., 2011). The measured egg sizes of this species were as follows: average diameter, $100.53 \pm 1.73$ $\mu$m, with minimum, 88 $\mu$m and maximum, 112 $\mu$m.

Four peaks were recorded in the population dynamics of *E. velox* from April to December: a single, relatively weak one in the middle of July, and 3 much more

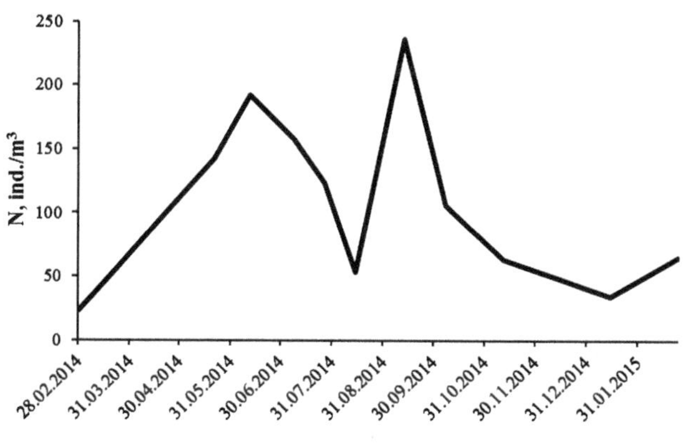


Fig. 8. Seasonal changes in the density of *Eurytemora lacustris* (Poppe 1887) nauplii of the first stage (NI).

TABLE IV

Interannual changes of *Eurytemora lacustris* (Poppe, 1887) density ($N$, ind./m$^3$) and percentual representation in the zooplankton (%) in two Belarusian lakes

| Year | Lake Volchin | | Year | Lake Vechelye | |
|---|---|---|---|---|---|
| | $N$, ind./m$^3$ | % | | $N$, ind./m$^3$ | % |
| 1950 | 4400[a] | – | 1977 | 3500[b] | – |
| 1992 | 1632 | 1.83 | 2002 | 2877 | 2.68 |
| 2002 | 1293 | 1.08 | 2005 | 647 | 0.50 |
| 2011 | 435 | 0.80 | 2011 | 216 | 0.63 |
| 2014 | 762 | 2.31 | 2014 | 1178 | 6.04 |
| 2015 | 910 | 1.78 | 2015 | 2172 | 6.60 |

[a]Data from Petrovich (1956).
[b]Data from Mitrakhovich & Lyakhnovich (1979).

prominent peaks from August to October. The density dynamics of separate stages allows to state that *E. velox* is a polycyclic species (fig. 9). The period of reproduction is extended and takes place in the combined season of summer-autumn. The process of reproduction thus takes place continuously, with overlapping generations.

The individual fecundity was 2-81 eggs, with an average number of 39 eggs per egg sac. This corresponds to the literature data from other freshwater habitats (from 18.2 to 49 eggs, with on average 41.5 eggs) (Viyushkova, 1971, 1985). The sizes of *E. velox* eggs vary in the range of 82-92 $\mu$m, with an average size of 87.1 $\mu$m.

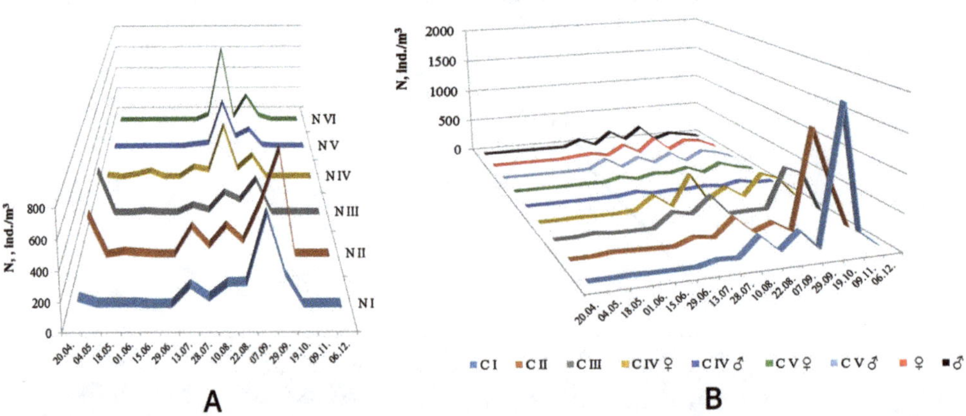


Fig. 9. Seasonal dynamics in the density of: A, the naupliar stages NI-NVI ($N$, ind./m$^3$); and, B, the copepodite stages CI-CVI; of *Eurytemora velox* (Lilljeborg, 1853).

## The status of the populations according to long-term data

Despite interannual fluctuations and the growth of numbers observed in the last 2 years, available long-term materials indicate a gradual decrease in *E. lacustris* density in both inhabited lakes (table IV). This process is more advanced in Lake Volchin, where in about 20 years (1992-2015) the density of these crustaceans reduced approximately to 50%, and calculated since 1950, it reduced to less than 25%. The reasons for this reduction are the species' preferences for dwelling in the hypolimnion, the cold water stenothermy of the species, its high demands in water quality, and the lack of any resting developmental stages, which together prevent *E. lacustris* from inhabiting other mesotrophic lakes, which would otherwise be fitting.

Such decrease in numbers of populations under the influence of eutrophication and pollution can lead to extinction of the species from the fauna of Belarus, something that is already observed in some lakes of the nearby countries, e.g., Lithuania, Latvia, Poland and Germany (Arbačiauskas & Kalytytė, 2010; Maier et al., 2011).

In Belarus, *E. velox* inhabits water bodies with a wide range of trophic niches. There are no long-term observations for this species. The link between the water quality of the studied points and the density of the alien *Eurytemora* is illustrated in fig. 10. It reflects an increase in crustacean density with a decline in water quality and the preference of *E. velox* for moderately polluted (i.e., the IIIrd quality class) waters.

*E. velox* is eurybiotic, so it successfully survives in unfavourable conditions at different stages of development (Lowndes, 1930; Reid & Saunders, 1986). Its number increases in reservoirs with low water quality, which allows us to predict its further spreading, and the subsequent colonization of new habitats.

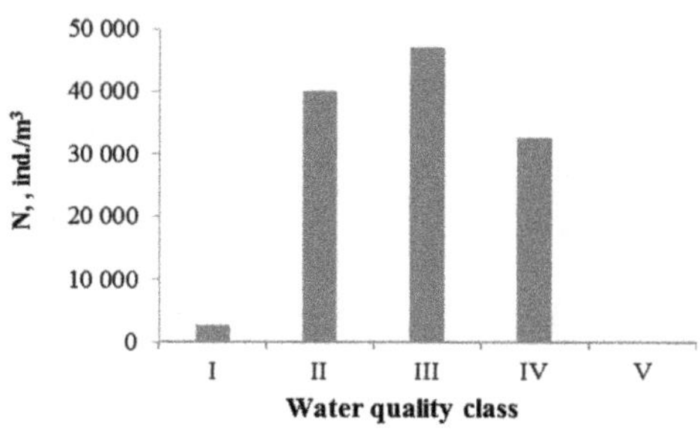


Fig. 10. The dependence of the density of *Eurytemora velox* (Lilljeborg, 1853) ($N$, ind./$m^3$) on water quality.

ACKNOWLEDGEMENTS

This work has been performed with a partial support from the Belarusian Republican Foundation for Fundamental Research: contracts B18MSIC-016 and B17MY-033.

The authors thank Iola Ejsmont-Karabin, Arseniy Volnisty, Andrej Nesterovich and Polina Ignatova for their assistance in the article's preparation.

## APPENDIX

TABLE A-I

The content of some ions (in mg/l) in the water of lakes with *Eurytemora lacustris* (Poppe, 1887)

| Date | Lake | | $PO_4^{3-}$ | $NO_3^-$ | $Fe^{2+}$ | $NH_4^+$ | $Cl^-$ | $Ca^{2+}$ | $Mg^{2+}$ |
|---|---|---|---|---|---|---|---|---|---|
| 24 July 2013 | Volchin | Surface | 0.10 | 1.5 | 0.09 | 0.05 | 0.09 | 89.0 | 14.0 |
| | | Bottom | 0.03 | 2.6 | 0.07 | 0.20 | 0.10 | 88.0 | 15.0 |
| 27 July 2013 | Vechelye | Surface | 0.03 | 1.4 | 0.03 | 0.12 | 0.21 | 98.0 | 14.0 |
| | | Bottom | 0.04 | 1.9 | 0.12 | 0.10 | 0.18 | 97.0 | 15.0 |

TABLE A-II

Hydrochemical parameters of water (ion content in mg/l) at the surveyed *Eurytemora velox* (Lilljeborg, 1853) sites (*N* corresponds to the numbers in fig. 2)

| *N* | *T* (°C) | pH | TDS ($\mu$S/cm) | $O_2$ | $Cl^-$ | $NO_3^-$ | $Fe^{2+}$ | $Ca^{2+}$ | $Mg^{2+}$ | $NH_4^+$ | $PO_4^{3-}$ |
|---|---|---|---|---|---|---|---|---|---|---|---|
| 1 | 21.0 | 7.9 | 130 | 5.9 | 0.06 | 1.8 | 0.28 | 135 | 6.0 | 0.63 | 0.18 |
| 2 | 21.0 | 8.5 | 206 | 9.0 | 0.14 | 2.2 | – | 167 | 7.0 | – | 0.07 |
| 3 | 21.1 | 7.7 | 214 | 6.4 | 0.07 | 1.2 | – | 197 | 11.0 | – | 0.06 |
| 4 | 23.2 | 7.9 | 153 | 6.9 | 0.09 | 0.8 | 0.06 | 144 | 3.0 | 0.68 | 0.12 |
| 5 | 21.8 | 7.8 | 192 | 6.9 | 0.19 | 1.0 | – | 196 | 13.0 | – | 0.64 |
| 6 | 19.2 | 7.4 | 156 | 4.2 | 0.1 | 3.5 | 0.38 | 143 | 8.0 | 0.66 | 0.00 |
| 7 | 21.8 | 7.7 | 187 | 6.6 | – | – | – | – | – | – | – |
| 8 | 23.7 | 7.7 | 211 | 6.5 | 0.14 | 2.2 | 0.23 | 167 | 7.0 | 0.53 | 0.07 |
| 9 | 21.8 | 7.8 | 205 | 6.8 | 0.09 | 2.3 | 0.67 | 189 | 9.0 | 0.69 | 0.26 |
| 10 | 22.0 | 7.6 | 171 | 4.3 | 0.1 | 2.6 | 1.21 | 141 | 8.0 | 0.5 | 0.16 |
| 11 | 23.0 | 7.8 | 195 | 5.0 | 0.07 | 1.7 | 0.94 | 188 | 9.0 | 0.76 | 0.04 |
| 12 | 23.3 | 7.5 | 254 | 6.7 | 0.05 | 2.2 | 2.07 | 237 | 6.0 | 1.28 | 0.48 |
| 13 | 22.8 | 7.4 | 149 | 6.5 | 0.15 | 0.00 | 2.26 | 109 | 11.0 | 2.5 | 0.05 |
| 15 | 23.2 | 7.6 | 184 | 7.2 | 0.19 | 1.0 | 2.03 | 171 | 9.0 | 1.28 | 0.64 |
| 16 | 24.3 | 7.5 | 182 | 5.6 | 0.09 | 2.3 | 0.44 | 172 | 11.0 | 0.62 | 0.26 |
| 17 | 24.2 | 7.5 | 187 | 7.2 | 0.06 | 0.5 | 1.82 | 168 | 5.0 | 1.22 | 0.04 |

## REFERENCES

ALEKSEEV, V. R. & A. SOUISSI, 2011. A new species within the *Eurytemora affinis* complex (Copepoda: Calanoida) from the Atlantic coast of USA, with observations on eight morphologically different European populations. Zootaxa, **2767**: 41-56.

ARBAČIAUSKAS, K. & D. KALYTYTĖ, 2010. Occurrence and interannual abundance variation of glacial relict calanoids *Limnocalanus macrurus* and *Eurytemora lacustris* in Lithuanian lakes. Acta Zoologica Lituanica, **20**: 61-67.

BORUTSKY, E. V., 1960. Guides to the identification of free-living freshwater Copepoda of the USSR and the adjacent countries by fragments in intestines of fishes: 1-117. (Academy of Sciences of the USSR Publishing House, Moscow).

BORUTSKY, E. V., L. A. STEPANOVA & M. S. KOS, 1991. Guides to the identification of freshwater Calanoida of the USSR: 1-504. (Science, Leningrad).

COSTELLO, M. J., C. EMBLOW & R. WHITE, 2001. European register of marine species: a checklist of the marine species in Europe and a bibliography of guides to their identification: 1-463. (Collection Patrimoines Naturels; Muséum national d'Histoire naturelle, Paris).

DODSON, S. I., D. A. SKELLY & C. E. LEE, 2010. Out of Alaska: morphological diversity within the genus *Eurytemora* from its ancestral Alaskan range (Crustacea, Copepoda). Hydrobiologia, **653**: 131-148.

ILLIES, J. (ed.), 1978. Limnofauna Europaea: 1-532. (Gustav Fischer Verlag, Stuttgart, New York).

KARATAYEV, A. Y. & E. N. MAKRITSKAYA, 1999. Zooplankton of the Narochansky region lakes. In: Results and prospects of hydroecological researches: materials of the international conference on water ecosystems: 108-113. (Belarusian State University, Minsk).

KOS, M. S., 1977. Species of the genus *Eurytemora* (Copepoda, Calanoida) northern part of the Pacific Ocean: taxonomy, distribution, variability. In: Researches of the seas fauna. Sea plankton (taxonomy and faunistiks): issue of scientific works: 20-53. (Leningrad).

KUZIN, B. S., 1972. Rybinsk reservoir and its life: 1-364. (Science, Leningrad).

LOWNDES, A. G., 1930. Some fresh-water calanoids: direct observation and indirect deduction. J. of Ecology, **18**: 151-155.

MAIER, G., B. SPETH, W. ARP, M. BAHNWART & P. KASPRZAK, 2011. New records of the rare glacial relict *Eurytemora lacustris* (Poppe, 1887) (Copepoda; Calanoida) in atypical lake habitats of northern Germany. J. Limnol., **70**: 145-148.

MITRAKHOVICH, P. A. & V. P. LYAKHNOVICH, 1979. Characteristic of *Eurytemora lacustris* (Poppe) population in the lake Vechelye of the Belarusian SSR. Bulletin of the BGU of V. I. Lenin, **2**: 39-44.

NORTHKOTE, T. G. & J. HAMMAR, 2006. Feeding ecology of *Coregonus albula* and *Osmerus eperlanus* in the limnetic waters of Lake Mälaren, Sweden. Boreal Environment Research, **11**: 229-246.

PETROVICH, P. G., 1956. Zooplankton species composition of fishery lakes of the BSSR western regions. Scientific Notes BGU of V. I. Lenin, **26**: 3-39.

POLISHCHUK, V. V., 1987. Zooplankton of river estuaries, flowing into the Kiev reservoir. Hydrobiological Journal, **23**: 70-81.

QUINTANA, X. D., F. A. XOMIN & R. MORENO-AMICH, 1998. Nutrient and plankton dynamics in a Mediterranean salt marsh dominated by incidents of flooding. J. of Plankton Research, **20**: 2109-2127.

REID, J. W. & J. F. SAUNDERS, 1986. The distribution of *Mesocyclops aspericornis* (Von Daday) in South America. J. of Crustacean Biol., **6**: 820-824.

SAZHINA, L. I., 1985. Nauplii of numerous species of pelagic Copepoda of the world ocean. Key to the identification: 1-237. (Scientific Thought, Kiev).

THIÉRY, A., D. DEFAYE & C. MARTIN, 1995. Consequences of exceptional precipitation on the brine shrimp *Artemia* and copepod populations in the saltworks of Sète-Villeroy (Languedoc, France). Crustaceana, **68**: 904-909.

VEZHNAVETS, V. V., 2009. Invasive species of Crustacea in water ecosystems of Belarus. The X congress of Hydrobiological Society at Russian Academy of Sciences: theses of reports of international scientific and practical conference: 65-66. (Dalnauka, Vladivostok).

VEZHNAVETS, V. V., 2004. Species of the genus *Eurytemora* (Copepoda, Calanoida) in water bodies of Belarus. In: Dynamics of fauna biological diversity, problems and prospects of steady use and protection of fauna of Belarus: theses of reports IX zoological scientific conference: 187-189. (Institute of zoology of NAN of Belarus, Minsk).

VEZHNAVETS, V. V., G. A. GALKOVSKAYA & V. V. ARAPOV, 1996. Pelagic zone fauna of lakes of the Belarusian Poozerje. In: Conservation of biodiversity of the Belarusian Poozerje: theses of regional scientific and practical conference reports: 101-102. (Vitebsk State University, Vitebsk).

VEZHNAVETS, V. V. & A. G. LITVINOVA, 2015a. Dimensional characteristic of *Eurytemora lacustris* (Poppe, 1887) in the lakes of Belarus. Reports of National Academy of Sciences of Belarus, **59**: 78-82.

VEZHNAVETS, V. V. & A. G. LITVINOVA, 2015b. Seasonal changes of daily vertical distribution of *Eurytemora lacustris* (Poppe, 1887) in the lake of Vechelye. In: Zoological readings — 2015: Proceedings of the international scientific-practical conference: 53-55. (Grodno State University, Grodno).

VYUSHKOVA, V. P., 1971. About Copepoda fecundity in the Volgograd reservoir. Hydrobiological J., **50**: 1580-1582.

VYUSHKOVA, V. P., 1985. Main species of zooplankton of the Volgograd and Saratov reservoirs. Hydrobiological J., **21**: 10-14.

WOLSKA, M. & W. G. PIASECKI, 2007. Seasonality of zooplankton changes phenomena observed in the estuarine part of the Oder River. Limnological Review, **2**: 117-121.

First received 6 September 2019.
Final version accepted 29 November 2019.

[When citing this volume, please refer to *Crustaceana* 93 (2020), 241-547]

*Proceedings of the* Eurytemora *conference, St. Petersburg, 2019*

# ABOUT THE SYSTEMATICS OF PALAEARCTIC *EURYTEMORA* (COPEPODA, CALANOIDA) BASED ON MORPHOLOGICAL ANALYSIS, WITH FOCUS ON *EURYTEMORA GRACILICAUDA* AKATOVA, 1949

BY

ELENA FEFILOVA[1,5]), NATALIA SUKHIKH[2]), EKATERINA ABRAMOVA[3]) and ILYA VELEGZHANINOV[1,4])

[1]) Institute of Biology, Komi Science Centre, Ural Division, Russian Academy of Sciences, Syktyvkar, 167982 Russia

[2]) Zoological Institute of the Russian Academy of Sciences, St. Petersburg, Russia

[3]) Lena Delta Nature Reserve, Tiksi, Sakha Republic, Russia

[4]) Polytechnical Institute, Vyatka State University, Kirov, Russia

## ABSTRACT

We studied the morphology of *Eurytemora* from inland waters at the shores of the White and Pechora seas and from the Lena River delta, and revealed a ubiquitous presence of *Eurytemora gracilicauda* Akatova, which results we confirmed with genetic data. We found this species for the first time in the Pechora Sea basin. In the White Sea basin, this species was previously described as *E. brodskyi* Kos, with the name that we suggest is a junior synonym of *E. gracilicauda*. *E. gracilicauda* differs from the co-living species: *E. lacustris* (Poppe), *E. arctica* Wilson M. S. & Tash, and *E. raboti* Richard, by the structure and armament of the caudal rami and the fifth legs (P5) of males and females. The caudal rami of both males and females were elongated. The female caudal rami showed a surface covered by spines. The male caudal rami were bare, or with rare spines on the sides. The appendages of the female P5 were also elongated: the length of the inner spine on the distal exopod segment was $2.27 \pm 0.12$ times as long as the outer spine; the exopod of the male right P5 had a specific trigonal ledge with a short spinule, and the coxopods (both or at least one) had groups of spinules. The morphometric parameters of the females were quite stable, while those of the male showed high variability within and between populations (CV = 11.5-43.5%). Similarities and differences of *E. gracilicauda* and three allochoric *Eurytemora* species were analysed, and the results presented herein.

## RÉSUMÉ

Nous avons étudié la morphologie de *Eurytemora* des eaux intérieures des rives de la mer Blanche et de la mer de Petchora ainsi que du delta de la Lena, et nous avons mis en évidence la présence constante de *Eurytemora gracilicauda* Akatova, et ces résultats ont été confirmés par les données génétiques. Nous avons trouvé cette espèce pour la première fois dans le bassin de la mer de

[5]) Corresponding author; e-mail: fefilova@ib.komisc.ru

 | DOI:10.1163/9789004465565_007

Petchora. Dans le bassin de la mer Blanche, cette espèce était auparavant décrite comme *E. brodskyi* Kos, que nous suggérons être un synonyme junior de *E. gracilicauda*. *E. gracilicauda* diffère des espèces *E. lacustris* (Poppe), *E. arctica* Wilson M.S. & Tash, et *E. raboti* Richard, par la structure et l'armature des rames furcales et par les cinquièmes pattes (P5) des mâles et des femelles. Les rames furcales des mâles et des females sont allongées. Les rames furcales de la femelle ont une surface couverte d'épines ; Les rames furcales du mâle sont glabres, ou avec de rares épines sur les côtés. Les appendices de la P5 femelle sont également allongées : l'épine interne du segment distal de l'exopodite est 2,27 ± 0,12 fois aussi longue que l'épine externe, l'exopodite de la P5 mâle droite a un rebord triangulaire particulier avec une spinule courte, et les coxopodites (les deux ou au moins un) présentent des groupes de spinules. Les paramètres morphométriques des femelles sont tout à fait stables, tandis que ceux des mâles montrent une variabilité au sein d'une même population et entre les populations (CV = 11,5-43,5%). Les similarités et les différences entre *E. gracilicauda* et les trois espèces allochoriques de *Eurytemora* ont été analysées, et les résultats présentés ici.

## INTRODUCTION

The first descriptions of the species of *Eurytemora* (Copepoda, Calanoida, Temoridae) published in the previous century do not contain information about their morphological variability (e.g., Williams, 1906; Akatova, 1949; Borutsky, 1949). This issue remains challenging, at least in the Russian literature, and in identification tables for the genus (Borutsky et al., 1991; Stepanova, 2010; Kos, 2016) as this complicates the diagnostics of *Eurytemora* spp. by morphological parameters. The main challenge in estimating the morphometric variability is, that *Eurytemora* spp. have the ability to easily colonize new habitats, to quickly disperse, and to change the limits of their habitat. These changes complicate the research on composition of the local fauna and might produce co-existence of numerous intraspecific morpho- and haplotypes, as well as interspecific hybrids (Lee & Frost, 2002). Recent descriptions of these copepod species include information on the variability of their principal morphometric parameters, fine morphological characters, and they are often presented complemented by data on molecular-genetic variability for species and populations, as well as by observations on their ecology (Lee, 1999; Lee & Frost, 2002; Dodson et al., 2010; Alekseev & Souissi, 2011; Moon et al., 2016; Sukhikh et al., 2016a). Such approach allows to review the systematics of individual species or groups.

In our research, we focused on morphologically similar individuals collected from geographically remote inland water bodies at the Arctic shore, which were identified previously as different species. *Eurytemora* cf. *brodskyi* Kos, 1993 was reported from White Sea rock pools, western Palaearctic (Kos, 1993; Sukhikh et al., 2016a). Morphologically similar to the previous one, *Eurytemora* sp. was collected in the basin of the Pechora Sea (Fefilova & Kononova, 2019) and *E. gracilicauda* Akatova, 1949 from the Lena River delta (Abramova et al., 2017). Moreover, the recent study of the White Sea population of *Eurytemora* (Sukhikh et al., 2016a) revealed some similarity between *E.* cf. *brodskyi* and *E.* cf. *americana* Williams, 1906. The latter is a wide spread non-Arctic species (Moon et al., 2016). All these examples show, that species identification within the genus *Eurytemora* represents a great challenge.

The aim of the present studies is to find a reliable species determination among *Eurytemora* representatives inhabiting brackish coastal water bodies of the Palaearctic, including the variability (i.e., stability and changeability) of diagnostically important features.

## MATERIAL AND METHODS

We studied *Eurytemora* material from zooplankton samples collected in three water bodies of European and Siberian Russia: a rock pool of Medyanka Island in Kandalaksha Bay of the White Sea (WS) (66°18′22″N 33°50′31″E) in summer 2014; a lake on Kashin Island in Korovinskaya Bay of the Pechora Sea (PS) (68°14′35″N 53°52′05″E) in August 2016 and 2017; and lakes in the Lena River Delta (LD) (72°22′12″N 126°31′12″E) in August 2017. Samples from the White Sea region were fixed in formaldehyde solution, all other samples were fixed in 96% ethanol and have been stored at −20°C.

The species identification guides by E. Borutsky et al. (1991) and M. Kos (2016) were used. For morphological analyses of the antennules, trunk somites, caudal rami and legs 5, structures of each specimen were used.

Specimens were examined, and drawings were made, using a Leica DM 4000 B microscope with a drawing tube under a magnification of 1000×. Photos were taken with the camera of an ASUS Phone ZE520KL under a magnification of 400×. The final version of each figure was optimized using the programs Adobe Photoshop CS3 Extended and Xara Photo & Graphic Designer 6.

For detailed comparative morphological analysis 22 specimens of *Eurytemora* of both sexes were used. Ten morphometric characters (fig. 1) and three ratios of females, as well as morphometry and quantitative morphological characters of males were taken into account (fig. 1). To estimate the variability of morphological characters of *Eurytemora*, coefficients of variation (CV) were calculated as:

$$\mathrm{CV} = \frac{\sigma}{\mu} * 100\%,$$

where $\sigma$ = standard deviation, $\mu$ = mean. If CV $< 10\%$, it was considered that a character had low variation and was stable, if the CV $> 10\%$, it was considered that a character showed considerable variation.

The morphological analysis of samples from PS and WS revealed four species, identified as *E. gracilicauda*, *E. lacustris* (Poppe, 1887), *E. arctica* Wilson M. S.

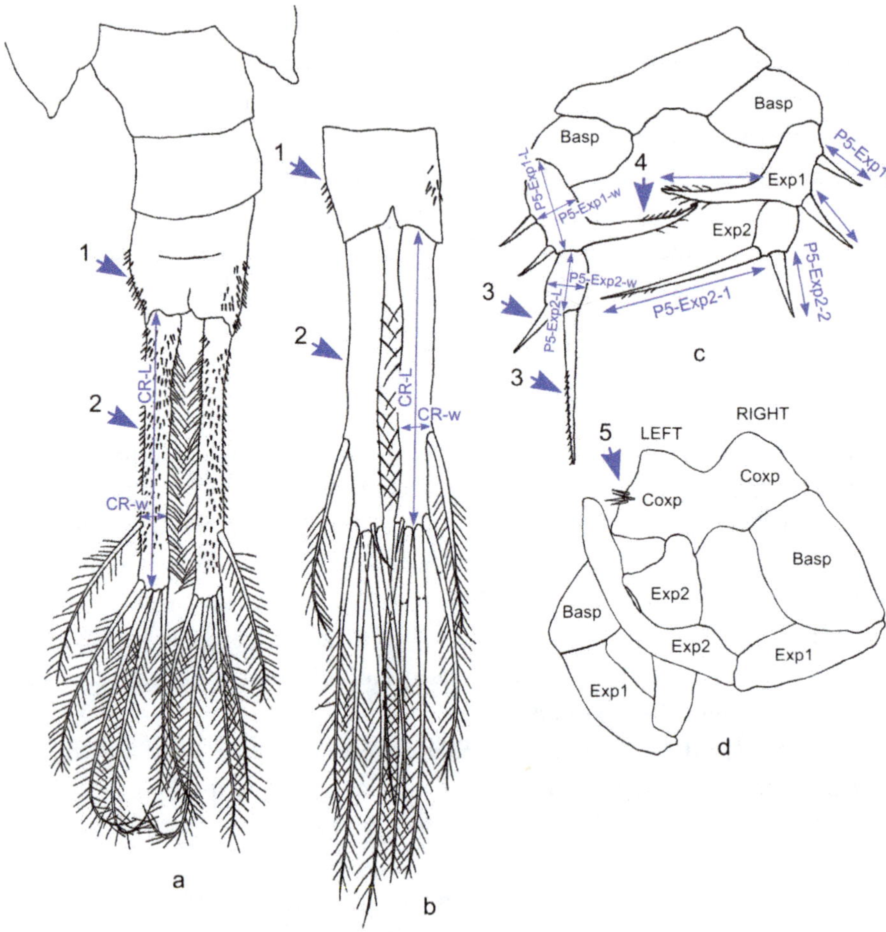


Fig. 1. Morphological characters of *Eurytemora* spp. used in this study: a, female caudal rami; b, male caudal rami; c, P5 of female; d, P5 of male: left and right; 1, spinules on anal somite; 2, spinules on caudal rami; 3, spinules on spines on distal segment of female P5; 4, setules on appendix on first segment of P5 of female; 5, spinules on Coxp of male; CR-L, length of caudal rami; CR-w, width of caudal rami; P5-Exp1-L and P5-Exp1-w, length and width of Exp1 of female, respectively; P5-Exp2-L and P5-Exp2-w, length and width of Exp2 of female, respectively; P5-Exp1-3, length of spine (longest) on Exp1 of female; P5-Exp1-APN, length of appendix on Exp1 of female; P5-Exp2-1, length of inner spine on Exp2 of female; P5-Exp2-2, length of outer spine on Exp2 of female.

& Tash, 1966, and *E. raboti* Richard, 1897. For the molecular identification, some specimens were used belonging to the first three species, but not *E. raboti*.

For the genetic analysis, 10 specimens of *Eurytemora* of both sexes were used. Two *E. gracilicauda* from LD and two from PS were analysed with COI and 18SrRNA markers. Three *E. lacustris* from PS and three *E. arctica* from PS were analysed only with COI gene DNA extractions, and PCR amplifications were performed in the Center for Collective Use facilities «Molecular Biology» of the Institute of Biology of the Komi Science Center, Ural Branch, Russian Academy of Sciences (RAS). In brief, DNA was isolated with a 6% Chelex-100 solution. The individual copepods were placed in 10 $\mu$l of the 6% Chelex-100 solution, 1 $\mu$l proteinase-K was added, and then samples were heated to 55°C for 30 min followed by 99°C for 15 min. Next, the samples were centrifuged at 5000 × *g* for 5 min and the supernatant was taken for amplification. The primer pair 18A1 mod (5′CTGGTTGATCCTGCCAGTCATATGC3′) and 1800 mod (5′GATCCTTCCGCAGGTTCACCTACG3′) was used for amplification of the fragment of the nuclear 18S rRNA gene (1800 bp) (Raupach et al., 2009). The following temperature conditions of the polymerase chain reaction (PCR) were used: initial denaturation at 95°C for 5 min, the following 35 cycles: 90°C for 30 s, 55°C for 30 s, 72°C for 60 s, and final elongation at 72°C for 5 min. The standard LCO1490/HCO2198 (Folmer et al., 1994) and modified EuF1 (5′CGTATGGAGTTGGGACAAGC3′) and EuR2 (5′CAAAATAAGTGTTGGTATAAAATTGGA3′) (Winkler et al., 2011) primers were used to amplify a fragment each (710 bp and 611 bp respectively) of the mitochondrial cytochrome c oxidase 1 subunit (COsI). The following temperature conditions were used in the PCR: initial denaturation at 95°C for 5 min, the following 5 cycles: 90°C for 30 s, 45°C for 60 s, 72°C for 90 s, then 27 cycles: 90°C for 30 s, 55°C for 45 s, 72°C for 60 s, and final elongation at 72°C for 5 min. The best results were obtained using the standard COI primers, so that the 710 bp amplicons were used for further data analysis.

The resulting products from the two genes were purified from the gel using Qiaquick PCR purification kit (Qiagen, Valencia, CA, U.S.A.). Sequencing was performed in both directions on an automated sequencer ABI 3130 (Applied Biosystems, Foster City, CA, U.S.A.) using BigDye v. 3.1. kits and the same primers that were used for amplification. Nucleotide sequences were aligned using the BioEdit 7.0.3 program (Hall, 1999) with the Clustal W algorithm (Thompson et al., 1994) and manually edited by hand. Sequencing was performed at the Research Resource Center for Molecular and Cell Technologies, Scientific Park of St. Petersburg State University on a 3500xL Genetic Analyzer with BigDye Terminator v3.1 Cycle Sequencing Kit. Previously analysed (Sukhikh et al., 2016a, b) COI mtDNA and 18S rRNA sequences (accession numbers in GenBank: HM474030-HM474035; MN256854-MN270893-MN270898) were used to compare with the obtained data.

The following abbreviations are used throughout the text and figures: Coxp, coxopod; Basp, basipodite; P5, legs 5; Exp1, proximal segment of P5 exopod; Exp2, distal segment of P5 exopod; and WS for Medyanka Island of Kandalaksha Bay of the White Sea; PS for Kashin Island of Korovinskaya Bay of the Pechora Sea; and LD for the Lena River Delta.

TABLE I

Results of molecular-genetic analysis of specimens of *Eurytemora* spp. from lakes of two regions: PS and LD, studied in 2016 and 2017

| No. | Sampling location | Sex | Species (according to results of DNA barcoding) | Accession numbers in GenBank | |
|---|---|---|---|---|---|
| | | | | COI | 18S |
| 1 | PS | ♀ | *E. gracilicauda* Akatova, 1949 | MN256855 | MN270894 |
| 2 | PS | ♀ | *E. lacustris* (Poppe, 1887) | MN256864 | |
| 3 | PS | ♀ | *E. lacustris* | MN256865 | |
| 4 | PS | ♂ | *E. lacustris* | MN256866 | |
| 5 | PS | n.det. | *E. gracilicauda* | MN256862 | MN270895 |
| 6 | LD | n.det. | *E. gracilicauda* | MN256863 | MN270896 |
| 7 | LD | ♀ | *E. arctica* Wilson M. S. & Tash, 1966 | MN256860 | |
| 8 | LD | ♀ | *E. arctica* | MN256861 | |
| 9 | LD | ♂ | *E. arctica* | MN256859 | |
| 10 | LD | ♂ | *E. gracilicauda* | MN256856 | MN270897 |

LD, Lena River Delta; PS, Pechora Sea; n.det., undetermined.

## RESULTS

### Molecular-genetic identification of *Eurytemora*

DNA barcoding of *Eurytemora* specimens from two sampling locations (PS and LD) revealed three species: *E. gracilicauda E. lacustris* and *E. arctica* (table I). Comparison of the COI mtDNA sequences with those previously published in GenBank and analysed in Sukhikh et al. (2016a, b) revealed that *E. gracilicauda* from LD and PS corresponded to *E.* cf. *brodskyi*. 18SrRNA sequences of these individuals were identical to *Eurytemora* from the White Sea, which was identified in Sukhikh et al. (2016a) as *E.* cf. *brodskyi*, not any nucleotide differences were observed among three these populations. The nucleotide diversity in 18SrRNA among *E. lacustris* and populations from LD, PS and WS was 2.5% in each case.

The COI sequences of the individuals from the Lena and Pechora rivers were also very similar to each other and to *Eurytemora* from the White Sea. The nucleotide differences in the COI gene between *E. lacustris* and populations from LD, PS and WS were 24.1%. Differences between *E. arctica* and populations from LD, PS and WS were 25.9%, between *E. arctica* and *E. lacustris* they were 22.6%. The comparison of differences among populations from LD, PS and WS were 0.4-1.4%, thus suggesting that *Eurytemora gracilicauda* and WS *E. brodskyi* are actually one and the same species. All obtained COI and 18SrRNA sequences were uploaded to GenBank, accession numbers are given in table I.

## Morphological identification of *Eurytemora*

Morphological analysis of PS and LD material revealed four species: *Eurytemora gracilicauda*, *E. lacustris*, *E. arctica* and *E. raboti*. *E. gracilicauda* and *E. lacustris* were the only species in the PS sample, while the LD sample contained all four mentioned *Eurytemora* species. *Eurytemora* from the WS rock pools was identified as *E. brodskyi* because of morphological similarity with the first description and because this description was done based on specimens from the same location of the White Sea.

Morphological and genetic comparison of three populations showed that PS and LD *Eurytemora gracilicauda* and WS *E. brodskyi* are actually the same species. At the same time, we should say that the type specimen of neither *E. gracilicauda* nor *E. brodskyi* was found for comparison with the present populations. Nevertheless, we considered *E. brodskyi*, described by Kos in 1993, as a junior synonym of *E. gracilicauda*, which was described by Akatova in 1949. Accordingly, we are going to use the name *E. gracilicauda* for this species from all sampling locations, including WS.

The various *Eurytemora* species, co-occuring in the PS and LD, are clearly different from each other by the morphology of females and males: the structure of the last thoracic and genital somites, the length and the integument of the caudal rami, the armature of segments in the middle part of the geniculate antennules, and the P5. The female P5 of *E. gracilicauda* shows that the length of the outer spine on the Exp2 is greater than the length of Exp2, and the long and slightly curved "annexe" [= large, inwardly directed, pointed processes of the segment proper] on the Exp1 (fig. 2a, c, e, f, in green). Males of this species have a specific ledge on the Exp2 of the right P5, armed with a short seta (fig. 2b, g, in green). One, or only the left coxopodite, has a group of spinules (fig. 2b, d, g, in green). *E. arctica*, which is very similar to *E. gracilicauda* in the morphology of trunk somites and caudal rami of females and males, has a specific long spine on the Exp1 of the left male P5 (fig. 2l, in blue) and a relatively short outer spine on the Exp2 of the female; the annexe on the Exp1 of the *E. arctica* female is sickle-curved and swollen at the base (fig. 2k, in blue). The P5 of *E. lacustris* males is typically armed on the left P5 Exp1 (fig. 2h, orange colour). Females of *E. raboti* have a wide proximal part in the annexe on the Exp1 of P5 and at the same time a short outer spine on the Exp2 (fig. 2j, in red).

## Morphological variability

We did not analyse the variability of the studied features of all *Eurytemora* species discussed herein, but the morphological variability of these characters has been studied for *E. gracilicauda*, and the results are presented here. Also, the

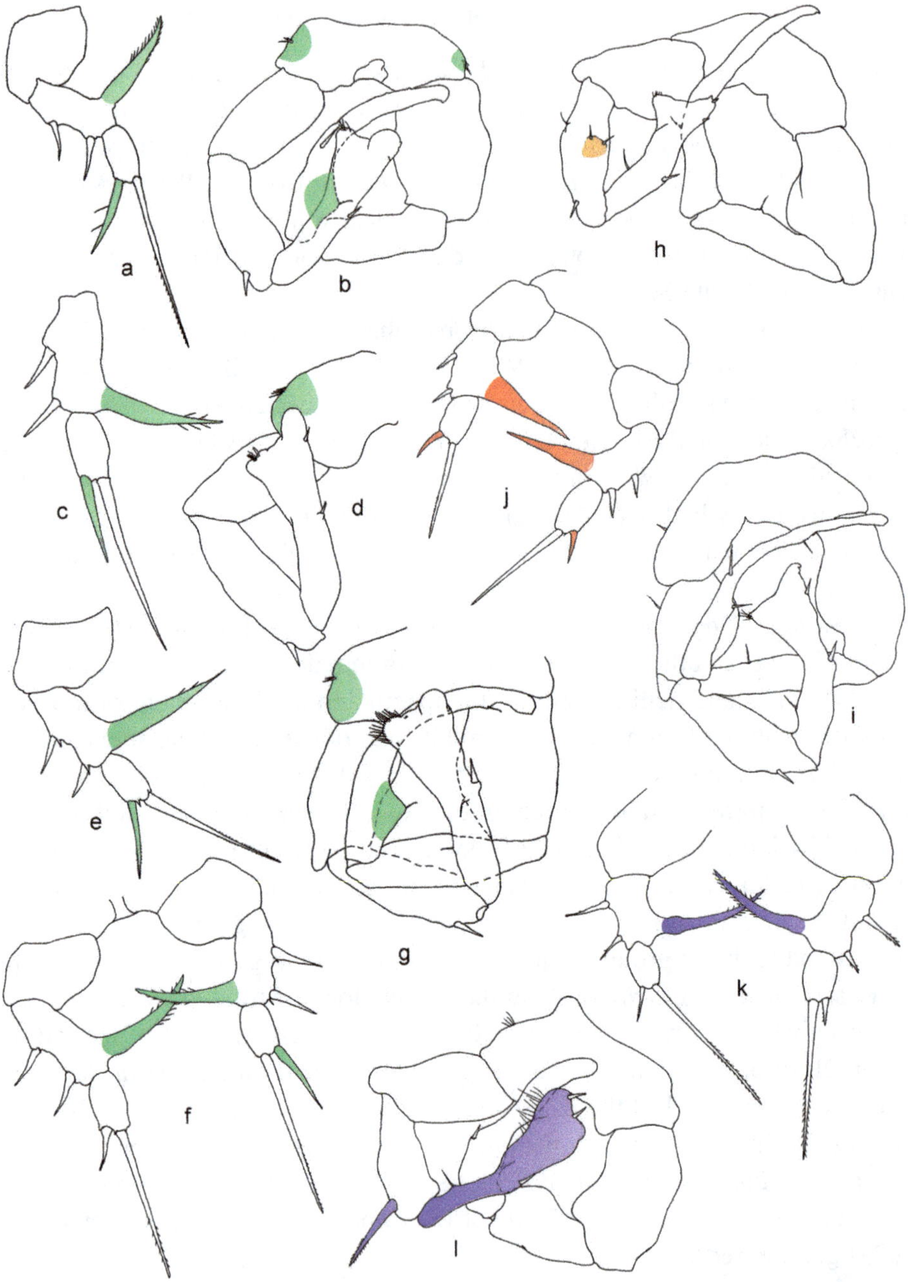


Fig. 2. Structure of P5 of females (a, c, e, f, j, k) and males (b, d, g, h, i, l) of the studied species of *Eurytemora*. a, b, *E. gracilicauda* Akatova, 1949 from WS; c, d, *E. gracilicauda* from PS; e-g, *E. gracilicauda* from LD; h, *E. lacustris* (Poppe, 1887) from PS; i, *Eurytemora* sp. from LD; j, *E. raboti* Richard, 1897 from LD; k, l, *E. arctica* Wilson M. S. & Tash, 1966 from LD. Colours highlight important parts for diagnostic features, as explained in the text. WS, White Sea; LD, Lena River Delta.

systematic diagnosis of *E. gracilicauda* was complemented in the present study with data on the variability of its morphological structures and morphometry. We studied the morphological features in three populations of *E. gracilicauda*, i.e., from PS, LD and WS. The following characteristics were shared: (1) caudal rami of females are longer than those of the male, their surface and the lateral sides of the anal somite are covered by small spines (table II, figs. 3d, 4a, b, 5a); (2) the last thoracic somite of females has lateral, triangular paddles, while males of all populations examined have none (figs. 3a, 4b); (3) the genital double-somite of females is without spines and the middle part of the male geniculate antennules (8-12 segments) has short spines of approximately equal length (figs. 3c, 4g). The shape of the P5 in both males and females was similar in all studied individuals of *E. gracilicauda* (figs. 2a-g, 3e-f, 4c, e, 5b, c). In females, the outer spine on Exp2

TABLE II

Morphometry and quantitative morphological characters (see fig. 1), ratios (CR-L/CR-w, P5-Exp1-L/P5-Exp1-w, P5-Exp2-L/P5-Exp2-w, P5-Exp2-1/P5-Exp2-2) of three studied populations of *Eurytemora gracilicauda* Akatova, 1949

| | WS | PS | LD | All |
|---|---|---|---|---|
| Morphometric characters and ratios, females: $\mu \pm \sigma$ | | | | |
| Number of specimens | 6 | 1 | 9 | 16 |
| CR-L (mm) | $0.251 \pm 0.009$ | 0.341 | $0.378 \pm 0.008$ | $0.328 \pm 0.016$ |
| CR-w (mm) | $0.025 \pm 0.0007$ | 0.03 | $0.034 \pm 0.002$ | $0.031 \pm 0.001$ |
| CR-L/CR-w | $10.13 \pm 0.430$ | 11.37 | $11.28 \pm 0.65$ | $10.82 \pm 0.39$ |
| P5-Exp1-L (mm) | $0.045 \pm 0.0008$ | 0.063 | $0.066 \pm 0.001$ | $0.056 \pm 0.003$ |
| P5-Exp1-w (mm) | $0.023 \pm 0.0008$ | 0.023 | $0.028 \pm 0.0009$ | $0.025 \pm 0.0009$ |
| P5-Exp1-L/ P5-Exp1-w | $1.99 \pm 0.05$ | 2.74 | $2.37 \pm 0.09$ | $2.21 \pm 0.09$ |
| P5-Exp1-3 (mm) | $0.020 \pm 0.0007$ | 0.018 | $0.030 \pm 0.001$ | $0.025 \pm 0.001$ |
| P5-Exp1-APN (mm) | $0.057 \pm 0.004$ | 0.063 | $0.084 \pm 0.006$ | $0.070 \pm 0.005$ |
| P5-Exp2-L (mm) | $0.030 \pm 0.002$ | 0.033 | $0.040 \pm 0.003$ | $0.035 \pm 0.002$ |
| P5-Exp2-w (mm) | $0.018 \pm 0.0002$ | 0.020 | $0.023 \pm 0.001$ | $0.020 \pm 0.001$ |
| P5-Exp2-L/ P5-Exp2-w | $1.78 \pm 0.09$ | 1.65 | $1.72 \pm 0.05$ | $1.71 \pm 0.05$ |
| P5-Exp2-1 (mm) | $0.096 \pm 0.003$ | 0.092 | $0.113 \pm 0.003$ | $0.105 \pm 0.003$ |
| P5-Exp2-2 (mm) | $0.047 \pm 0.003$ | 0.042 | $0.049 \pm 0.004$ | $0.048 \pm 0.002$ |
| P5-Exp2-1/ P5-Exp2-2 | $2.06 \pm 0.06$ | 2.19 | $2.41 \pm 0.20$ | $2.27 \pm 0.12$ |
| Morphometric and quantitative morphological characters and ratios of males: $\mu \pm \sigma$ | | | | |
| Number of specimens | 3 | 1 | 2 | 6 |
| CR-L (mm) | $0.212 \pm 0.004$ | 0.269 | $0.255 \pm 0.005$ | $0.236 \pm 0.011$ |
| CR-w (mm) | $0.023 \pm 0.001$ | 0.023 | $0.026 \pm 0.003$ | $0.024 \pm 0.001$ |
| CR-L/CR-w | $9.44 \pm 0.65$ | 11.70 | $10.11 \pm 1.19$ | $10.04 \pm 0.56$ |
| Left P5 | $2.33 \pm 0.33$ | 3 | $3.00 \pm 0.00$ | $2.86 \pm 0.59$ |
| Right P5 | $3.33 \pm 1.45$ | – | $3.00 \pm 0.00$ | $3.20 \pm 0.80$ |

CR-L/CR-w, ratio of length of caudal rami to width of caudal rami; P5-Exp1-L/P5-Exp1-w, ratio of length of Exp1 to width of Exp1 of female; P5-Exp2-L/P5-Exp2-w, ratio of length of Exp2 to width of Exp2 of female; $\sigma$, standard deviation; $\mu$, mean; WS, White Sea; PS, Pechora Sea; LD, Lena River Delta.

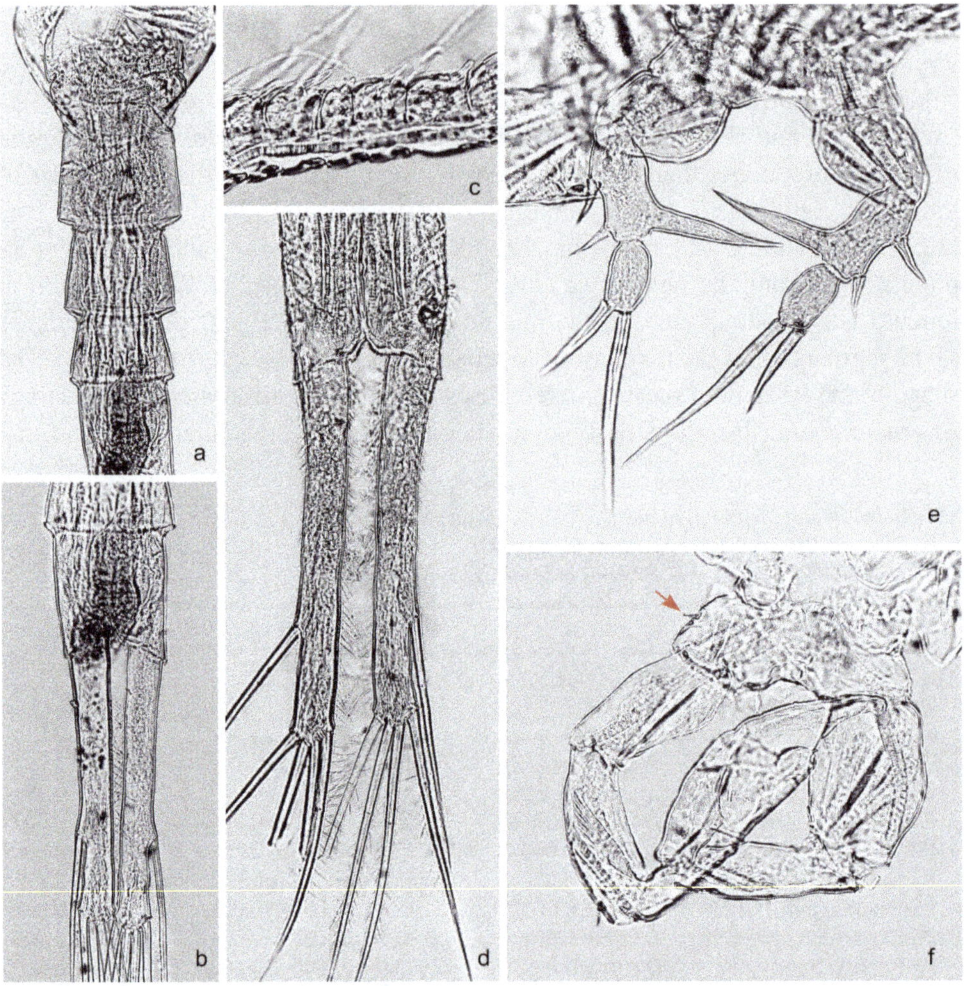


Fig. 3. *Eurytemora gracilicauda* Akatova, 1949 from WS: a-c, f, male; d, e, female; a, genital somite; b, d, caudal rami; c, middle part of geniculate antennule; e, f, P5. WS, White Sea.

of P5 was longer than the Exp2, but more than twice as short as the inner spine (figs. 2a, c, e, f, 3e, 4c, 5b, table II).

Females and males of *E. gracilicauda* from WS had the shortest and narrowest caudal rami, and the shortest segments and appendices of P5 (table II). One of the males from this population had a smooth anal somite and smooth caudal rami. The caudal rami of the other individuals had rare spines on the lateral sides of the middle part. The male anal somite had no spines (fig. 3a). Only the left, or both coxopods of the P5 of males had lateral groups of small spines or setules (figs. 2b, 3f). The total number of spines on the exopod and endopod of the P5 of males varied from six to eight.

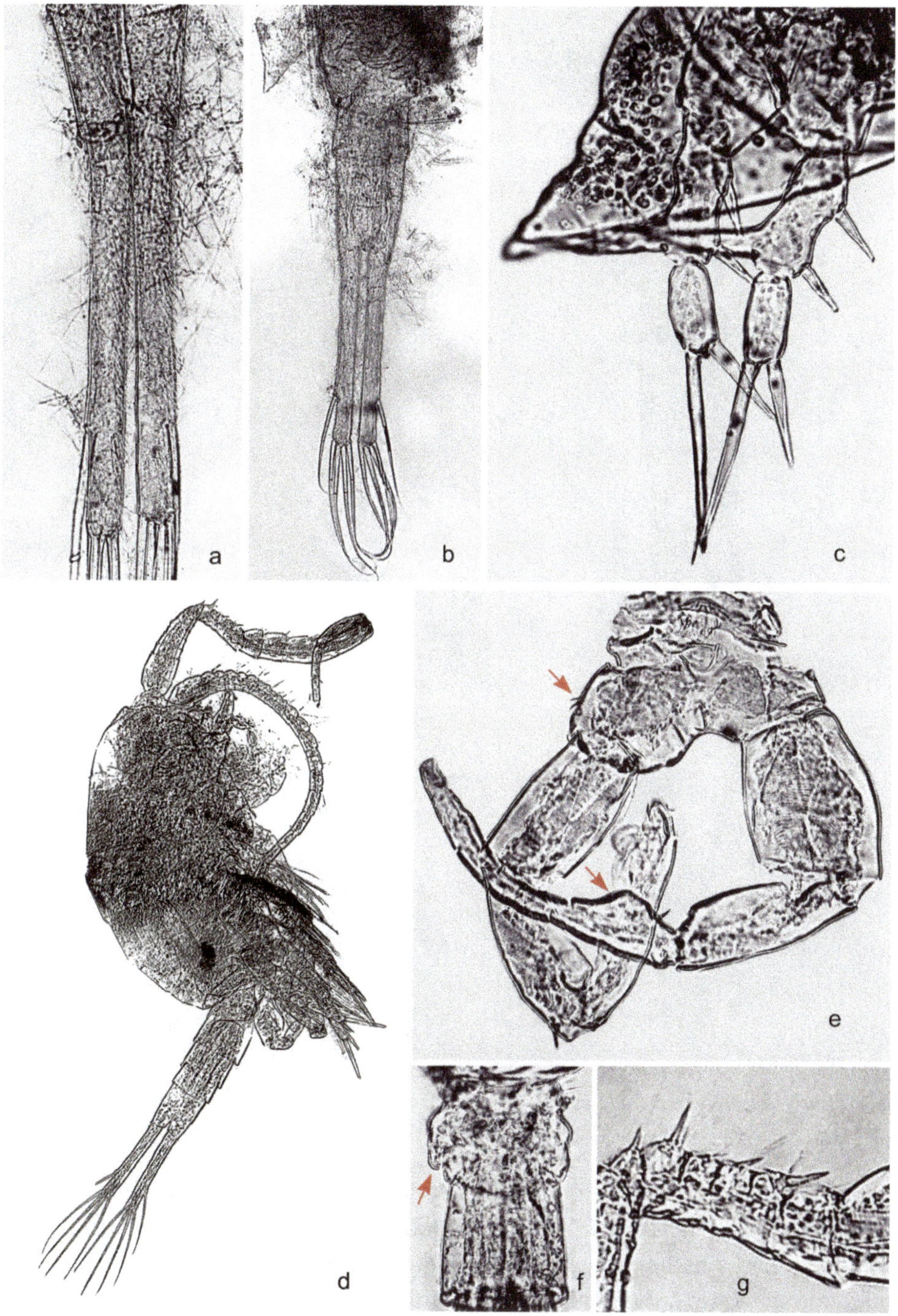


Fig. 4. *Eurytemora gracilicauda* Akatova, 1949 from PS: a-c, female; d-g, male; a, caudal rami; b, abdomen; c, e, P5; d, overall view; f, genital somite; g, middle part of geniculate antennule. PS, Pechora Sea.

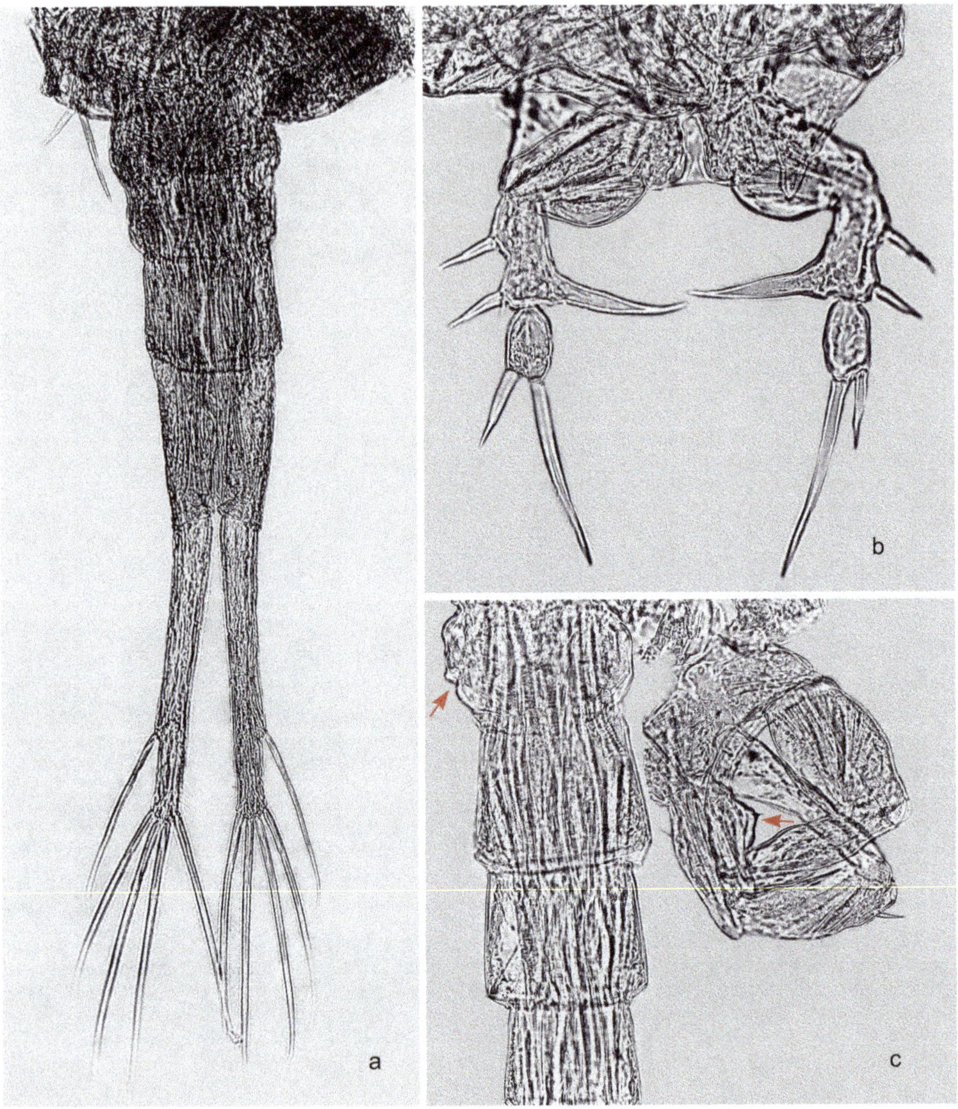


Fig. 5. *Eurytemora gracilicauda* Akatova, 1949 from LD: a, b, female; c, male; a, abdomen and caudal rami; b, P5; c, genital somite and P5. LD, Lena River Delta.

*E. gracilicauda* from PS had longer caudal rami and segments of the female P5 compared to *E. gracilicauda* from WS (table II). At least one male from the PS population had small spinules in the middle of the outer sides of the caudal rami and at the same time, the genital somite without annexes and at least one male had smooth caudal rami and a short, finger-shaped appendix on one side of the genital somite (fig. 4f). Only the left or both coxopods of the P5 of males had groups of spinules on the outer side (figs. 2d, 4e).

Females from LD had the most slender caudal rami and the longest segments of the P5 compared to females from the other sites; moreover, they had the longest appendix on the exopod 2 of the P5 (table II), and in some females it was crescent (figs. 2f, 5b). The caudal rami and anal somite of males were without spinules and the genital somite bore a finger-shaped appendix on one side (fig. 5c). The coxopod of the left P5 of males had a group of spinules on the outer side (fig. 2g). The number of spinules and setae on the P5 of males was six.

The spinules on the spines and the annexe on the exopod segments of the female P5 were variable characters in all studied populations of *E. gracilicauda*; they could be present on each annexe, but could also be absent. According to our data, the morphometric characters and ratios of females of *E. gracilicauda* are rather stable, but the majority of male characters concerning structure of caudal rami and P5 are very variable (table III). The calculated variation coefficients show, that the *E. gracilicauda* population of LD displays a higher variability (CV: 6.58 ± 2.15%) than the population from WS (CV: 3.88 ± 0.48%; table III).

TABLE III

Coefficients of variation (%) of morphometry and quantitative morphological characters (see fig. 1, table I), ratios (see table I) of three studied populations of *Eurytemora gracilicauda* Akatova, 1949

| | WS | LD | All |
|---|---|---|---|
| Morphometric characters and ratios, females | | | |
| CR-L | 3.6 | 2.4 | 4.9 |
| CR-w | 2.8 | 3.0 | 3.2 |
| CR-L/CR-w | 4.2 | 5.8 | 3.6 |
| P5-Exp1-L | 1.8 | 1.5 | 5.4 |
| P5-Exp1-w | 3.7 | 3.2 | 3.6 |
| P5-Exp1-L/P5-Exp1-w | 2.5 | 3.8 | 4.1 |
| P5-Exp1-3 | 3.5 | 6.9 | 4.0 |
| P5-Exp1-APN | 7.0 | 7.1 | 7.1 |
| P5-Exp2-L | 6.6 | 7.5 | 5.7 |
| P5-Exp2-w | 1.1 | 4.3 | 5.0 |
| P5-Exp2-L/P5-Exp2-w | 5.1 | 2.9 | 2.9 |
| P5-Exp2-1 | 3.1 | 2.6 | 2.8 |
| P5-Exp2-2 | 6.4 | 8.1 | 4.3 |
| P5-Exp2-1/P5-Exp2-2 | 2.9 | 8.3 | 5.3 |
| Morphometric and quantitative morphological characters, and ratios of males | | | |
| CR-L | 1.9 | 2.0 | 4.7 |
| CR-w | 4.3 | *11.5* | 4.2 |
| CR-L/CR-w | 6.9 | *11.7* | 5.4 |
| Left P5 | *14.2* | *43.5* | *20.6* |
| Right P5 | *43.5* | 0.00 | *25.0* |

Italics indicates the most variable characters. WS, White Sea; LD, Lena River Delta.

## DISCUSSION

We found that three remote *Eurytemora* populations of the Russian Arctic coast, which were identified earlier as different species (*E. brodskyi* and *E. gracilicauda*) (Kos, 1993, 2016; Abramova & Zhulay, 2016; Sukhikh et al., 2016a; Abramova et al., 2017), are morphologically and genetically very similar, and thus represent one single species, *E. gracilicauda*. Taking into account only the stable morphological characters, the description of WS *E. brodskyi* (cf. Kos, 2016) corresponds to the complemented diagnosis of species *E. gracilicauda*. The same is true for *E. kurenkovi* Borutzky, 1961 from the Okhotsk Sea (Borutzky, 1961), which was recognized earlier as a junior synonym of *E. gracilicauda* (cf. Borutzky et al., 1991), and *E. gracilicauda occidentalis* Fefilova, 2008 from Vaigach Island (Fefilova, 2008). It is interesting that *E. gracilicauda occidentalis* is more similar to the studied population from WS than to the ones from LR and PS by the morphometry of the caudal rami (CR-L/CR-w approx. 8.5-9.1), being geographically closer to the population from PS. More likely, genetically *E. gracilicauda occidentalis* is also closer to European WS and PS populations and more distant to the Siberian LS populations. The latter is probably more similar to Far East *E. gracilicauda* populations and in our study showed 1.4% nucleotide substitution differences towards European populations. This level of difference corresponds to the subspecies level according to other Copepoda studies (Bucklin et al., 1994; Sukhikh et al., 2016b). As a result, we can assume that subspecies *E. gracilicauda occidentalis* inhabits European waters and the Asian waters are inhabited by at least one other subspecies. The unification of *E. gracilicauda* with *E. americana* and *E. arctica* based on 26 morphological parameters of Alaskan *Eurytemora*, was reported by Dodson et al. (2010). However, these 26 morphological characters are not indicative for these species' (*E. gracilicauda*, *E. americana* and *E. arctica*) diagnostics, and therefore we suggest that this unification might not be adequate. In contrast, our genetic and morphological studies revealed undoubtedly differentiation of the two species *E. gracilicauda* and *E. arctica*. The most important morphological parameter is the structure of P5 in males and females, dividing *E. americana* from *E. gracilicauda* and *E. arctica* according to the literature (Williams, 1906; Brylinski, 2009; Heron, 2009; Kos, 2016; Moon et al., 2016). Unfortunately, there is no description of *E. americana* with the full dimensions of the female P5, but it is clear that its relative size (P5-Exp1-L/P5-Exp1-w and P5-Exp2-L/P5-Exp2-w) differs from what we calculated for *E. gracilicauda*. In addition, segments of female P5 of *E. americana* have setae instead of spines. The male P5 of *E. gracilicauda* also differs in the shape and armature from the P5 of *E. americana*.

Some of the characters that we suggested as variable for *E. gracilicauda* demonstrated clear variability also in other populations of the species, as well as in the other *Eurytemora* species. For example, *E. gracilicauda* described in (Kos, 2016) has an either smooth, or spinulate anal segment in males, while the caudal rami are always smooth.

The genital somite of the males of E. *lacustris* (cf. Kos, 2016) and *E. americana* (cf. Kos, 2016; Moon et al., 2016) may be armed by finger-shaped annexes, but this appendix is not a constant character of this species, like it is observed in *Eurytemora grimmi* (Sars, 1897) (cf. Kos, 2016). None of the studied females of *E. arctica* and *E. lacustris* had any annexes at the basipodite of P5, while Alaskan *E. americana* and *E. lacustris* from Lake Ladoga have a seta on the outer site of that structure (Borutzky et al., 1991; Kos, 2016). We did not expect the morphometric parameters of *E. gracilicauda* to be stable within the population. On the contrary, we suggest at least the length and width of the caudal rami to be variable, because size dimensions of copepods essentially depend on factors such as food concentration and temperature (e.g., Klein Breteler & Gonzalez, 1988).

Possibly, some other environmental factors determine the variability of parameters of *E. gracilicauda*. Salinity can be suggested as one of the driving factors, because most *Eurytemora* species occur in inland waters, communicating with the sea, where salinity varies strongly. For example, in the rock pools of the White Sea with *E. gracilicauda*, salinity ranges from 0 to 16‰ during the summer (Polyakova & Starkov, 2019). Kos (2016) reported *E. gracilicauda* from a wide range of water bodies, from fresh to brackish, with a salinity of 6 psu. In Alaska, *E. gracilicauda* was found in water with a salinity of 5.7 psu (Heron, 2009).

We did not find a clear relationship between the morphology of the studied *E. gracilicauda* populations with their geographical location. Three populations demonstrated a trend of increasing the length of the caudal rami and the segment of P5, moving from the east Palaearctic to the west, but the species from Vaigach Island (Fefilova, 2008) was not in line with this trend. So, additional research is required to reveal potential ecological drivers of *E. gracilicauda* variability.

The question about the origin of some populations of *E. gracilicauda* in the Palaearctic also remains open. This species was described from western Siberia in 1949 (Akatova, 1949), from the far East in the 1960s (Kos, 2016), Kamchatka (Borutzky, 1961), and Alaska (Heron, 1964). Recent studies found this species in the Lena River Delta (Abramova & Zhulay, 2016), and on Vaigach Island (Fefilova, 2008). The White Sea population of *E.* cf. *brodskyi* (cf. Kos, 1993) and the Pechora Sea population of *Eurytemora* sp., according to our data, also belong to the species *E. gracilicauda* (cf. Fefilova & Kononova, 2019). Pelletier & Carstens (2018) concluded that the genetic diversity of small invertebrates, including arthropods, depends mainly on the geographical isolation of populations, rather than on environmental conditions.

## ACKNOWLEDGEMENTS

The work was done within the framework of the State Tasks of the Animals Ecology Department of the Institute of Biology, Komi SC UrD RAS (AAAA-A17-117112850235-2), UrD RAS Complex Program (AAAA-A18-118011390005-9) (to EF), was conducted in accordance with the national initiative AAAA-A19-119020690091-0 (to NS), and financially supported by the RFBR grants: 18-44-110017 p_a (to EF, IV), 20-34-70020, 19-04-00217_a (to NS). For this study, the Federal Collection of the Zoological Institute of the Russian Academy of Sciences (St. Petersburg, Russia) was used.

## REFERENCES

ABRAMOVA, E., I. VISHNYAKOVA, J. BOIKE, A. ABRAMOVA, G. SOLOVYEV & F. MARTYNOV, 2017. Structure of freshwater zooplankton communities from tundra waterbodies in the Lena River Delta, Russian Arctic, with a discussion on new records of glacial relict copepods. Polar Biology, **40**: 1629-1643. DOI:10.1007/s00300-017-2087-2.

ABRAMOVA, E. N. & I. A. ZHULAY, 2016. The occurrence of the new zooplankton species in the Lena River Delta. Proceedings of the Zoological Institute of RAS, **320**(4): 437-487. [In Russian.]

AKATOVA, N. A., 1949. Zooplankton of the Kolyma River and its basin. Uchenye Zapiski LGU (Ser. Biologicheskaya), **126**(21): 341-367. [In Russian.]

ALEKSEEV, V. R. & A. SOUISSI, 2011. A new species within the *Eurytemora affinis* complex (Copepoda: Calanoida) from the Atlantic coast of USA, with observations on eight morphologically different European populations. Zootaxa, **2767**: 41-56.

BORUTZKY, E. V., 1961. New calanoid copepods of the Far East. Sbornik Trudov Zoologicheskogo Museya, **8**: 3-19. [In Russian.]

BORUTZKY, E. V., L. A. STEPANOVA & M. S. KOS, 1991. Guidance of determinaton of Calanoida of fresh water of USSR: 1-503. (Nauka Press, St.-Petersburg, Russia). [In Russian.]

BRYLINSKI, J.-M., 2009. The pelagic copepods in the Strait of Dover (eastern English Channel). A commented inventory 120 years after Eugene Canu. Cah. Biol. Mar., **50**: 251-260.

BUCKLIN, A. & T. C. LAJEUNESSE, 1994. Molecular genetic variation of *Calanus pacificus* (Copepoda: Calanoida): preliminary evaluation of genetic structure and subspecific differentiation based on mt-DNA sequences. Calif. Cooperative Oceanic Fish. Invest. Rep., **35**: 45-51.

CHU, K. H., P. F. TAM, C. H. FUNG & Q. C. CHEN, 1997. A biological survey of ballast water in container ships entering Hong Kong. Hydrobiologia, **352**: 201-206.

DODSON, S. I., D. A. SKELLY & C. E. LEE, 2010. Out of Alaska: morphological diversity within the genus *Eurytemora* from its ancestral Alaskan range (Crustacea, Copepoda). Hydrobiologia, **653**: 131-148. DOI:10.1007/s10750-010-0351-3.

FEFILOVA, E., 2008. New species of the genus *Moraria* (Copepoda, Harpacticoida) and a new subspecies of the genus *Eurytemora* (Calanoida) from islands of the Barents Sea. Russian Zoological Journal, **87**(4): 393-402. [In Russian.]

FEFILOVA, E. & O. KONONOVA, 2019. Diversity of planktonic fauna of the Pechora River Delta. Contemporary Problems of Ecology, **12**(3): 254-264. DOI:10.1134/S1995425519030041.

FOLMER, O., M. BLACK, W. HOEH, R. LUTZ & R. VRIJENHOEK, 1994. DNA primers for amplification of mitochondrial Cytochrome C oxidase subunit I from diverse metazoan invertebrates. Molecular Marine Biology and Biotechnology, **3**(5): 294-299.

HERON, G. A., 1964. Seven species of *Eurytemora* (Copepoda) from northwestern North America. Crustaceana, **7**(3): 199-211. DOI:10.1098/rsbl.2017.0566.

KOS, M. S., 1993. A new species of the genus *Eurytemora* of Chupa Lip (White Sea), in Issledovaniya Fauny Morei. In: O. A. SKARLATO (ed.), Study of the sea fauna: marine plankton: systematics, ecology, distribution, **45**(53): 30-37. (Zoological Institute RAS, St.-Petersburg, Russia). [In Russian.]

KOS, M. S., 2016. Calanoid copepods of the families Stephidae and Temoridae of the seas of Russia and adjacent waters. In: N. V. VYSHKVARTZEVA & G. N. BUZHINSKAJA (eds.), Proceedings of the zoological institute of the Russian academy of sciences, **179**: 1-108. (Zoological Institute RAS, St.-Petersburg, Russia). [In Russian.]

LEE, C. E., 1999. Rapid and repeated invasions of freshwater by the saltwater copepod *Eurytemora affinis*. Evolution, **53**: 1423-1434.

LEE, C. E. & B. W. FROST, 2002. Morphological stasis in the *Eurytemora affinis* species complex (Copepoda: Temoridae). Hydrobiologia, **480**: 111-128.

MOON, S. Y. & B. A. V. MARAN, 2016. New extension range and complementary description of *Eurytemora americana* (Calanoida: Temoridae) in northern Korea. Marine Biodiversity Records, **9**(1). DOI:10.1186/s41200-016-0076-8.

PELLETIER, T. A. & B. C. CARSTENS, 2018. Geographical range size and latitude predict population genetic structure in a global survey. Biol. Lett., **14**: 20170566.

POLYAKOVA, N. & A. STARKOV, 2019. *Eurytemora* sp. in the rock pools of the White Sea islands. In: Use of molecular-genetic and morphological methods to study the taxonomy, phylogeny, biogeography, and ecology of *Eurytemora* species: 33-34. (Book of Abstracts of International Scientific Conference, Saint Petersburg, Russia).

RAUPACH, M. J., C. MAYER, M. MALYUTINA & J.-W. WÄGELE, 2009. Multiple origins of deep-sea Asellota (Crustacea: Isopoda) from shallow waters revealed by molecular data. Proc. Biol. Sci., **276**(1658): 799-808.

STEPANOVA, L. A., 2010. Calanoids (Calaniformes). In: V. R. ALEKSEEV (ed.), Determinant of zooplankton and zoobenthos of fresh water of European Russia, **1**. Zooplankton: 284-327. (Nauka, St. Petersburg, Russia). [In Russian.]

SUKHIKH, N. M., V. CASTRIC, N. V. POLYAKOVA, S. SOUISSI & V. R. ALEKSEEV, 2016a. Isolated populations of *Eurytemora americana* Williams (Crustacea, Copepoda) in the White Sea rock pools — Postglacial relicts or anthropogenic invasions? Russian Journal of Biological Invasions, **7**(4): 396-404.

SUKHIKH, N. M., A. SOUISSI, S. SOUISSI, G. WINKLER, V. CASTRIC, A. C. HOLL & V. R. ALEKSEEV, 2016b. Genetic and morphological heterogeneity among populations of *Eurytemora affinis* (Crustacea: Copepoda: Temoridae) in European waters. C.R. Biol., **339**: 197-206.

THOMPSON, J. D., D. J. HIGGINS & T. J. GIBSON, 1994. CLUSTAL W: improving the sensitivity of progressive multiple sequence alignment through sequence weighting, position-specific gap penalties and weight matrix choice. Nucleic Acids Res., **22**: 4673-4680.

WILLIAMS, L. W., 1906. Notes on marine Copepoda of Rhode Island. Amer. Nat., **40**: 639-660.

WINKLER, G., S. SOUISSI, C. POUX & V. CASTRIC, 2011. Genetic heterogeneity among *Eurytemora affinis* populations in Western Europe. Mar. Biol., **158**: 1841-1856.

First received 5 September 2019.
Final version accepted 19 December 2019.

[When citing this volume, please refer to *Crustaceana* 93 (2020), 241-547]

*Proceedings of the* Eurytemora *conference, St. Petersburg, 2019*

# STOCHASTIC PHENOTYPIC VARIATION: EMPIRICAL RESULTS AND POTENTIAL USE IN *EURYTEMORA* RESEARCH (COPEPODA, CALANOIDA)

BY

DMITRY LAJUS[1,3]), NATALIA SUKHIKH[2]) and VICTOR ALEKSEEV[2])

[1]) Saint-Petersburg State University, 16 linia V.O. 7-29, 199034 St. Petersburg, Russia

[2]) Zoological Institute of the Russian Academy of Sciences, Universitetskay nab. 1., 199034 St. Petersburg, Russia

## ABSTRACT

This study analyses the potential of stochastic phenotypic variation for investigating the population biology of *Eurytemora*. Stochastic variation is the third component of phenotypic variance, standing on equal footing with genotypic variation and phenotypic plasticity. This is a manifestation of developmental instability and usually increases under stress. In morphological traits, stochastic variation is most often studied using fluctuating asymmetry (FA) of bilateral traits. Here, using data on the FA of nine populations of three *Eurytemora* species from Europe and North America, we found no correlation between FA and temperature, salinity or tidal amplitude. Invasive American *E. carolleeae* in the Gulf of Finland (Baltic Sea) had lower FA than the same species in its native Chesapeake Bay, or than *E. affinis* in its native Gulf of Finland. This pattern may be caused by global warming, which brought Chesapeake Bay temperatures beyond *E. carolleeae*'s optimal conditions, but made the Gulf of Finland a more suitable environment. Stochastic variation in life history traits is technically more difficult to study, but it may provide important information on fitness. In particular, it manifests in bet-hedging, a risk-spreading strategy beneficial in unpredictable environments. As resting eggs are common in *Eurytemora*, bet-hedging can be considered a genus strategy. Understanding how stochastic variation contributes to total phenotypic variance may help to interpret changes under unpredictable environmental conditions. Therefore, studies of stochastic phenotypic variation may supply information about the population biology of *Eurytemora* and other copepods.

## RÉSUMÉ

Cette étude analyse le potentiel de variation phénotypique stochastique pour étudier la biologie des populations d'*Eurytemora*. La variation stochastique est le troisième composant de la variance phénotypique, avec la variation génotypique et la plasticité phénotypique. C'est une manifestation de l'instabilité développementale et elle augmente habituellement avec le stress. Pour les caractères morphologiques, la variation stochastique est le plus souvent étudiée en utilisant l'asymétrie fluctuante (FA) des caractères bilatéraux. Ici, en utilisant les données de la FA de neuf populations

[3]) Corresponding author; e-mail: dlajus@gmail.com

 | DOI:10.1163/9789004465565_008

de trois espèces d'*Eurytemora* d'Europe et d'Amérique du Nord, nous n'avons trouvé aucune corrélation entre la FA et la température, la salinité ou l'amplitude des marées. L'espèce invasive américaine *E. carolleeae* du Golfe de Finlande (Mer Baltique) avait une FA inférieure à celle de la même espèce de Chesapeake Bay, ou à celle d'*E. affinis* dans son Golfe de Finlande d'origine. Ce modèle peut être causé par le réchauffement global qui a apporté à Chesapeake Bay des températures situées au-delà des conditions optimales de *E. carolleeae*, mais qui a conféré au Golfe de Finlande un environnement plus favorable. La variation stochastique des traits d'histoire de vie est techniquement plus difficile à étudier, mais elle peut fournir des informations importantes sur la « fitness ». En particulier, elle se manifeste par la minimisation des risques (bet-hedging), une stratégie de dispersion des risques avantageuse dans les environnements imprévisibles. Comme les œufs de diapause sont communs chez *Eurytemora*, le bet-hedging peut être considéré comme une stratégie du genre. Comprendre comment la variation stochastique contribue à la variance phénotypique totale peut aider à interpréter les changements dans des conditions environnementales imprévisibles. En conséquence, les études de la variation phénotypique stochastique peuvent apporter des informations sur la biologie des populations d'*Eurytemora* et d'autres copépodes.

## INTRODUCTION

Currently, marine organisms face serious challenges caused by fisheries, pollution, biological invasions and climate change. In many cases, human-driven ecosystem changes occur so unpredictably that they negate responses based on genetic diversity, and so rapidly that they exceed the adaptive capacities of populations. Due to extensive river runoff and the proximity of human settlements, anthropogenic pressure on estuaries is especially high. Changes resulting from multiple factors cause both direct effects on individual species and indirect effects via interaction with other species in this complex ecosystem. Hence, forecasting ecosystem conditions and proactive management require comprehensive scientific information addressing different aspects of population biology. *Eurytemora* species are typical estuary inhabitants. Some invading species compete with native organisms, including those in their own genera. They play an important role in trophic chains by actively consuming phytoplankton and providing food for numerous fish species.

A variety of research techniques in population biology deal with phenotypic variation. The ability to regulate phenotype is a basic property pertaining to any individual organism. Traditionally, variation in phenotypic traits is considered to be caused by two factors: genetic diversity, a function of parental genotypes, and phenotypic plasticity, a function of environmental heterogeneity (Falconer & Mackay, 1996). The present concept of phenotypic variation is rooted in the studies of Charles Darwin, who differentiated between indefinite and definite phenotypic variation. Focusing on the evolution of organisms, and postulating heredity as a key mechanism of evolution, Darwin stated that "Any uninherited variation is unimportant for us" (Darwin, 1886: 9). Over the years, focus on evolution predetermined scientific interest in indefinite, or inherited variation, which eventually transformed into genotypic variation. Definite, or uninherited variation, as environmental or non-genotypic variation came to be called, received much less attention.

Nevertheless, soon after Darwin's concept first appeared, some researchers found serious limitations in the focus on indefinite variation. At the end of the 19$^{th}$ century, Pearson empirically analysed variation in replicated structures of plants and fungi, such as spines on leaves, segments in capsules, seed pods, gill indices, stigmatic fronds, leaf indices, veins in leaves, and members of leaf whorls. In choosing these structures, Pearson assumed his specimens shared a similar genetic basis and developed under the same environmental conditions. This minimized both known types of variation. Empirical results, however, showed unexpectedly high levels of residual variation. The nature of this variation was unclear, and Pearson wrote: "We can measure this variation, we can describe it quantitatively, but its cause stays a mystery like the cause of gravitation" (Pearson, 1911: 530).

One problem with Pearson's study was that environmental heterogeneity cannot be fully ignored. Three decades later, Astaurov (1930, 1974) selected traits more rigorously. He analysed variation in the number of chaetae among homologous segments of polychaetes and myriapods, compared left and right halteres in inbred lines of *Drosophila*, and studied data in the literature on identical (monozygotic) twins. Astaurov also observed sizable residual variation. He concluded that "... [any] process of formation in an organism has some independent variation, which can be attributed neither to genotypic differences, nor to the direct effect of the environment" (Astaurov, 1974: 103). Similarly, Waddington (1957) differentiated between canalization and developmental stability. Canalization is an organism's ability to reproduce a consistent phenotype under different environmental conditions. Its converse is phenotypic plasticity. In contrast, developmental stability refers to a developing organism's ability to produce a consistent phenotype in a given environment. Therefore, stochastic phenotypic variation is a manifestation of developmental instability.

Waddington discovered stochastic phenotypic variation independently of Astaurov and Pearson. In the mid-20th century his work was often cited in studies of one kind of stochastic variation, fluctuating asymmetry (FA) (Mather, 1953; Tebb & Thoday, 1954; Thoday, 1958; Beadmore, 1960; Reeve, 1960). This attribute was suggested as an effective measure of environmental stress (Leary & Allendorf, 1989; Zakharov, 1989; Parsons, 1990; Graham et al., 1993), and the numerous publications on FA that appeared during the 1990s (Lajus et al., 2019) focused on that practical application. FA's potential as a third component of total phenotypic variation, on equal footing with genotypic and environmental variation, was not considered (Lajus et al., 2003a).

Using empirical data on *Eurytemora* and information obtained about other organisms, our goal is to discuss the potential of stochastic phenotypic variation studies to improve our understanding of the population biology of *Eurytemora* and other marine crustaceans.

## DIVERSE TYPES OF STOCHASTIC VARIATION

### Fluctuating asymmetry

FA represents random deviations from prefect symmetry (Ludvig, 1932). Most often it is measured on traits that manifest bilateral, or mirror symmetry, and the analytical methodology is well developed (e.g., Palmer & Strobeck, 2003). But other types of symmetry exist — such as rotational, radial, dihedral, and translational symmetry — and the concept can be expanded to fractals (symmetry of scale) (Graham et al., 1993). Graham and co-authors (Graham et al., 2010) reviewed different types of symmetry in detail, with examples from the plant and animal kingdoms. FA techniques can be applied on all structures that manifest any type of symmetry, as a measure of the extent to which average individuals depart from perfect symmetry.

Statistically, ideal fluctuating bilateral asymmetry occurs when measured differences between the left and right structures are symmetrically distributed, around zero. Yet, symmetry is often broken, as in directional asymmetry or antisymmetry (Graham et al., 2010). Directional asymmetry takes place when either the left or right structure is systematically larger than its counterpart, i.e., the mean of the left minus the right distribution differs from zero. Antisymmetry occurs when one structure is larger than the other, but the disparity occurs unpredictably — it can appear on either the right, or the left side (Timofeeff-Ressovsky, 1934). These data appear as a bimodal distribution with a mean equal to zero. Once identified, directional symmetry and antisymmetry can be statistically corrected to reveal bilateral FA (Graham et al., 1998).

FA changes due to variation in environmental and genetic factors. It can increase due to non-optimal temperature (Leary et al., 1992; Benderlioglu & Dow, 2017; Kozlov & Zverev, 2018), parasite infection (Reimchen, 1997; Pojas & Tabugo, 2015), pollutants (Valentine et al., 1973; Ostbye et al., 1997; Lajus et al., 2015a) and other factors (see Graham et al., 2010, for a review).

In total phenotypic variance, the contribution of FA varied from 10 to 40% for metric (measured) traits, and from 50 to 70% for meristic (counted) traits (Lajus et al., 2003a). This estimate, however, cannot be extrapolated to all bilateral traits because, for FA studies, researchers usually choose traits with relatively high asymmetry to reduce the contribution of measurement error. It is important to note that the magnitude of stochastic variation may differ by an order of magnitude or even more for similar morphological traits (Lajus & Alekseev, 2000). Also, traits with high stochastic variation show a strong tendency towards high non-stochastic variation, caused by genetic variation and environmental heterogeneity (Lajus, 2001).

### Morphological abnormalities

Traditionally, analysis of morphological abnormalities had been used to assess the effects of stress (e.g., see Hamdoun & Epel's, 2007 review of embryos). However, this technique has limitations. It works better on early developmental stages because malformations that affect viability and cause mortality in early life stages decline in frequency in older individuals. Thus, malformation frequency in adults is usually so low that detection may require large sample sizes. Furthermore, abnormalities are caused not only by stochastic variation, but also by genetic or environmental influences. This problematizes interpretation, so that judging developmental instability based on data about morphological abnormalities must proceed with caution (Møller & Swaddle, 1997). Whereas studying abnormalities requires clearly defining a norm, which is not easy to do, the norm for FA is easily defined as perfect symmetry.

### Life history and other traits

Stochastic variation in morphological structures is of interest as a proxy for population fitness, and life history traits, as components of fitness, relate directly to a population's vitality. However, stochastic variation of life history traits is difficult or even impossible to study in wild populations. To obtain information equivalent to that provided by the analysis of symmetrical morphological structures, one must estimate variation among genetically identical individuals, or in replicate measurements of the same individual, under standardized laboratory conditions. Unfortunately, there is no symmetry in life history traits. A few studies of stochastic variation in life history traits were performed on the life span of mice, *Mus musculus* Linnaeus, 1758 (cf. Williams & Pelton, 1966) and of the nematode, *Caenorhabditis elegans* (Maupas, 1900) (cf. Kirkwood & Finch, 2002); on the production of resting eggs by *Daphnia magna* Straus, 1820 (cf. Lajus & Alekseev, 2004); and on a number of traits related to the marbled crayfish, *Procambarus virginalis* Lyko, 2017 reproduction (Vogt et al., 2008).

Data on the stochastic variation of physiological and behavioural types of traits were reviewed by Vogt (2015). A few of his examples relate to crustaceans, particularly the marbled crayfish, which is a convenient model for the study of stochastic variation because it often produces clonal progeny. Clonal marbled crayfish juveniles, kept individually under standard laboratory conditions, increase in variability with age. At age 71 days, the body weight coefficient of variation was about 10%; at age 351 days it had grown almost five-fold (Vogt et al., 2008).

A different technique was used on blue mussels to measure stochastic variation in behavioural traits by analysing within-individual variation in behavioural responses to the same experimental stimulus (Lajus & Sukhotin, 1998; Sukhotin et al., 2003) The animals were placed in a tank with cyclical changes in salinity. Salinity was recorded when mussels closed (as salinity decreased) or opened their shells (as salinity increased). This allowed calculating the repeatability of this behavioural response, which was an inverse index of stochastic variation. Roughly 40% of the total phenotypic variance of this trait was attributed to stochastic variation. Vogt and co-authors (Vogt et al., 2008) suggested that stochastic variation in behaviour probably developed from very small behavioural differences in self-reinforcing circuitries, involving metabolism and neuroendocrine feedback.

## MATERIAL AND METHODS

For this study we used the same methods described in our previous research on *Eurytemora* (Lajus et al., 2015b), which was predominantly devoted to variation in body shape and provided limited information on FA. The present study focuses exclusively on FA variation. We also analysed a number of abiotic parameters that can potentially influence FA, and estimated their contribution to total stochastic variation.

At European and eastern U.S.A. locations, between March 2005 and June 2011, samples were collected from surface water (1-2 m deep) using a 100 $\mu$m plankton net, and these were preserved in 95% ethanol. Species identification was done using the morphological differences described in earlier papers (Alekseev & Souissi, 2011; Sukhikh et al., 2013; Sukhikh & Alekseev, 2013), and was supported via identification of the mitochondrial CO1 gene. In total, we analysed 231 specimens from nine populations of three species (see table I for details).

Measurements were performed only on females. Using a dissection microscope (Olympus SZX2, Tokyo, Japan) with 5-$\mu$m resolution, we analysed the six morphological traits shown in fig. 1. All traits were measured on both the left and right sides of the body. The magnitude of FA was calculated with the following formula:

$$\mathrm{FA} = |L - R|/0.5(L + R)$$

where $L$ and $R$ are the left and right values of the trait, respectively (Lajus & Alekseev, 2000). To normalize the obtained distribution, we used the Box-Cox

TABLE I

Characteristics of samples of *Eurytemora* species, sampling locations, time periods, and fluctuating asymmetry

| Species | Sampling locations | Geographical coordinates | Sampling date | $N$ | Salinity, range (PST) | Temperature, range (°C) | Tide amplitude, range (m) | FA |
|---|---|---|---|---|---|---|---|---|
| *E. caspica* Sukhikh & Alekseev, 2013 | Caspian Sea | 45°48′N 49°38′E | June 2011 | 19 | 0-10 | 4-26 | 0 | $0.103 \pm 0.5828$ |
| *E. carolleeae* Alekseev & Souissi, 2011 | Chesapeake Bay | 39°23′N 76°03′W | April 2008 | 13 | 0.5-30 | 5-25 | 0.2-3.5 | $-0.474 \pm 0.5706$ |
| *E. carolleeae* | Gulf of Finland | 59°24′N 28°11′E | July 2008, August 2009 | 45 | 0-8 | 0-20 | 0 | $-1.410 \pm 0.3476$ |
| *E. affinis* (Poppe, 1880) | Elbe estuary | 53°53′N 09°08′E | March 2006 | 17 | 0-30 | 0-20 | 0.1-4.3 | $-0.891 \pm 0.6527$ |
| *E. affinis* | Seine estuary | 49°N 00°W | May 2008, July 2008 | 17 | 2.5-25 | 3-21 | 0-8.0 | $1.482 \pm 0.7013$ |
| *E. affinis* | Gulf of Riga | 57°04′N 23°04′E | August 2008 | 28 | 0.5-7 | 3-20 | 0 | $0.868 \pm 0.5538$ |
| *E. affinis* | Gulf of Finland | 59°24′N 28°11′E | July 2008, August 2009 | 44 | 0-8 | 0-20 | 0 | $0.244 \pm 0.4038$ |
| *E. affinis* | Vistula Lagoon | 54°65′N 20°23′E | October 2007 | 30 | 2-4 | 0-23 | 0 | $1.443 \pm 0.5683$ |
| *E. affinis* | Loire estuary, Gironde estuary | 47°17′N 02°01′W; 45°31′N 01°57′W | July 2009, April 2006, March 2005 | 18 | 3.3-30 | 8-24 | 0-5.9 | $-1.153 \pm 0.5590$ |

Abiotic data were obtained from Podgornyi (2018) and Kotta et al. (2018, https://tides4fishing.com), as well as Kimmel et al. (2006, https://upload.wikimedia.org/wikipedia/commons/0/09/Tide_final4.jpg). The species include *Eurytemora caspica* Sukhikh & Alekseev, 2013, *E. carolleeae* Alekseev & Souissi, 2011 and *E. affinis* (Poppe, 1880).

$N$, number of specimens; FA, fluctuating asymmetry.

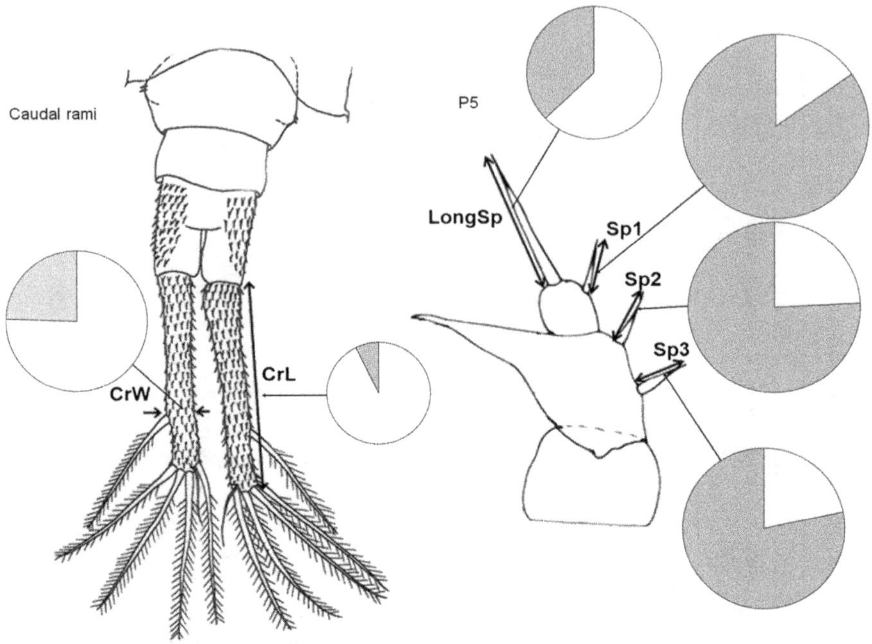


Fig. 1. Morphological traits used to analyse fluctuating asymmetry in *Eurytemora* (based on pictures from Sukhikh & Alekseev, 2013). CrL, CrW (on caudal rami), LongSp, Sp1,Sp2, Sp3 on protopodite 5 (P5). The average total standardized variance for across all 231 individuals is shown by circles, where the encircled area is proportional to the magnitude of total variance. Grey sectors indicate the stochastic component, and the white sector indicates the non-stochastic component.

transformation (Box & Cox, 1964):

$$(\mathrm{FA} + \lambda 1)^{\lambda 2}$$

with $\lambda 1$ ranging from 0.005 to 0.02, and $\lambda 2$ from 0.2 to 0.6 for different traits ($\lambda 1$ and $\lambda 2$ values minimize departure from the normal distribution). Palmer & Strobeck (1986) recommended this transformation for FA studies. The composite FA index of an individual is the average of the standardized FAs of its individual traits (Palmer & Strobeck, 2003). To assess the heterogeneity of FA among samples, we used a one-way ANOVA with consequent application of a Tukey HSD post-hoc test.

To estimate the contribution of stochastic variance to the total, we used a procedure described in our earlier papers (e.g., Lajus et al., 2003b, 2015b). Briefly, this includes log-transformation of all data, standardizing data by individual size with consequent use of residuals, testing the normality of the data distribution, as well as the distribution of L-R values.

To characterize abiotic conditions, we used minimum and maximum monthly average temperatures, salinity and tide amplitudes. We also calculated the yearly averages and range of variation of these environmental factors, i.e., in total, we analysed twelve abiotic parameters. To study their association with FA, we used Spearman's coefficient of correlation, as the distributions of most abiotic factors were not normal.

For statistical comparison of samples by magnitude of FA, we used a Student's t-test. Statistica 7.0 and Microsoft Excel 2010 software were employed in statistical analysis.

## RESULTS

Tests for directional asymmetry (performed using a $t$-test of standardized values) and antisymmetry (with analysis of kurtosis) showed one and three cases of statistically significant ($p<0.05$) effects, respectively, which become non-significant after sequential Bonferroni correction (Rice, 1989). Table I provides the average values of the composite FA index for the nine samples, obtained by averaging indices for the six traits. They differ somewhat from data provided in our previous paper (Lajus et al., 2015b); here, a different methodology (Box-Cox transformation) allowed us to normalize the $|L - R|$ distributions and apply parametric statistics such as ANOVA.

Among all nine samples, the highest FA was observed in samples of *Eurytemora affinis* (Poppe, 1880) from the Seine estuary, the Vistula Lagoon and the Gulf of Riga; the lowest FA was found in *E. carolleeae* Alekseev & Souissi, 2011 from the Gulf of Finland and populations from the Loire and Gironde River estuaries (France) (table I). One-way ANOVA showed a statistically significant effect for the factor "Sample" on FA (df $=$ 8, $p$ $=$ 0.00005). The post hoc Tukey HSD test showed that *E. carolleeae* from the Gulf of Finland showed a significantly lower FA than *E. affinis* from the Seine estuary ($p$ $=$ 0.0043), the Gulf of Riga ($p$ $=$ 0.0113), the Gulf of Finland ($p$ $=$ 0.0813) and the Vistula Lagoon ($p$ $=$ 0.0002). In addition, the Vistula Lagoon sample showed a significantly higher FA than samples from the French rivers ($p$ $=$ 0.030). Other differences were not statistically significant ($p>0.1$).

Total phenotypic variance, and the combined contribution of the six traits' FA to total variance, were measured as an average among nine samples (fig. 1). Standardized total variance among traits ranged about three-fold, with the lowest variance on the length of the caudal rami and the highest on the Sp1 of P5. The observed stochastic variation contributed, on average, 51% for all six traits. The stochastic component and its contribution to total variance were also largest on the Sp1 (84.6%) and smallest on the length of the caudal rami (7.2%). Given that the length of the caudal rami is almost nine-fold as large as that of the Sp1, this confirms earlier findings showing that the magnitude of stochastic variance is inversely related to the size of the morphological trait (Soule, 1967; Lajus, 2001; Lajus et al., 2003c).

No significant correlation was found between FA and the twelve abiotic variables characterizing temperature, salinity and tidal regime in the studied locations.

## DISCUSSION

In this section we discuss both the empirical results of our study of stochastic variation manifested in morphological asymmetries, and potential applications of our analysis to other types of stochastic phenotypic variation in *Eurytemora*.

### Morphological traits

FA contributes significantly to total phenotypic variance, accounting for more than half of total variance, which is typical of these structures in studies of other copepods (Lajus & Alekseev, 2000). Here, we only analysed the observed stochastic variation, which is higher than true variation because it includes any error from inconsistent measurements (Palmer & Strobeck, 2003). Most likely, the magnitude of measurement error in *Eurytemora* is close to that found in our previous study of similar structures on the copepod *Acanthocyclops signifer* Mazepova, 1952 (cf. Lajus & Alekseev, 2000): there, it accounted for about a third of the observed stochastic variation. Since the measurement error did not differ across samples, it was unlikely to affect the results from comparing FA across samples. Patterns of relationship between variance and trait size were similar to those found in other studies (Soule, 1967, 1982; Lajus, 2001).

Due to the absence of an expected association between FA and environmental factors, some researchers, in repeated discussions, have doubted the utility of FA to measure stress (e.g., Bjorsten et al., 2000a, b; Møller, 2000; Van Dongen & Lens, 2000). In terms of this discussion, we point out that developmental programmes are intended to control phenotypic variation in growth, which is energetically costly (Koehn & Bayne, 1989; Sommer, 1996). Under stress, larger portions of an organism's available energy must be allocated to important functions like maintenance, growth, and reproduction. This reduces the resources available for developmental control and results in increased stochastic variation. Therefore, there are direct physiological links between stochasticity and stress. A lack of response in a stressed population's FA does not suggest that FA is an inappropriate measure of stress. Rather, it indicates that the tested factors do not detectably affect the development of those particular traits under those particular conditions (Lajus, 2014).

The absence of a significant correlation between FA and key hydrological variables in this study can be explained by the intrinsically high sampling error of FA estimates: FA measures have only one degree of freedom (Palmer & Strobeck, 2003). Another explanation is the presence of other factors with a stronger effect on FA than the controlled environmental variables, for instance, the genetic properties of populations. FA is reported to increase towards the edge of a species' distribution range, where populations are often more homozygous (Kat, 1982; White & Searle, 2008), or due to directional selection resulting in reduced genetic variation (Alibert & Auffray, 2003; Leamy & Klingenberg, 2005). Among environmental factors, the most attention has probably been paid to chemical pollution (see reviews by Graham et al., 2010; Lajus, 2014; and a number of empiric studies, e.g., Tracy et al., 1995; Tan-Kristanto et al., 2003; Ambo-Rappe et al., 2007, 2008; Lajus et al., 2015a). As *Eurytemora* inhabits estuarine and coastal areas, they often encounter various pollutants. This subject is being actively studied (e.g., Legrand et al., 2017; Tian et al., 2017; Berasategui et al., 2018; Presta et al., 2019) via effects that can be assessed using FA techniques. It should also be noted for samples that differ in number (from 13 to 45), sampling error is higher when sample size is lower: for these samples the probability of finding a significant effect decreases.

It is of special interest to compare three samples: native *E. carolleeae* in Chesapeake Bay, invasive *E. carolleeae* in the Gulf of Finland (which likely originated from Chesapeake Bay) and native *E. affinis* in the Gulf of Finland (Sukhikh et al., 2013). The invasive American population living in the Gulf of Finland is characterized by the lowest FA compared with the native populations, the FAs of which do not differ significantly from each other, although native *E. caroleeae* from the Chesapeake Bay shows a lower FA than native *E. affinis* from the Gulf of Finland. Note that the six morphological traits used in this study (fig. 1) did not reveal statistically significant differences between the two *E. carolleeae* samples: under more comprehensive analysis involving sixteen traits, these differences appeared (Lajus et al., 2015b).

The difference in FA between genetically different forms under the same conditions — i.e., samples from the Gulf of Finland — suggests that the form with the lower FA is better adapted to this environment. The Baltic Sea is known as a sea of invaders (Leppakoski et al., 2002), and the eastern part of the Gulf of Finland, with more than forty invasive species (Lehtiniemi et al., 2016), is especially ridden with them. These ecosystems have been heavily disturbed by various anthropogenic pressures (see reports of the Helsinki Commission (www.helcom.fi); Lajus et al., 2015d; Berezina et al., 2016), and intensive navigation has caused most introductions (Panov et al., 2003). Anthropogenic effects facilitate changes in ecological niches, which may be readily occupied by invaders. *E. carolleeae* was first reported in the Gulf of Finland in 2007 (Alekseev et al., 2009; Sukhikh et al., 2013). In trying to forecast the future of the invader in the Gulf of Finland as it competes with the native *E. affinis*, we noted its high reproductive potential features almost double the clutch size of the native population (Sukhikh et al., 2019). In addition to an advantageous life history, analysis of developmental stability shows that local environments in the Gulf of Finland are more favourable for the invader than for the native species. This explains why the invasion has been successful so far, and suggests that *E. carolleeae* has the potential to completely displace the local species. Given that climate change will most likely result in increased temperature in the Gulf of Finland, the invader should be better adapted than the native species due to the higher average temperatures in its native range.

Data in the published literature comparing the FA of invasive and native species within the same range are limited, but exemplify situations in which the invader FA is both lower (African invaders vs European native honeybees *Apis mellifera* Linnaeus, 1758 (cf. Schneider et al., 2003)) and higher (invasive European vs native North American wood borer beetle *Tetropium fuscum* (Fabricius, 1787) (cf. Goczal et al., 2018)) than that of the native species. FA analysis suggests that the American *Eurytemora* is lucky to find favourable conditions in the Gulf of Finland that facilitate its chance for success.

In fact, FA data show that *E. carolleeae* experiences more stressful conditions in its native habitat than in the Gulf of Finland. One can assume that this is because the climate is changing faster than the rate at which organisms are able to adapt. In their native Chesapeake Bay, rising water temperatures can exceed the optimal range of *E. carolleeae*, resulting in an increased FA. The same thing may happen to *E. affinis* in the Gulf of Finland. There, it was reported that the invasive *E. carolleeae* occurs at all temperatures from 15.8°C to 24.5°C, whereas the native population disappears at temperatures below 17.1°C and above 24.5°C (Sukhikh et al., 2019). Reduced genetic variation due to the bottleneck affect can cause an increase in FA (Alibert & Auffray, 2003; Leamy & Klingenberg, 2005). However, in this case environmental factors may be the main driver(s) as there is no evidence of reduced genetic variation in the invading species (Sukhikh et al., 2019).

The relatively high FA of *Eurytemora* from the other six locations of our study may also be explained by quick environmental shifts causing the dis-adaptation of local species. In these areas, invasive *E. carolleeae* may be more successful because the optimal range of environmental conditions is becoming disadvantageous to local forms. So far, invasive forms were found in four of the six study locations: the Elbe and the Loire/Gironde estuaries were the exceptions.

The bilateral structures we used for measuring FA are also opportune for taxonomic analyses, having high trait diversity and well developed statistical methods despite their high sampling error (see above). Structures possessing other types of symmetry — translational symmetry in particular — usually have more than two replicates and may have a lower sampling error. They are, however, rarely used, and only on the FA analysis of plants (Tan-Kristanto et al., 2003; Ambo-Rappe et al., 2007). *Eurytemora* and other crustaceans have structures that are good candidates for such studies. For instance, their antennular segments possess clear translational symmetry. After correcting for normal size changes from a segment's base to its end, FA can be analysed successfully. Setae on antennular segments are potentially suitable structures that deserve special attention when sample size is limited, because each individual antennula has multiple segments.

We did not find studies of developmental abnormalities in *Eurytemora* or in copepods in general. However, studies were performed on other crustaceans, particularly on *Daphnia*, because of their use as a test object in ecotoxicology. Due to the effects of different chemicals and bacteria, an increased frequency of embryo abnormalities was found, including undeveloped antennae, curved shell spines, curved post-abdomen spines, malformed carapaces and eyes, broken membranes spilling yolk, and stunted appendages with missing setae (Toumi et al., 2015; Mushegian et al., 2016; Cui et al., 2017). This technique can be used on *Eurytemora* as well.

## Life history traits and bet-hedging

Stochastic variation is usually considered to be non-adaptive, as it is a consequence of the inability of developmental control mechanisms to produce the perfect phenotype. This might be true for morphological traits. In life history traits that are directly related to fitness, however, stochastic variation may be adaptive. Such a trait is bet-hedging, a risk-spreading strategy equivalent to not placing all your eggs in one basket. The concept is borrowed from plant biology. Originally, bet-hedging was used to explain the seed bank phenomenon, a reservoir of ungerminated seeds in the soil that mitigates the risk to plants from unfavourable conditions. If a drought kills germinated plants, ungerminated seeds remaining in the seed bank will have a fitness advantage compared to plants whose seeds sprouted immediately. Simons & Johnson (1997) first linked bet-hedging to developmental instability. Stochastic phenotypic variation may possess evolutionary advantages and adaptive significance as a bet-hedging strategy, because it increases the probability of survival under unpredictable environmental changes. As developmental instability has low, but significant heritability (Whitlock, 1998; Cadee, 2000; Van Dongen, 2000; Leamy & Klingenberg, 2005), the capacity for bet-hedging can be subject to selection and, thus, adaptively respond to environmental changes.

Bet-hedging is of special relevance to animals producing resting eggs, the ecological analogue to dormant seeds. Like plants, animal populations with resting eggs need to achieve a trade-off between adjusting hatching to environmental cues and leaving some eggs unhatched if photoperiods or temperature signals do not coincide with favourable environmental conditions (Simons, 2011). In *Eurytemora*, this strategy was considered in a study of the production patterns of diapausing eggs (Glippa et al., 2013). Although the percentage of diapausing eggs was low (about 7% in the Seine River estuary), they ensure population survival under unpredictable environmental conditions. It is important to note that individual variation in resting egg production depended very little on environmental conditions: in *Daphnia*, more than 98% of the non-genetic component of phenotypic variation in this trait was explained by stochastic variation (Lajus & Alekseev, 2004).

As a manifestation of stochastic variation that increases in stressful conditions, variation resulting in bet-hedging should also increase under stress. Such patterns can increase the adaptive significance of this strategy, as bet-hedging should play a more important role under more stressful environmental conditions. In fact, Vogt (2015) suggests that bet-hedging may increase the invasive potential of a population.

Behavioural and physiological traits have interesting potential for studies of stochastic variation in *Eurytemora* and other copepods, although so far we are not aware of such research. Because clones are not common among copepods, within-individual variation in response to the same stimuli and cyclical traits are promising candidates for such analysis. These could replicate behavioural responses to environmental factors (light, salinity, temperature) in experiments, or analysis of motion patterns like jumps, which are actively studied in copepods, including *Eurytemora* (Borg et al., 2012; Michalec et al., 2015, 2017).

## CONCLUSION

Studies of stochastic phenotypic variation in *Eurytemora* may provide important information on different aspects of their population biology. Data on FA show that, in the Chesapeake Bay and the Gulf of Finland, native forms likely experience stress from suboptimal environmental conditions, probably due to quick changes in climate that exceed the adaptive potential of *Eurytemora* populations. When North American *E. carolleeae* appeared in the Gulf of Finland, they likely found more favourable conditions, resulting in a decrease in FA. Bet-hedging may be an important strategy allowing copepods to respond adequately to environmental fluctuations. When the environment is fundamentally unpredictable, the only way for populations to adapt is to generate intrinsic stochasticity. In this modern epoch characterized by increased anthropogenic pressure resulting in unpredictable environmental changes, including greater probability of competition from invasive species, organisms using bet-hedging strategies may gain the advantage. As stochastic variation is heritable, it may be subjected to selection. Along with bet-hedging, other manifestations of stochastic variation in life history and other traits represent a responsiveness to selection and may be used as stress indicators. Therefore, studies of stochastic phenotypic variation may disclose innovative information on various aspects of the population biology of *Eurytemora*, and the conclusions of this study may be applicable to other copepods.

## ACKNOWLEDGEMENTS

This study was supported by a grant from the Russian Science Foundation 19-14-00092 for D.L.; N.S. and V.A. were supported by a grant from the Russian Foundation for Basic Research 20-04-00035; N.S. was also supported by the Russian Foundation for Basic Research 19-04-00217 and by the national initiative AAAA-A19-119020690091-0. Karen Alexander helped with improving the English. The Federal Collection of the Zoological Institute of the Russian Academy of Sciences (St. Petersburg, Russia) was used for this study.

## REFERENCES

ALEKSEEV, V. R., N. I. ABRAMSON & N. M. SUKHIKH, 2009. Introduction of sibling species to the ecosystem of the Baltic Sea. Dokl. Biol. Sci., **429**(5): 694-697.

ALEKSEEV, V. R. & A. SOUISSI, 2011. A new species within the *Eurytemora affinis* complex (Copepoda: Calanoida) from the Atlantic coast of USA, with observations on eight morphologically different European populations. Zootaxa, **2767**: 41-56.

ALIBERT, P. & J.-C. AUFFRAY, 2003. Genomic coadaptation, outbreeding depression, and developmental instability. In: M. POLAK (ed.), Developmental instability: causes and consequences: 116-134. (Oxford University Press, New York, NY).

AMBO-RAPPE, R., D. L. LAJUS & M. J. SCHREIDER, 2007. Translational fluctuating asymmetry and leaf dimension in seagrass, *Zostera capricorni* Aschers in a gradient of heavy metals. Environ. Bioindic., **2**: 99-116.

AMBO-RAPPE, R., D. L. LAJUS & M. J. SCHREIDER, 2008. Increased heavy metal and nutrient contamination does not increase fluctuating asymmetry in the seagrass *Halophila ovalis*. Ecol. Indic., **8**: 100-103.

ASTAUROFF, B. L., 1930. Analyse der erblichen Störungsfälle der bilateralen Symmetrie im Zusammenhang mit der selbstständigen Variabilität ähnlicher Strukturen. Zeitschrift für Induktive Abstammungs- und Vererbungslehre, **55**(3): 183-262.

ASTAUROV, B. L., 1974. [Study of hereditary deviations from bilateral symmetry, associated with variation of metameric structures.] In: P. F. ROKITSKII (ed.), Nasledstvennost' i razvitie: 54-109. (Nauka Press, Moscow). [Translation from German into Russian of the paper Astaurov (1930).]

BEARDMORE, J. A., 1960. Developmental stability in constant and fluctuating temperatures. Heredity, **14**(3/4): 411-422.

BENDERLIOGLU, Z. & E. DOW, 2017. Low temperatures during ontogeny increase fluctuating asymmetry and reduce maternal aggression in the house mouse, *Mus musculus*. Ethology, **123**(6-7): 442-452.

BERASATEGUI, A. A., F. BIANCALANA, A. FRICKE, M. D. FERNANDEZ-SEVERINI, R. UIBRIG, M. S. DUTTO, J. MARCOVECCHIO, D. CALLIARI & M. S. HOFFMEYER, 2018. The impact of sewage effluents on the fecundity and survival of *Eurytemora americana* in a eutrophic estuary of Argentina. Estuarine, Coastal and Shelf Science, **211**: 208-216.

BEREZINA, N. A., Y. I. GUBELIT, Y. M. POLYAK, A. N. SHAROV, V. A. KUDRYAVTSEVA, V. A. LUBIMTSEV, V. A. PETUKHOV & T. D. SHIGAEVA, 2017. An integrated approach to the assessment of the eastern Gulf of Finland health: a case study of coastal habitats. J. Mar. Syst., **171**: 159-171. DOI:10.1016/j.jmarsys.2016.08.013.

BJORKSTEN, T. A., K. FOWLER & A. POMIANKOWSKI, 2000a. What does sexual trait FA tell us about stress? Trends Ecol. Evol., **15**: 163-166.

BJORKSTEN, T. A., K. FOWLER & A. POMIANKOWSKI, 2000b. Reply from T. Bjorksten, K. Fowler and A. Pomiankowski. Trends Ecol. Evol., **15**: 331. DOI:10.1016/S0169-5347(00)01926-1.

BORG, C. M. A., E. BRUNO & T. KIØRBOE, 2012. The kinematics of swimming and relocation jumps in copepod nauplii. PLoS One, **7**(10): e47486. DOI:10.1371/journal.pone.0047486.

CADEE, N., 2000. Genetic and environmental effects on morphology and fluctuating asymmetry in nestling barn swallows. J. Evol. Biol., **13**: 359-370.

CUI, F., T. CHAI, L. QIAN & C. WANG, 2017. Effects of three diamides (chlorantraniliprole, cyantraniliprole and flubendiamide) on life history, embryonic development and oxidative stress biomarkers of *Daphnia magna*. Chemosphere, **169**: 107-116.

DARWIN, CH. R., 1886. On the origin of species by means of natural selection or the preservation of favoured races in the struggle for life (6th ed., with additions and corrections to 1872): 1-458. (John Murray, London).

FALCONER, D. S. & F. C. MACKAY, 1996. Introduction to quantitative genetics (4th ed.): i-xii, 1-438. (Longman Scientific & Technical, New York, NY).

GLIPPA, O., V. R. ALEKSEEV & S. SOUISSI, 2013. Effects of photoperiod on egg production in *Eurytemora affinis* Poppe, 1880 (Copepoda: Calanoida) from the Seine estuary (France). Italian J. Zool., **80**(4): 518-525.

GOCZAŁ, J., R. ROSSA, A. NAWROCKA, J. D. SWEENEY & A. TOFILSKI, 2018. Developmental costs of biological invasion: the exotic wood borer *Tetropium fuscum* (Coleoptera: Cerambycidae) is more asymmetric and smaller in invaded area. Envir. Entomol., **47**(4): 982-989.

GRAHAM, J. H., J. M. EMLEN, D. C. FREEMAN, L. J. LEAMY & J. KIESER, 1998. Directional asymmetry and the measurement of developmental instability. Biol. J. Linn. Soc., London, **64**: 1-16.

GRAHAM, J. H., D. C. FREEMAN & J. M. EMLEN, 1993. Developmental stability: a sensitive indicator of populations under stress. In: W. G. LANDIS, J. HUGHES & M. A. LEWIS (eds.), Environmental toxicology and risk assessment. ASTM STP, **1179**: 136-158. (American Society for Testing and Materials, Philadelphia, PA).

GRAHAM, J. H., S. RAZ, H. HEL-OR & E. NEVO, 2010. Fluctuating asymmetry: methods, theory, and applications. Symmetry, **2**: 466-540.

HAMDOUN, A. & D. EPEL, 2007. Embryo stability and vulnerability in an always changing world. Proc. Natl. Acad. Sci. USA, **104**(6): 1745-1750.

KAT, P. W., 1982. The relationship between heterozygosity for enzyme loci and developmental homeostasis in peripheral populations of aquatic bivalves (Unionidae). Am. Nat., **119**: 824-832.

KIMMEL, D. G., W. D. MILLER & M. R. ROMAN, 2006. Regional scale climate forcing of mesozooplankton dynamics in Chesapeake Bay. Estuar. Coasts, **29**(3): 375-387.

KIRKWOOD, T. B. L. & C. E. FINCH, 2002. Ageing: the old worm turns more slowly. Nature, **419**(6909): 794-795.

KOEHN, R. K. & B. L. BAYNE, 1989. Towards a physiological and genetic understanding of the energetics of the stress response. Biol. J. Linn. Soc., London, **37**: 157-171.

KOTTA, J., V. LAURINGSON, G. MARTIN, M. SIMM, I. KOTTA, K. HERKÜL & H. OJAVEER, 2008. Gulf of Riga and Pärnu Bay. In: U. SCHIEWER (ed.), Ecology of Baltic coastal waters. Ecological Studies (Analysis and Synthesis), **197**: 217-243. (Springer, Berlin).

KOZLOV, M. V. & V. ZVEREV, 2018. Temperature and herbivory, but not pollution, affect fluctuating asymmetry of mountain birch leaves: results of 25-year monitoring around the copper-nickel smelter in Monchegorsk, northwestern Russia. Science of the Total Environment, **640-641**: 678-687.

LAJUS, D. L., 2001. Variation patterns of bilateral characters: variation among characters and among populations in the White Sea herring, *Clupea pallasi marisalbi* (Berg) (Clupeidae, Teleostei). Biol. J. Linn. Soc., London, **74**: 237-253.

LAJUS, D. L., 2014. Developmental instability and fluctuating asymmetry. [Vignette in:] M. C. NEWMAN, Fundamentals in ecotoxicology — the science of pollution (4$^{th}$ ed.): 253-259. (Taylor & Francis, Boca Raton, FL).

LAJUS, D. L. & V. R. ALEKSEEV, 2000. Components of morphological variation in Baikalian endemial cyclopid *Acanthocyclops signifer* complex from different localities. Hydrobiologia, **417**: 25-35.

LAJUS, D. L. & V. R. ALEKSEEV, 2004. Phenotypic variation and developmental instability of life-history traits: a theory and a case study on within-population variation of resting eggs formation in *Daphnia*. J. Limnol., **63**: 37-44.

LAJUS, D. L., M. CIOSTEK, M. DRASZANOWSKA & T. SYWULA, 2003c. Geographic and ontogenetic patterns of chaetotaxy variation in glacial relict *Saduria entomon* (Crustacea, Isopoda): inter-population, inter-individual and intra-individual variations (fluctuating asymmetry). Ann. Zool. Fenn., **40**: 411-419.

LAJUS, D. [L.], YE. GLAZKOVA, D. SENDEK, V. KHAITOV & YU. LAJUS, 2015d. Dynamics of fish catches in the eastern Gulf of Finland (Baltic Sea) and downstream of the Neva River during the 20th century. Aquat. Sci., **77**(3): 411-425. DOI:10.1007/s00027-014-0389-9.

LAJUS, D. L., P. V. GOLOVIN, A. O. YURTSEVA, T. S. IVANOVA, A. S. DORGHAM & M. V. IVANOV, 2019. Fluctuating asymmetry as an indicator of stress and fitness in stickleback: analysis of publications and testing cranial structures. Evolutionary Ecology Research, **20**: 83-106.

LAJUS, D. L., J. H. GRAHAM & A. V. KOZHARA, 2003a. Developmental instability and the stochastic component of total phenotypic variance. In: M. POLAK (ed.), Developmental instability: causes and consequences: 343-363. (Oxford University Press, New York, NY).

LAJUS, D. [L.], M. KATOLIKOVA, P. STRELKOV & H. HUMMEL, 2015c. Fluctuating and directional asymmetry of the blue mussel (*Mytilus edulis*): improving methods of morphological analysis to explore species performance at the northern border of its range. Symmetry, **7**(2): 488-514. DOI:10.3390/sym7020488.

LAJUS, D. L., R. KNUST & O. BRIX, 2003b. Fluctuating asymmetry and other parameters of morphological variation of eelpout *Zoarces viviparus* from different parts of distributional range. Sarsia, **88**: 247-260.

LAJUS, D. [L.], N. SUKHIKH & V. ALEKSEEV, 2015b. Cryptic or pseudocryptic: can morphological methods inform copepod taxonomy? An analysis of publications and a case study of the *Eurytemora affinis* species complex. Ecol. Evol., **5**(12): 2374-2385. DOI:10.1002/ece3.1521.

LAJUS, D. L. & A. A. SUKHOTIN, 1998. A new approach for the assessment of stochastic variation: analysis of behavioural response in blue mussel (*Mytilus edulis* L.). Helgoländer Wissenschaftliche Meeresuntersuchungen, **52**: 141-145.

LAJUS, D. [L.], A. YURTSEVA, G. BIRCH & D. BOOTH, 2015a. Fluctuating asymmetry as a pollution monitor: the Australian estuarine smooth toadfish *Tetractenos glaber* (Teleostei: Tetraodontidae). Marine Pollution Bulletin, **101**(2): 758-767. DOI:10.1016/j.marpolbul.2015.09.038.

LEAMY, L. J. & C. P. KLINGENBERG, 2005. The genetics and evolution of fluctuating asymmetry. Annu. Rev. Ecol. Evol. Syst., **36**: 1-21.

LEARY, R. F. & F. M. ALLENDORF, 1989. Fluctuating asymmetry as an indicator of stress: implications for conservation biology. Trends Ecol. Evol., **4**: 214-217.

LEARY, R. F., F. W. ALLENDORF & K. L. KNUDSEN, 1992. Genetic, environmental, and developmental causes of meristic variation in rainbow trout. Acta Zool. Fenn., **191**: 79-95.

LEGRAND, E., C. BOULANGÉ-LECOMTE, G. RESTOUX, G. TRÉMOLET, A. DUFLOT & J. FORGET-LERAY, 2017. Individual and mixture acute toxicity of model pesticides chlordecone and pyriproxyfen in the estuarine copepod *Eurytemora affinis*. Environmental Science and Pollution Research, **24**(6): 5976-5984.

LEHTINIEMI, M., A. ANTSULEVICH, J. KOTTA, A. MAXIMOV, H. OJAVEER & M. ORLOVA, 2016. Non-indigenous species. In: M. RAATEOJA & O. SETÄLÄ (eds.), The Gulf of Finland assessment. Reports of the Finnish Environment Institute, Helsinki, **2016**: 264-275.

LEPPAKOSKI, E., S. GOLLASCH, P. GRUSZKA, H. OJAVEER, S. OLENIN & V. PANOV, 2002. Baltic — a sea of invaders. Canadian Journal of Fisheries and Aquatic Sciences, **59**(7): 1175-1188.

LUDVIG, W., 1932. Das Rechts-Links Problem im Tierreich und beim Menschen: 1-496. (Springer, Berlin).

MATHER, K., 1953. Genetical control of stability in development. Heredity, **7**(3): 297-336.

MICHALEC, F. G., S. SOUISSI & M. HOLZNER, 2015. Turbulence triggers vigorous swimming but hinders motion strategy in planktonic copepods. J. Roy. Soc. Interface, **12**: 20150158.

MICHALEC, F.-G., I. FOUXON, S. SOUISSI & M. HOLZNER, 2017. Zooplankton can actively adjust their motility to turbulent flow. Proceedings of the National Academy of Sciences of the United States of America, **114**(52): E11199-E11207. DOI:10.1140/epje/i2015-15108-2.

MØLLER, A. P., 2000. Symmetry, size and stress. Trends Ecol. Evol., **15**: 330.

MØLLER, A. P. & J. P. SWADDLE, 1997. Asymmetry, developmental stability and evolution: i-xiii, 1-291. (Oxford University Press, Oxford, UK).

MUSHEGIAN, A. A., E. BURCKLEN, T. M. M. SCHÄR & D. EBERT, 2016. Temperature-dependent benefits of bacterial exposure in embryonic development of *Daphnia magna* resting eggs. Journal of Experimental Biology, **219**(6): 897-904.

OSTBYE, K., S. A. OXNEVAD & L. A. VOLLESTAD, 1997. Developmental stability in perch (*Perca fluviatilis*) in acidic aluminium-rich lakes. Can. J. Zool., **75**: 919-928.

PALMER, A. R. & C. STROBECK, 1986. Fluctuating asymmetry: measurement, analysis, patterns. Annu. Rev. Ecol. Syst., **17**: 391-421.

PALMER, A. R. & C. STROBECK, 2003. Fluctuating asymmetry analyses revisited. In: M. POLAK (ed.), Developmental instability: causes and consequences: 279-319. (Oxford University Press, New York, NY).

PANOV, V. E., D. E. BYCHENKOV, N. A. BEREZINA & A. M. MAXIMOV, 2003. Alien species introductions in the eastern Gulf of Finland: current state and possible management options. Proc. Estonian Acad. Sci., (Biol. Ecol.) **52**(3): 254-267.

PARSONS, P. A., 1990. Fluctuating asymmetry: an epigenetic measure of stress. Biol. Rev., **65**: 131-145.

PEARSON, K., 1911. Grammatika nauki: 1-655. (Shipovnik Press, St.-Petersburg). [Translation into Russian of the second English edition of "The Grammar of Science".]

PODGORNY, K. A., 2018. A mathematical model for studying ecosystems Vistula Gulf of the Baltic Sea. Part 1. Theoretical foundations and model structure, methodology for preparing the initial data for performing calculations. Atlantichefii Nauchno-Issledovatelsky Institute Rybnogo Khoziaistva i Okeanografii: 1-271. (AtlantNIRO, Kaliningrad).

POJAS, R. G. & S. R. M. TABUGO, 2015. Fluctuating asymmetry of parasite infested and non-infested *Sardinella* sp. from Misamis Oriental. AACL Bioflux, **8**(1): 7-14.

PRESTA, M. L., F. L. CAPITANIO & M. S. HOFFMEYER, 2019. Spatio-temporal dynamics of mesozooplankton in the subantarctic Beagle Channel: the case of Ushuaia Bay. Regional Studies in Marine Science, **28**: 100546.

REEVE, E. C. R., 1960. Some genetic tests on asymmetry of sternopleural chaeta number in *Drosophila*. Genet. Res., **1**: 151-172.

REIMCHEN, T. E., 1997. Parasitism of asymmetrical pelvic phenotypes in sticklebacks. Can. J. Zool., **75**: 2084-2094.

RICE, W. R., 1989. Analysing tables of statistical tests. Evolution, **43**: 223-225.

SCHNEIDER, S. S., L. J. LEAMY, L. A. LEWIS & G. DEGRANDI-HOFFMAN, 2003. The influence of hybridization between African and European honeybees, *Apis mellifera*, on asymmetries in wing size and shape. Evolution, **57**(10): 2350-2364. DOI:10.1111/j.0014-3820.2003.tb00247.x.

SIMONS, A. M., 2011. Modes of response to environmental change and the elusive empirical evidence for bet hedging. Proc. Biol. Sci., **278**(1712): 1601-1609. DOI:10.1098/rspb.2011.0176.

SIMONS, A. M. & M. O. JOHNSTON, 1997. Developmental instability as a bet-hedging strategy. Oikos, **80**: 401-406.

SOMMER, C., 1996. Ecotoxicology and developmental stability as in situ monitor of adaptation. Ambio, **25**: 374-376.

SOULÉ, M. E., 1967. Phenetics of natural populations. II: asymmetry and evolution in a lizard. Am. Nat., **101**: 141-160.

SOULÉ, M. E., 1982. Allomeric variation. I. The theory and some consequences. Am. Nat., **120**: 751-764.

SUKHIKH, N. M. & V. R. ALEKSEEV, 2013. *Eurytemora caspica* sp. nov. from the Caspian Sea — one more new species within the *E. affinis* complex (Copepoda: Calanoida). Proc. Zool. Inst. RAS, **317**(1): 85-100.

SUKHIKH, N. M., A. SOUISSI, S. SOUISSI & V. R. ALEKSEEV, 2013. Invasion of *Eurytemora* sibling species (Copepoda: Temoridae) from North America into the Baltic Sea and European Atlantic coast estuaries. J. Nat. Hist., London, **47**(5-12): 753-767.

SUKHIKH, N. [M.], S. SOUISSI, N. V. SCHIZAS & V. [R.] ALEKSEEV, 2019. Life in sympatry: coexistence of native *Eurytemora affinis* and invasive *Eurytemora carolleeae* in the Gulf of Finland (Baltic Sea). Oceanologia, **61**(2): 227-238. DOI:10.1016/j.oceano.2018.11.002.

SUKHOTIN, A. A., D. L. LAJUS & P. A. LESIN, 2003. Influence of age and size on activity and stress resistance in marine bivalve *Mytilus edulis*. Journal of Experimental Marine Biology and Ecology, **284**(1/2): 129-144.

TAN-KRISTANO, A., A. HOFFMANN, R. WOODS, P. BATTERHAM, C. COBBETT & C. SINCLAIR, 2003. Translational asymmetry as a sensitive indicator of cadmium stress in plants: a laboratory test with wild-type and mutant *Arabidopsis thaliana*. New Phytol., **159**: 471-477.

TEBB, G. & J. M. THODAY, 1954. Stability in development and relational balance of X-chromosomes in *Drosophila melanogaster*. Nature, **174**: 1109-1110.

THODAY, J. M., 1958. Homeostasis in a selection experiment. Heredity, **12**(4): 401-415.

TIAN, Y., C. BOULANGÉ-LECOMTE, A. BENAMAR, N. GIUSTI-PETRUCCIANI, A. DUFLOT, S. OLIVIER, C. FREDERICK, J. FORGET-LERAY & F. PORTET-KOLTALO, 2017. Application of a crustacean bioassay to evaluate a multi-contaminated (metal, PAH, PCB) harbor sediment before and after electrokinetic remediation using eco-friendly enhancing agents. Science of the Total Environment, **607-608**: 944-953.

TIMOFEEFF-RESSOVSKY, N. V., 1934. Über der Einfluss des genotypischen Milieus und der Aussenbedingungen auf die Realisation des Genotypes. Genmutation vti bei *Drosophila funebris*. Nachr. Acad. Wiss. Göttingen, (II, Math.-Phys.) **1**: 53-106.

TOUMI, H., M. BOUMAIZA, M. MILLET, C. M. RADETSKI, B. I. CAMARA, V. FELTEN & J.-F. FERARD, 2015. Investigation of differences in sensitivity between 3 strains of *Daphnia magna* (crustacean Cladocera) exposed to malathion (organophosphorous pesticide). Journal of Environmental Science and Health, (Part B, Pesticides, Food Contaminants, and Agricultural Wastes) **50**(1): 34-44.

TRACY, M., D. C. FREEMAN, J. M. EMLEN, J. H. GRAHAM & R. A. HOUGH, 1995. Developmental instability as a biomonitor of environmental stress: an illustration using plants and macroalgae. In: F. M. BUTTERWORTH, L. D. CORKUM & J. GUZMÁN-RINCÓN (eds.), Biomonitors and biomarkers as indicators of environmental change: 313-337. (Plenum Press, New York, NY).

VALENTINE, D. W., M. E. SOULÉ & P. SAMOLOW, 1973. Asymmetry analysis in fishes: a possible indicator of environmental stress. Fish. Bull., **71**: 357-370.

VAN DONGEN, S., 2000. The heritability of fluctuating asymmetry: a Bayesian hierarchical model. Ann. Zool. Fenn., **37**: 15-23.

VAN DONGEN, S. & L. LENS, 2000. Symmetry, size and stress. Trends Ecol. Evol., **15**: 330.

VOGT, G., 2015. Stochastic developmental variation, an epigenetic source of phenotypic diversity with far-reaching biological consequences. Journal of Biosciences, **40**(1): 159-204.

VOGT, G., M. HUBER, M. THIEMANN, G. VAN DEN BOOGAART, O. J. SCHMITZ & C. D. SCHUBART, 2008. Production of different phenotypes from the same genotype in the same environment by developmental variation. J. Exp. Biol., **211**: 510-523.

WADDINGTON, C. H., 1957. The strategy of the genes: i-ix, 1-262. (George Allen & Unwin, London).

WHITE, T. A. & J. B. SEARLE, 2008. Mandible asymmetry and genetic diversity in island populations of the common shrew, *Sorex araneus*. J. Evol. Biol., **21**: 636-641.

WHITLOCK, M. C., 1998. The repeatability of fluctuating asymmetry: a revision and extension. Proceedings of the Royal Society of London, (B, Biological Sciences) **265**: 1429-1431.

WILLIAMS, R. J. & R. B. PELTON, 1966. Individuality in nutrition: effects of vitamin A-deficient and other deficient diets on experimental animals. Proc. Natl. Acad. Sci. USA, **55**: 126-134.

ZAKHAROV, V. M., 1989. Future prospects for population phenogenetics. Soviet Scient. Rev., (sect. F) **4**: 1-79.

First received 5 September 2019.
Final version accepted 22 January 2020.

[When citing this volume, please refer to *Crustaceana* 93 (2020), 241-547]

*Proceedings of the* Eurytemora *conference, St. Petersburg, 2019*

# VARIABILITY OF MANDIBLE SHAPE IN THE FRESHWATER GLACIAL RELICT *EURYTEMORA LACUSTRIS* (POPPE, 1887) (COPEPODA, CALANOIDA, TEMORIDAE)

BY

ŁUKASZ SŁUGOCKI[1,2,3])

[1]) Department of Hydrobiology, Institute of Biology, University of Szczecin, Wąska 13, 71-415 Szczecin, Poland

[2]) Center of Molecular Biology and Biotechnology, University of Szczecin, Wąska 13, 71-415 Szczecin, Poland

## ABSTRACT

The mandibles of copepods are used to classify the feeding modes of certain species, because these appendages have specific shapes. The number, shape, location, and structure of the mandibular teeth are traits that are related to the feeding preferences of copepods. Therefore, the mandibles have been used as parameters in order to determine the role of the corresponding species in the trophic network. In this study, we aimed to determine the morphological variability of the mandible shape in a rare calanoid, *Eurytemora lacustris* (Poppe, 1887). The size of the mandibles of *E. lacustris* is closely related to its body size; however, considering the proportions of the width of the mandibles to that body size, young individuals have relatively larger mandibles. Despite the large body size of males, the females are characterized by larger mandibles than those of males. The teeth of males were found to be smaller, because of intensive abrasion. Moreover, the season and habitat affected the size of the mandibles and also the values of the edge index (EI). The food conditions in the studied lakes fluctuated within seasons, which led to changes in the shape of the dental crowns. Using this index (EI), *E. lacustris* was classified as an omnivorous species; however, the range of calculated values varied considerably. We note that the values of the EI decrease along with maturation, which suggests that with increase in age, the copepod becomes more herbivorous; however, this contradicts the paradigm that copepods become omnivorous during maturation. All those morphological differences, furthermore, indicate the high phenotypic plasticity of the species, which was found to be related to maturity, sex, phenology, and the environmental conditions of the habitat.

## RÉSUMÉ

Les mandibules des copépodes sont utilisées pour classer les modes d'alimentation de certaines espèces, car ces appendices ont des formes spécifiques. Le nombre, la forme, la position et la structure des dents mandibulaires sont des caractères en relation avec les préférences alimentaires des copépodes. Ainsi, les mandibules ont été utilisées comme paramètres afin de déterminer le rôle

[3]) e-mail: lukasz.slugocki@usz.edu.pl

 | DOI:10.1163/9789004465565_009

des espèces correspondantes dans le réseau trophique. Dans cette étude, nous avions pour objectif de déterminer la variabilité morphologique de la forme de la mandibule chez un calanoïde rare, *Eurytemora lacustris* (Poppe, 1887). La taille des mandibules de *E. lacustris* est étroitement liée à la taille de son corps ; cependant, si on considère le rapport de la largeur des mandibules à la taille du corps, les individus jeunes ont des mandibules relativement plus grandes. Malgré la grande taille du corps des mâles, les femelles sont caractérisées par de plus grandes mandibules que les mâles. Les dents des mâles ont été trouvées plus petites, en raison d'une abrasion intensive. De plus, la saison et l'habitat affectent la taille des mandibules et aussi, la valeur de l'indice de tranchant (EI : edge index). Les conditions de nutrition dans les lacs étudiés variaient au cours des saisons, conduisant à des changements dans la forme des couronnes dentaires. En utilisant cet index (EI), *E. lacustris* a été classée comme espèce omnivore ; cependant, l'échelle des valeurs calculées variait considérablement. Nous notons que les valeurs de EI diminuent tout au long du vieillissement, ce qui suggère qu'avec l'âge, le copépode devient plus herbivore ; cependant, cela contredit le paradigme suivant lequel les copépodes deviennent plus omnivores avec le temps. Toutes les différences morphologiques, en outre, indiquent la haute plasticité phénotypique de l'espèce, qui s'est avérée être liée à la maturité, au sexe, à la phénologie, et aux conditions environnementales de l'habitat.

## INTRODUCTION

A typical habitat of *Eurytemora lacustris* (Poppe, 1887) is constituted by deep, clear-water lakes (Kasprzak et al., 2005). Extensive agriculture and nutrients of anthropogenic origin have increased the eutrophication of many inland waterbodies; hence, nutrient-rich waters have become a serious threat for populations of *E. lacustris* (cf. Maier et al., 2011; Karpowicz & Kalinowska, 2018; Sługocki & Czerniawski, 2018). Therefore, *E. lacustris* has been classified as an endangered species by some European countries (Maier et al., 2011; Spikkeland et al., 2016). Rapid changes in the trophic status and the effects of global climate change, may force organisms to quickly adapt to new conditions. If the species is not phenotypically fitted to cope with such new situations, it loses its chances of winning the competition with other species (Darwin, 1859). Organisms have to be successful in feeding, survival, and reproduction to thrive properly (Brun et al., 2016); therefore, in a rapidly changing environment, feeding is one of the major factors that determine the survival of any species (Litchman et al., 2013).

The number, shape, location, and construction of the teeth on the mandibles are traits that are related to the feeding preference of copepods. They have been playing an important role in the species' evolution by determining the role of copepods in the trophic network (Itoh, 1970; Ohtsuka & Onbé, 1989; Michels & Gorb, 2015). Therefore, knowledge about mandible variability can serve as a marker in quick assessment of changes in trophic relationships. Mandibles have species-specific shapes that are related to the feeding mode of a species; accordingly, they are used to classify copepods as carnivorous, herbivorous, or omnivorous, based on the calculation of the so-called edge index (EI) (Itoh, 1970). This index is useful for determining the trophic status of hydrobionts, because the morphology of the oral appendages and the contents of the stomachs of specimens are necessary tools for a comprehensive assessment and confirmation of the species' feeding preferences. In addition, the mandible could be used for identifying microfossils (Harvey et al., 2012) or for estimating the size of an ingested copepod found in the gut contents of a predator (Giesecke & González, 2004).

There are very few studies in the literature about the variability of the mandible in copepods. Usually, studies regarding mandibles are based on a small number of specimens; therefore, in some cases, calculations considering species differ considerably. Among the genus *Eurytemora*, the EI value was calculated for *Eurytemora affinis* (Poppe, 1880), and its value for *E. affinis* ranged from 522 (Revis et al., 1991) to 760 (Schnack, 1982). Furthermore, Revis et al. (1991) attempted to discover the morphology of the mandible in *E. affinis*. They observed differences between sexes but they did not demonstrate the actual calculations behind these observations. The elemental analysis conducted by Vezhnovets et al. (2012) revealed a relatively greater strength of the mandibular cutting edge in *E. lacustris* than in *Eurytemora velox* (Lilljeborg, 1853). Therefore, unlike other species of *Eurytemora*, it seems that *E. lacustris* teeth may be characterized by low abrasiveness. During maturation, differences are observed in the behaviour of copepods, which is also connected with morphological changes. According to Dussart & Defaye (2001), diet changes with the body size of the copepod, and if the species is omnivorous, predation starts at stage C3 or rarely C2. Therefore, a shift in the mandible shape is expected among maturity stages, which can be easily described by the EI.

Although there are many studies regarding biological and ecological aspects of *E. affinis* (e.g., Revis et al., 1991; Alekseev & Souissi, 2011; Sukhikh et al., 2013; Lajus et al., 2015), for the rare glacial relict *E. lacustris*, studies have mainly focused on the ecology of the species (e.g., Maier et al., 2011; Karpowicz & Kalinowska, 2018) and have rarely focused on their morphology (Sługocki et al., 2019). Therefore, this study will supplement the state of knowledge on *Eurytemora* species — in fact, a model aquatic crustacean (Engström et al., 2000; Winkler et al., 2008; Lee, 2016).

Accordingly, this study aims to determine the morphological variability of the mandible shape in the freshwater calanoid *E. lacustris*. The following hypotheses have been proposed: (i) the mandible shape differs along with the maturation; (ii) the shape of the mandible depends on sex; and (iii) the shape of the mandible depends on environmental conditions.

## MATERIAL AND METHODS

During daytime in the spring, summer, and autumn of 2018, a set of zooplankton samples was collected from six deep, dimictic lakes located in northwestern Poland; a total of 18 zooplankton samples were collected. Based on existing maps, the sampling stations were set up at the deepest point in each of the lakes (Jańczak, 1996). Table I shows the mean values of temperature, Secchi disc depth (SD), total dissolved solids (TDS), and chlorophyll *a* that were measured with a Hydrolab DS5 sensor (Sutron/ITT Hydromet, Sterling, VA, U.S.A.), in addition to the coordinates, maximum depth, volume, and area of the lakes (Choiński, 2006). At each site, the phytoplankton samples were collected from the epilimnion, metalimnion and hypolimnion. For water samples, a Van Dorn 3-liter sampler was used. According to the procedure reported by Utermöhl (1958), phytoplankton was counted on an inverted plankton microscope. Individuals were counted in a sedimentation chamber (2 ml). Biomass was then calculated from cell volume measurements (Pliński et al., 1984; Kawecka & Eloranta, 1994), and the calculation of the phytoplankton biomass from each zone was combined with other thermal zones.

Zooplankton samples were collected by dragging a net (mesh 100 $\mu$m; mouth diameter = 20 cm) from the bottom of the lake to the surface. Concentrated samples were poured into a 110-ml tube and fixed in a buffered 4% formalin solution. The specimens of *Eurytemora lacustris* were placed in a 2-ml plankton chamber using a Zeiss Primo Vert reversed microscope (Germany). The species was identified using taxonomic keys (e.g., Einsle, 1996; Błędzki & Rybak, 2016). Specimens have been stored in collections at the University of Szczecin. The values of the various mandible characteristics were measured from multiple digital images obtained using a Nikon Eclipse 50i microscope (Nikon, Tokyo, Japan) and the software ToupView (ToupTek Photonics, Hangzhou, P.R. China). The body size of the specimens was measured without the caudal setae.

To characterize the feeding modes of the copepods, we applied the mandible edge index proposed by Itoh (1970), where $w_i$ is the space between each pair of teeth, $W$ is the total width of the mandible blade, $h_i$ is the height of each tooth, $H$ is the total height of the cutting edge, and $N$ is the total number of teeth on the mandible blade (fig. 1). Filtering herbivorous species are characterized by an EI of <500, omnivorous species have an EI between 500 and 900, and carnivorous species have an EI of >900. Thus:

$$\text{Edge index} = \frac{\sum(\frac{w_i}{W} * \frac{h_i}{H} * 10^4)}{N}.$$

The relationships between size and mandible traits were verified using Pearson's correlation ($p < 0.05$). To determine whether sex, maturity, seasons, and lakes are

TABLE I

The variables for lakes with *Eurytemora lacustris* (Poppe, 1887) populations

| | Lake | | | | | |
|---|---|---|---|---|---|---|
| | Cieszęcino (53°55′44″N 16°49′27″E) | Drawsko (53°35′22″N 16°10′54″E) | Ińsko (53°26′39″N 15°32′34″E) | Lubie (53°27′23″N 15°54′14″E) | Siecino (53°36′12″N 16°01′13″E) | Żerdno (53°36′17″N 16°12′51″E) |
| Depth (m) | 38 | 82.2 | 42 | 46.2 | 44.2 | 36 |
| Volume ($km^3$) | 13 790 | 331 443 | 65 182 | 169 880 | 104 441 | 31 240 |
| Area (ha) | 102 | 1871.5 | 590 | 1439 | 729 | 205 |
| $T$ (°C) | 11.8 ± 5.3a | 12.3 ± 5.5a | 11.2 ± 5.8a | 10.8 ± 5.8a | 12.3 ± 5.5a | 12.7 ± 5.8a |
| SD (m) | 3 ± 1.4 | 2.4 ± 0.9 | 5.2 ± 0.7 | 1.7 ± 0.5 | 3.9 ± 0.6 | 2.6 ± 0.7 |
| TDS (g/l) | 0.131 ± 0.005a | 0.157 ± 0.015b | 0.094 ± 0.009c | 0.180 ± 0.009d | 0.122 ± 0.006e | 0.147 ± 0.008f |
| chl $a$ ($\mu$g/l) | 2.39 ± 1.97a | 3.99 ± 5.42a | 2.23 ± 1.79a | 8.5 ± 10.39b | 2.42 ± 1.68a | 4.4 ± 4,87a |

Mean and standard deviation values of: $T$, temperature; SD, Secchi disc visibility; TDS, total dissolved solids; and, chl $a$, chlorophyll $a$. Different letters indicate significant differences in values of variables between lakes ($p < 0.05$).

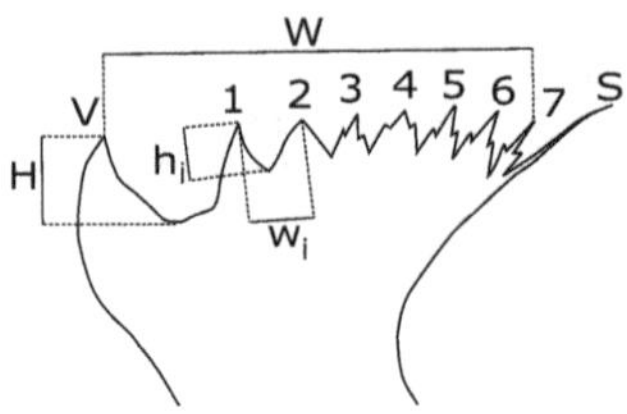


Fig. 1. Mandibular plate of *Eurytemora lacustris* (Poppe, 1887). V, ventral tooth; 1-7, teeth nos. 1 to 7; S, spine; W, width of mandibular plate; H, height of ventral tooth; $w_i$, width between teeth; $h_i$, height of teeth.

important factors for mandible shape, three mandible traits were used: mandible width, mean teeth height, and EI. The parametric one-way ANOVA and the post-hoc Duncan's test were used to determine significant differences in mandible traits according to maturity ($n = 303$, all samples), sex ($n = 278$ — specimens for whom sex determination was possible), lake ($n = 148$, adults only), and season ($n = 148$, adults only) ($p < 0.05$) (Statistica 12, StatSoft). In statistical analysis, specimens with jagged and sharp teeth were included. In each seasons females of *Eurytemora* had egg sacs and adult specimens were present in all lakes during the studied period. The lower numbers of measurements for lakes and season were used to eliminate effects of maturity stage among populations. To determine the best predictors for the values of characteristics of *E. lacustris* mandibles ($n = 148$, adults only) in relation to phytoplankton biomass, a multiple stepwise regression was used ($p < 0.05$).

## RESULTS

The lakes with *Eurytemora lacustris* populations are different in terms of their morphology and physico-chemical characteristics ($F = 332.03$; $p < 0.05$) (table I). In this study, we have considered both clean water lakes such as Ińsko, and turbid lakes such as Lubie. The biomass of different groups of phytoplankton fluctuated during different seasons (fig. 2). In general, diatoms were the most dominant group of phytoplankton and contributed to approx. 58% of the total biomass. The additional contributors to total biomass were Dinophyta (15%), Chrysophyceae (9%), Cyanoprokaryota (8%), Chlorophyta (6%), and Charophyta (3%). Other groups of phytoplankton were <1% of total biomass. In Cieszęcino, Drawsko, Lubie, Siecino and Żerdno, diatoms dominated the phytoplankton biomass, whereas Ińsko was dominated by Chrysophyceae. In Lubie, 99% of the phytoplankton biomass consisted of Dinophyta during the summer.

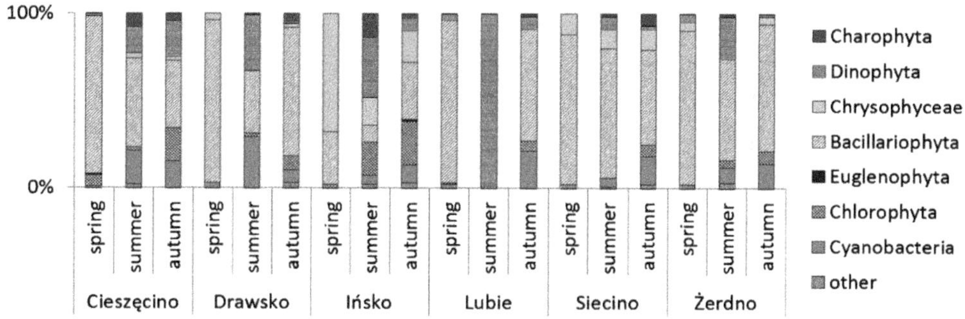


Fig. 2. Percentage contribution of particular groups of phytoplankton to the total biomass in the studied lakes in 2018.

## Mandible characteristics and maturity

It has been observed, that with the moulting of worn teeth (fig. 3A-B), the "new" teeth become sharp again (fig. 3C), but this was not observed among any adult individuals. Juvenile individuals (CII-CV) did not differ in terms of the number of teeth from adults (A). The feature that allows distinguishing adults from copepodid individuals is the bulge between ventral tooth and first central tooth (fig. 3D). Significant differences were noted between the height of adult *E. lacustris* teeth

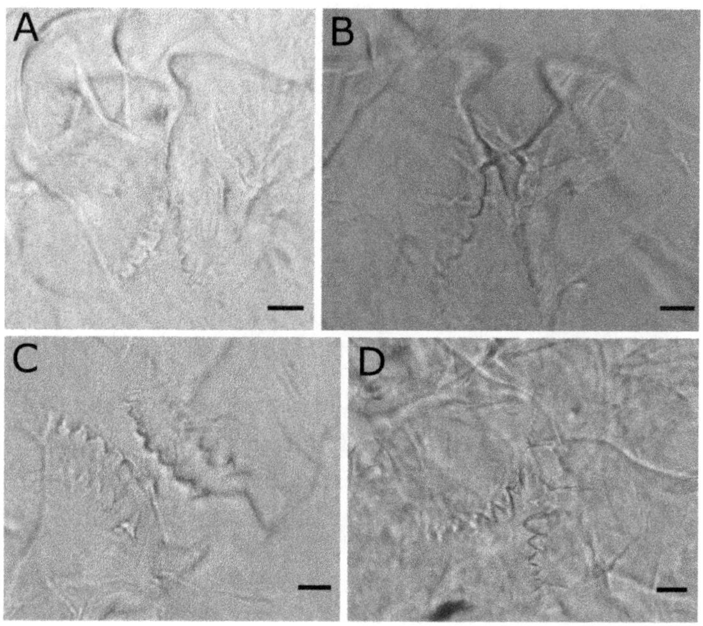


Fig. 3. Mandibles of *Eurytemora lacustris* (Poppe, 1887); especially the masticatory edges. A, male mandible with all teeth jagged; B, male mandible with some teeth jagged; C, mandibles during moulting; D, bulge between first two teeth of adult *Eurytemora*. Scale bar — 10 $\mu$m.

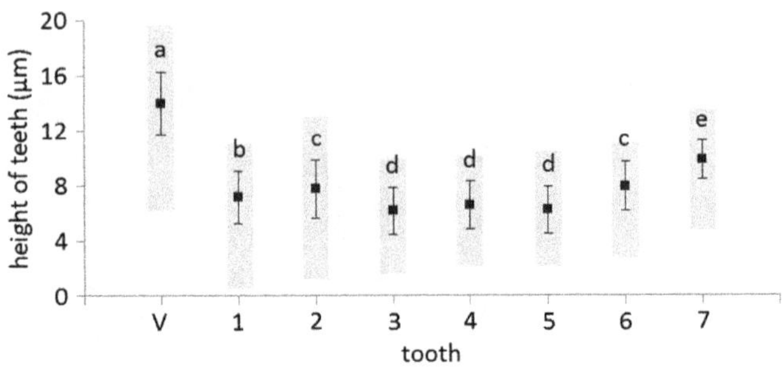


Fig. 4. Mean height (with SD) of mandibular teeth of adult *Eurytemora lacustris* (Poppe, 1887) and significant differences among them ($p < 0.05$). Grey boxes indicate ranges of values. Percentage of standard deviation in relation to mean value: V (ventral tooth), 16%; 1, 26%; 2, 27%; 3, 28%; 4, 26%; 5, 27%; 6, 22%; 7, 14%.

($F = 301.97$; $p < 0.05$) (fig. 4). The teeth located on the edges were characterized by the lowest diversity according to the mean high value.

A close relationship was established between the mandible width and the carapace length of *E. lacustris* ($r = 0.78$; $p < 0.001$) (fig. 5A). Less clear but significant tendencies are observed in the relationship between the size of individuals and the average tooth height ($r = 0.46$; $p < 0.001$) (fig. 5B). The EI, which considers the width between the teeth and the height of the teeth, is highly dispersed; however, it shows a statistically significant negative relation to the size of the individuals ($r = -0.15$; $p = 0.006$) (fig. 5C). The variability of EI is much greater than that observed by other researchers (fig. 6). Regarding ANOVA and post-hoc tests, the mean size of teeth ($F = 19.32$; $p < 0.05$) and width of mandibles ($F = 94.05$; $p < 0.05$) differ significantly among maturity stages (fig. 7A-C). Furthermore, mandible width as a percentage of carapace length ($6.1 \pm 0.6\%$) significantly drops as specimens become larger ($r = -0.57$; $p < 0.001$) (fig. 8).

## Differences between sexes

Females are characterized by significantly wider and on average higher teeth than males ($p < 0.05$) (fig. 7D-F). No significant differences between sexes were revealed for body size (body length). Despite the high EI values for females, the statistical test did not confirm significant differences in this parameter between sexes ($p > 0.05$).

## Lakes, seasons, and food base

In terms of the investigated traits of mandibles, the populations from the studied lakes significantly differ. The values of mean teeth height were significantly high

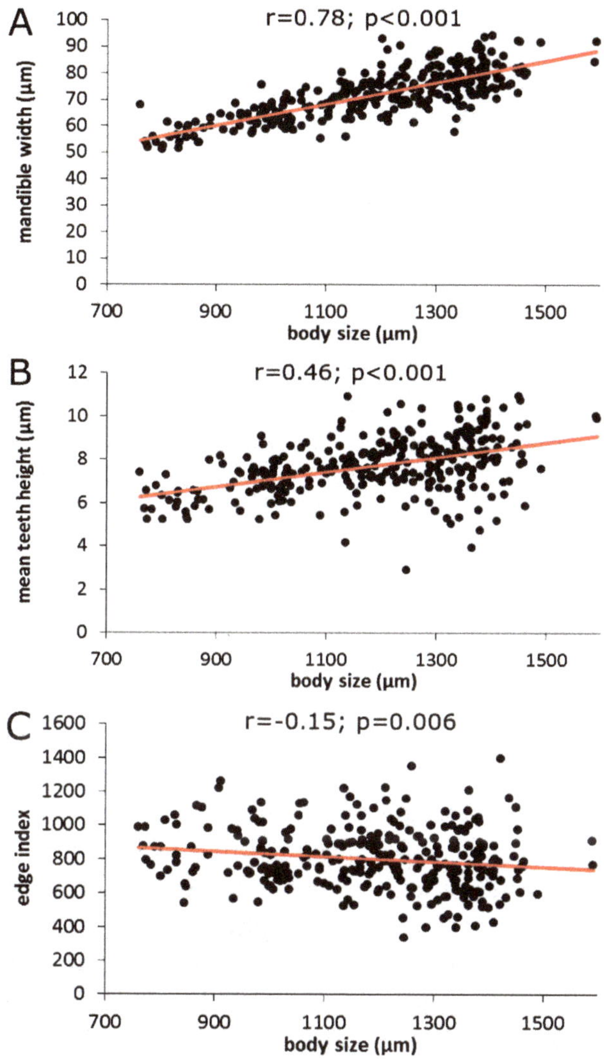


Fig. 5. Pearson correlations between mandible characteristics and body size of *Eurytemora lacustris* (Poppe, 1887) ($p < 0.05$). A, the relationship between mandible width and body size; B, the relationship between mean teeth height and body size; C, the relationship between edge index value and body size.

for Lubie, Ińsko and Siecino compared to other lakes ($F = 4.43$; $p < 0.05$) (fig. 7G-I). The width of mandibles was significantly higher for Lubie and the least for Ińsko ($F = 14.19$; $p < 0.05$). For Ińsko, the value of EI was the highest compared to other lakes ($F = 9.24$; $p < 0.05$).

In the summer, the width of mandibles was the largest and differed significantly between spring and autumn ($F = 19.45$; $p < 0.05$) (fig. 7J-L). According to mean teeth height, no significant differences were found; however, with regard

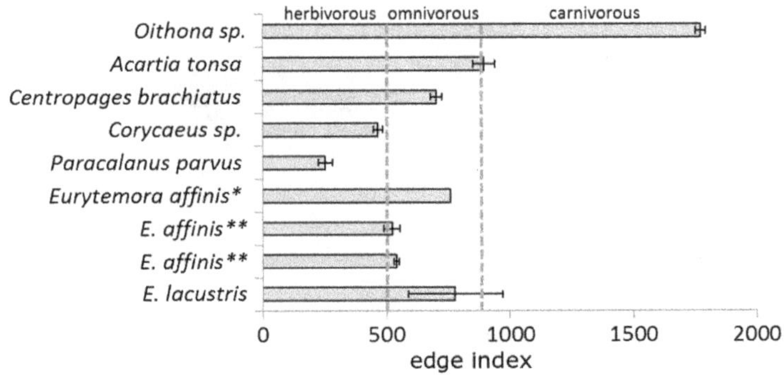


Fig. 6. Mean values (with SD) of the Edge Index (EI) calculated for adult specimens of some calanoid species, where 0-500 is for herbivorous, 500-900 is for omnivorous, >900 is for carnivorous. *Oithona* sp. ($n = 8$), *Acartia tonsa* Dana, 1849 ($n = 14$), *Centropages brachiatus* (Dana, 1849) ($n = 11$), *Corycaeus* sp. ($n = 7$), *Paracalanus parvus* (Claus, 1863) ($n = 15$) (all from Giesecke & González, 2004); *Eurytemora affinis* (Poppe, 1880)∗ (from Schnack, 1982); *E. affinis*** ($n = 5$ in each case: 5 specimens from the Gironde and 5 from the Westerschelde estuary, respectively) (from Revis et al., 1991); and *Eurytemora lacustris* (Poppe, 1887) ($n = 148$) (present study).

to EI values, significant differences between spring and summer were observed ($F = 7.90$; $p < 0.05$).

According to multiple regression, the groups of phytoplankton most significantly related to characteristics of *E. lacustris* were Cyanobacteria, Dinophyta, Charophyta, Chlorophyta, Bacillariophyta and Euglenophyta (table II). EI was negatively related with the biomass of Cyanobacteria ($\beta = -0.49$; $p < 0.001$), Charophyta ($\beta = -0.27$; $p < 0.05$) and positively with Chlorophyta ($\beta = 0.16$; $p < 0.01$), while teeth height was negatively correlated with biomass of Cyanobacteria ($\beta = -0.25$; $p < 0.01$), Bacillariophyta ($\beta = -0.17$; $p < 0.05$), Euglenophyta ($\beta = -0.24$; $p < 0.01$) and Charophyta ($\beta = -0.20$; $p < 0.05$). Note that mandible width was positively related with the biomass of Cyanobacteria ($\beta = 0.28$; $p < 0.01$), Dinophyta ($\beta = 0.28$; $p < 0.001$), and Charophyta ($\beta = 0.23$; $p < 0.01$) while negatively with Euglenophyta ($\beta = -0.17$; $p < 0.05$). Moreover, copepod size was negatively related with the biomass of Chlorophyta ($\beta = -0.41$; $p < 0.001$), Bacillariophyta ($\beta = -0.21$; $p < 0.01$) and positively related with Dinophyta ($\beta = 0.20$; $p < 0.01$) and Charophyta ($\beta = 0.22$; $p < 0.01$).

## DISCUSSION

The proportions of the mandible width and the height of teeth in copepods may depend on multiple factors, which may lead to changes in their feeding mode and can be described on the basis of the EI. In all studied individuals, the

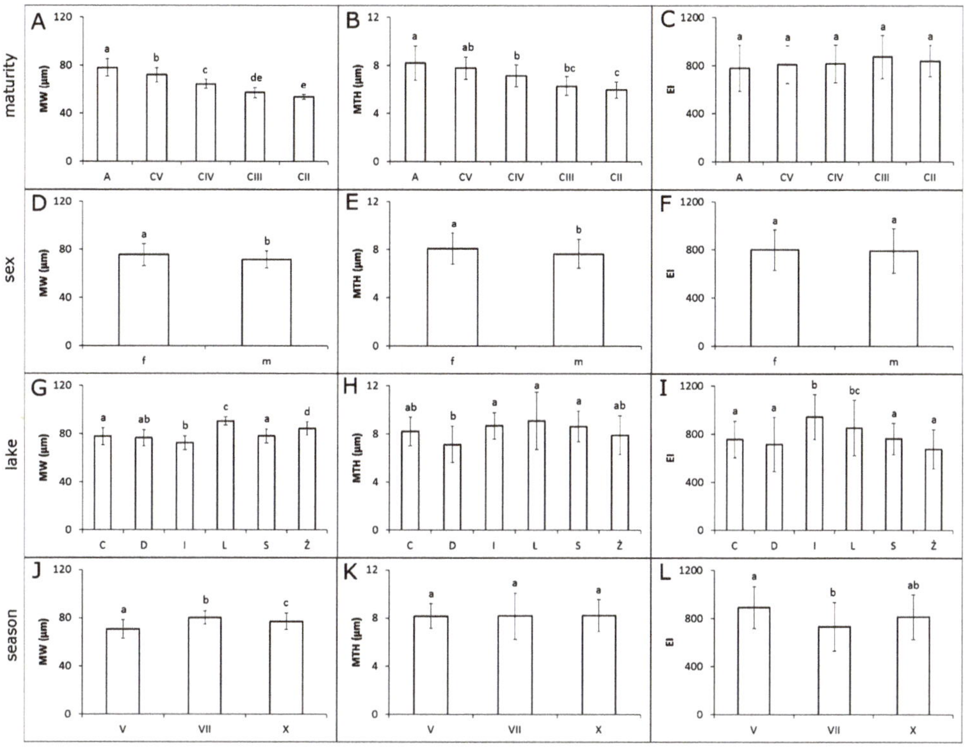


Fig. 7. Mean values (with standard deviation) and differences in mandible biometrics of *Eurytemora lacustris* (Poppe, 1887): MW, mean width; MTH, mean teeth height; and, EI, edge index; respectively, among: A-C, maturity stages; D-F, sexes; G-I, different lakes (C, Cieszęcino; D, Drawsko; I, Ińsko; L, Lubie; S, Siecino; Ż, Żerdno); and, J-L, seasons (V, spring; VII, summer; X, autumn). Different letters indicate significant differences ($p < 0.05$) in values of variables.

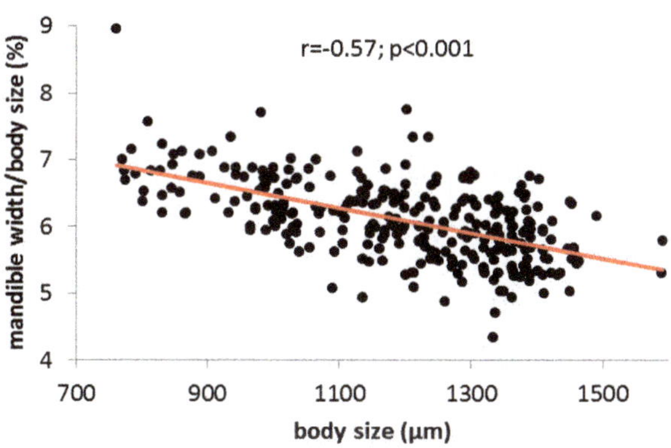


Fig. 8. Pearson correlation in *Eurytemora lacustris* (Poppe, 1887), between percentage of mandible width with regard to body size, plotted against body size.

TABLE II

Significances of the effects of different groups of phytoplankton on EI, mean height of teeth, mandible width, and copepod size (total body length) in *Eurytemora lacustris* (Poppe, 1887) based on multiple regression

| Phytoplankton taxa | Statistical parameter | EI | Teeth height | Mandible width | Copepod size |
|---|---|---|---|---|---|
| Cyanobacteria | $\beta$ | −0.49*** | −0.35** | 0.28** | |
| Chlorophyta | $\beta$ | 0.16* | | | −0.41*** |
| Bacillariophyta | $\beta$ | | −0.17* | | −0.21** |
| Euglenophyta | $\beta$ | | −0.24** | −0.17* | |
| Dinophyta | $\beta$ | | | 0.28*** | 0.20** |
| Charophyta | $\beta$ | −0.27*** | −0.20* | 0.23** | 0.22** |
| | $r$ | 0.36 | 0.21 | 0.37 | 0.33 |
| | $F$ (7.14) | 11.23 | 5.23 | 11.64 | 9.79 |
| | $p$ | <0.0001 | <0.0001 | <0.0001 | <0.0001 |

$^{*}p < 0.05$, $^{**}p < 0.01$, $^{***}p < 0.001$.

number of teeth was the same; however, a bulge between ventral and first central tooth was observed only among adult specimens. Grice & Gibson (1982) reported that the number of teeth in *Labidocera wollastoni* (Lubbock, 1857) varied from 4 to 6 during copepodid development. Jerling & Wooldridge (1994) described a diverse number of teeth in the calanoid *Paracartia longipatella* (Connell & Grindley, 1974) from 7 in the youngest copepodids (CI), to 10 in both CIII and adults. An equal number of teeth for all copepodid stages of *Eurytemora lacustris* in the present study could be informative for palaeoecological studies. Despite the lack of significant differences between the maturity stage and the EI value, a pattern is noticeable, indicating that the value of EI decreases with maturation. Jerling & Wooldridge (1994) reported differences among estuarine copepod maturity stages, where calculations for EI values of *P. longipatella* ranged from 675 among the youngest specimens (Cl), to 592 for CIII and 554 for adults. They observed a similar pattern for *Pseudodiaptomus hessei* (Mrázek, 1894). Both studies contradict (at least according to EI calculation) the popular assumption that copepods become more omnivorous with maturation (Rybak & Błędzki, 2010). Probably, the use of the EI is not applicable for assessing the feeding mode of immature stages of calanoids.

Although females did not differ in body size from males, females were characterized by larger mandibles (in width and teeth height). Females thus show adaptation for feeding on a broader spectrum of organisms than males, to provide a better nutrition for their offspring (therefore, also, the female body is stockier than the male body). The adult stage occurs earlier in males than in females (Dussart & Defaye, 2001); therefore, males are more susceptible for "jagging", i.e., abrasion, of the teeth than females. This could alternatively be explained by certain speculations, as an increase of metabolism (intense feeding), another diet, or an increased brittleness of the teeth due to the characteristics of mineralization. Svetlichny et al. (2012), who investigated two other calanoids (*Calanipeda aquaedulcis* Krichagin, 1873 and *Arctodiaptomus salinus* (Daday, 1885)), found that oxygen consumption doubled in males in comparison with females, in terms of both total and basal metabolism. In that case, males have a greater expenditure of energy that would be compensated by more intense feeding, which consequently accelerates the jagging of the teeth. The differences in the width of the jaws and the height of the teeth between the sexes could result from the adaptation of the female to a larger particle size spectrum, and the greater abrasion of the teeth in males, which may result from faster maturation in males, another diet, an increased brittleness of the teeth due to the characteristics of mineralization, and/or differences in metabolism.

In the summer, *Eurytemora* achieves a larger body size, which would explain the higher value of the width of the mandibles (which is strongly related to body size). Moreover, in the summer the abrasion of the mandibles was substantial; hence, no significant differences were found for the average teeth height between the seasons. Therefore, the EI values were lower in the summer period, which was influenced by the ratio of the mandible width to the average teeth height.

Both abiotic and biotic factors could affect populations of *Eurytemora* and could lead to differences in traits related to the mandible. In the most turbid lake (Lubie), the width of *Eurytemora* mandibles was the highest, and distinct from values of that trait from all other lakes. In contrast, in the most clear-water lake (Ińsko), the width of the mandibles was the lowest. This could be explained by the larger body size of *Eurytemora* in Lubie, resulting from the growth pattern of the copepods as related to the different conditions in the various lakes.

Populations of *E. lacustris* from Drawsko were characterized by the smallest teeth. This is because, in Drawsko, the mandibles of a few individuals of *Eurytemora* were jagged, which could be related to the food that was available for these populations. However, most calanoids are opportunistic, and food is selected by size and taste and depends on season, temperature, and possibly many other factors (Dussart & Defaye, 2001). Diatoms serve as a main source of food for *Eurytemora* (Vezhnovets et al., 2012), and these were occurring in almost every lake throughout the entire period. The high nutrition qualities of diatoms make calanoids improve their adaptations for the acquisition, retention, mastication and consumption as a valuable food. However, the intensity of feeding could have a significant impact on the formation and development of the shape of the mandible in the examined copepods. On the basis of regression analysis, it can be concluded which are the conditions in the lake that contribute to the development of algal groups that play a role in shaping the characteristics of *Eurytemora* mandibles. It seems that diatoms, with their strong silica cell walls, can significantly affect the state of the dental crown. It was earlier confirmed that *E. lacustris* mandibles are stronger than the mandibles of *E. velox* (cf. Vezhnovets et al., 2012), as the habitats of the latter species (eutrophicated waters) offer a large amount and diversity of available particles of food. When the feeding apparatus is losing its ability for crumbling diatoms, this could lead to a drop in feeding efficiency. Therefore, the progressive nicking of the mandibles of *Eurytemora lacustris* could have an impact on life history of this species. The trophic changes in the ambient waters may force *E. lacustris* to change its dietary preferences, which is herein demonstrated by the differences in mandible morphology among the populations of the studied lakes.

Furthermore, in adult specimens, the teeth lying in the central part of the mandible were characterized by the highest variability, which can be explained by their greatest abrasiveness through the frequent contact of these teeth with food particles. There is also an assumption that the physico-chemical conditions of the lakes affected the structure of the teeth; therefore, in some lakes the teeth were more susceptible to abrasion. The teeth on the mandibular gnathobasis of copepods consist of composites with silica-containing structures, connected to the regular chitinous exoskeleton with resilin (Michels & Gorb, 2015). According to Miller et al. (1980), silicon deficiencies can lead to increased effects of wear on insufficiently mineralized teeth of the marine copepod *Acartia tonsa* Dana, 1849. Therefore, the composition of the gnathobase could be related to some chemical elements that are offered by the lake, which could, if sufficiently available, increase their resistance to mechanical damage. Additionally, the representation of various chemical elements in the lakes could have an impact on the intensity of tooth abrasion. Wilken et al. (2011) have experimentally demonstrated that diatoms increase cellular silica concentrations under iron limitation and in consequence increase their frustule strength — what, so it seems, could lead to higher pressure on the mandibular teeth.

An EI of $<500$ indicates that the copepod is herbivorous (Itoh, 1970) and that calculation could be expected for *E. lacustris*. To date, EI calculations have been carried out for *E. affinis*. The calculations of Revis et al. (1991) of the EI for adult females of *E. affinis* showed a mean value of 540 ($n = 5$, SD $= 10$) for specimens collected from the Westerschelde estuary, and of 522 ($n = 5$, SD $= 33$) for specimens collected from the Gironde estuary. The values calculated for the Westerschelde and the Gironde were lower, however, than the results of Schnack (1982) (EI $= 760$) who classified *E. affinis* as an omnivore. The above-mentioned variations of the EI value for a species were greater than those observed by Giesecke & González (2004) for other calanoid species. In the present study, the variation of the EI value was higher than that observed by all other authors. The range of this value in our present study oscillated from 334 to 1396. Thus, *E. lacustris* could be classified as ranging from herbivorous to carnivorous.

Studies on gut contents show that *E. lacustris* is a typical herbivore, as the food pellets mostly contained pennate diatom algae, whereas what remained did not exceed 40 $\mu$m in length (Vezhnovets et al., 2012). In the present study, the gastrointestinal tracts in the populations of each lake were also mostly filled with pennate diatoms. Probably, the common ancestor for this trait in the genus *Eurytemora* that is related to the shape of the mandibles (i.e., sharp teeth), was not fitted to the actual feeding behaviour of *E. lacustris*. We must conclude that the initially omnivore ancestor of *E. lacustris*, became more herbivorous, despite its omnivore-like feeding apparatus. Consequently, the freshwater form *E. lacustris* that has become accustomed to feeding on easily accessible food (diatoms) in clear-water lakes, still has a feeding apparatus that enables it to acquire a diversified diet. Nevertheless, molecular studies of gut contents are necessary for verifying whether *Eurytemora* is exclusively a herbivore, or whether it occasionally preys on animals as well.

Based on the above results, it can be concluded, that with the increase in body size the size of mandibles and also that of their teeth increases. Giesecke & González (2004) reported a close relationship between the mandible width and the carapace length of the three calanoid species *Paracalanus parvus* (Claus, 1863), *Centropages brachiatus* (Dana, 1849) and *Acartia tonsa*. Compared with other calanoids (Giesecke & González, 2004), the relatively small size of *E. lacustris*' mandible, could have implications related to feeding and can result in preying on smaller particles. Copepodids have proportionally higher mandibles than adults, which suggests the adaptation of copepods for feeding on a large spectrum of particles right from the youngest copepodid instars, which provides better nutritional conditions. In large individuals, there are less sharp teeth on average, which suggests that teeth are not crucial for their survival; however, they are perhaps crucial for the effective feeding and rapid growth of new generations before reaching adulthood.

## ACKNOWLEDGEMENT

This research was supported by funds from the National Science Center granted on the basis of decision number DEC-2017/01/X/NZ8/00793.

## REFERENCES

ALEKSEEV, V. R. & A. SOUISSI, 2011. A new species within the *Eurytemora affinis* complex (Copepoda: Calanoida) from the Atlantic coast of U.S.A., with observations on eight morphologically different European populations. Zootaxa, **2767**: 41-56.

BRUN, P. G., M. R. PAYNE & T. KIØRBOE, 2016. A trait database for marine copepods. Earth System Science Data Discussions, **9**: 99-113.

DARWIN, C. R., 1859. On the origin of species by means of natural selection — or, the preservation of races in the struggle for life: 1-532. (Routledge & Kegan Paul, London). [Reprint 2004.]

DUSSART, B. H. & D. DEFAYE, 2001. Introduction to the Copepoda (2nd ed., revised and enlarged). Guides to the Identification of the Microinvertebrates of the Continental Waters of the World, **16**: i-viii: 1-344. (Backhuys Publishers, Leiden).

ENGSTRÖM, J., M. KOSKI, M. VIITASALO, M. REINIKAINEN, S. REPKA & K. SIVONEN, 2000. Feeding interactions of the copepods *Eurytemora affinis* and *Acartia bifilosa* with the Cyanobacteria *Nodularia* sp. Journal of Plankton Research, **22**: 1403-1409.

GIESECKE, R. & H. E. GONZÁLEZ, 2004. Mandible characteristics and allometric relations in copepods: a reliable method to estimate prey size and composition from mandible occurrence in predator guts. Revista Chilena de Historia Natural, **77**: 607-616.

GRICE, G. D. & V. R. GIBSON, 1982. The developmental stages of the calanoid copepod *Labidocera wollastoni* (Lubbock) with observations on its eggs. Cahiers de Biologie Marine, **23**: 215-225.

HARVEY, T. H., M. I. VÉLEZ & N. J. BUTTERFIELD, 2012. Exceptionally preserved crustaceans from western Canada reveal a cryptic Cambrian radiation. Proceedings of the National Academy of Sciences, **109**: 1589-1594.

ITOH, K., 1970. A consideration of feeding habits of planktonic copepods in relation to the structure of their oral parts. Bulletin of the Plankton Society of Japan, **17**: 1-10.

JERLING, H. L. & T. H. WOOLDRIDGE, 1994. Comparative morphology of the feeding appendages of four mesozooplankton species in the Sundays River estuary. African Zoology, **29**: 252-257.

KARPOWICZ, M. & K. KALINOWSKA, 2018. Vertical distribution of the relic[t] species *Eurytemora lacustris* (Copepoda, Calanoida) in stratified mesotrophic lakes. Biologia, Bratislava, **73**: 1197-1204.

KASPRZAK, P., C. REESE, R. KOSCHEL, M. SCHULZ, L. HAMBARYAN & J. MATHES, 2005. Habitat characteristics of *Eurytemora lacustris* (Poppe, 1887) (Copepoda, Calanoida): the role of lake depth, temperature, oxygen concentration and light intensity. International Review of Hydrobiology, **90**: 292-309.

KAWECKA, B. & P. V. ELORANTA, 1994. Outline of the ecology of algae in freshwater and terrestrial habitats: 1-252. (PWN, Warszawa). [In Polish.]

LAJUS, D., N. SUKHIKH & V. ALEKSEEV, 2015. Cryptic or pseudocryptic: can morphological methods inform copepod taxonomy? An analysis of publications and a case study of the *Eurytemora affinis* species complex. Ecology and Evolution, **5**: 2374-2385.

LEE, C. E., 2016. Evolutionary mechanisms of habitat invasions, using the copepod *Eurytemora affinis* as a model system. Evolutionary Applications, **9**: 248-270.

LITCHMAN, E., M. D. OHMAN & T. KIØRBOE, 2013. Trait-based approaches to zooplankton communities. Journal of Plankton Research, **35**: 473-484.

MAIER, G., B. SPETH, A. R. P. WOLFGANG, M. BAHNWART & P. KASPRZAK, 2011. New records of the rare glacial relict *Eurytemora lacustris* (Poppe, 1887)(Copepoda; Calanoida) in atypical lake habitats of northern Germany. Journal of Limnology, **70**: 145-148.

MICHELS, J. & S. N. GORB, 2015. Mandibular gnathobases of marine planktonic copepods — feeding tools with complex micro- and nanoscale composite architectures. Beilstein Journal of Nanotechnology, **6**: 674-685.

MILLER, C. B., D. M. NELSON, R. R. GUILLARD & B. L. WOODWARD, 1980. Effects of media with low silicic acid concentrations on tooth formation in *Acartia tonsa* Dana (Copepoda, Calanoida). The Biological Bulletin, **159**: 349-363.

OHTSUKA, S. & T. ONBÉ, 1989. Evidence of selective feeding on larvaceans by the pelagic copepod *Candacia bipinnata* (Calanoida: Candaciidae). Journal of Plankton Research, **11**: 869-872.

PLIŃSKI, M., J. PICIŃSKA & L. TARGOŃSKI, 1984. Method defining the biomass of marine phytoplankton by means of computers. Zeszyty Naukowe Wydziału Biologii i Nauk o Ziemi Uniwersytetu Gdańskiego, **10**: 129-155. [In Polish.]

REVIS, N., J. CASTEL & M. L. M. TACKX, 1991. Some reflections on the structure of the mandible plate of *Eurytemora affinis* (Copepoda, Calanoida). Hydrobiological Bulletin, **25**: 45-50.

RYBAK, J. I. & L. A. BŁĘDZKI, 2010. Słodkowodne skorupiaki planktonowe: klucz do oznaczania gatunków: 1-268. (Wydawnictwa Uniwersytetu Warszawskiego, Warszawa). [In Polish.]

SCHNACK, S. B., 1982. The structure of the mouth parts of copepods in Kiel Bay. Meeresforschung., **29**: 89-101.

SŁUGOCKI, Ł. & R. CZERNIAWSKI, 2018. Trophic state (TSISD) and mixing type significantly influence pelagic zooplankton biodiversity in temperate lakes (NW Poland). Peer J., **6**: e5731.

SŁUGOCKI, Ł., A. RYMASZEWSKA & L. KIRCZUK, 2019. Insights into the morphology and molecular characterisation of glacial relict *Eurytemora lacustris* (Poppe, 1887) (Crustacea, Copepoda, Calanoida, Temoridae). ZooKeys, **864**: 15-33.

SPIKKELAND, I., B. KINSTEN, G. KJELLBERG, J. P. NILSSEN & R. VÄINÖLÄ, 2016. The aquatic glacial relict fauna of Norway — an update of distribution and conservation status. Fauna Norvegica, **36**: 51-65.

SUKHIKH, N., A. SOUISSI, S. SOUISSI & V. ALEKSEEV, 2013. Invasion of *Eurytemora* sibling species (Copepoda: Temoridae) from North America into the Baltic Sea and European Atlantic coast estuaries. Journal of Natural History, **47**: 753-767.

SVETLICHNY, L., A. KHANAYCHENKO, E. HUBAREVA & L. AGANESOVA, 2012. Partitioning of respiratory energy and environmental tolerance in the copepods *Calanipeda aquaedulcis* and *Arctodiaptomus salinus*. Estuarine, Coastal and Shelf Science, **114**: 199-207.

UTERMÖHL, H., 1958. For improvement of quantitative phytoplankton methods. Verhandlungen des Internationalen Verein Limnologie, **9**: 1-38. [In German.]

VEZHNOVETS, V. V., I. Y. ZAIDYKOV, E. Y. NAUMOVA & E. A. SYSOVA, 2012. Biological peculiarities of two copepod species (Crustacea, Copepoda, Calanoida) as possible causes of changes in their geographical ranges. Russian Journal of Biological Invasions, **3**: 243-250.

WILKEN, S., B. HOFFMANN, N. HERSCH, N. KIRCHGESSNER, S. DIELUWEIT, W. RUBNER, L. J. HOFFMANN, R. MERKEL & I. PEEKEN, 2011. Diatom frustules show increased mechanical strength and altered valve morphology under iron limitation. Limnology and Oceanography, **56**: 1399-1410.

WINKLER, G., J. J. DODSON & C. E. LEE, 2008. Heterogeneity within the native range: population genetic analyses of sympatric invasive and noninvasive clades of the freshwater invading copepod *Eurytemora affinis*. Molecular Ecology, **17**: 415-430.

First received 29 August 2019.
Final version accepted 18 November 2019.

[When citing this volume, please refer to *Crustaceana* 93 (2020), 241-547]

*Proceedings of the* Eurytemora *conference, St. Petersburg, 2019*

# ABNORMALITIES IN SHAPE AND SIZE OF *EURYTEMORA AFFINIS* (COPEPODA, CALANOIDA) AND ITS EGGS UNDER DIFFERENT ENVIRONMENTAL CONDITIONS

BY

ANISSA SOUISSI[1,3]) and SAMI SOUISSI[2])

[1]) Univ. Littoral Côte d'Opale, UMR 1158 BioEcoAgro, TERRA Viollette, USC Anses, INRAe, Univ. Lille, Univ. Artois, Univ. Picardie Jules Verne, Univ. Liège, Yncréa, 62200 Boulogne-sur-Mer, France

[2]) Laboratoire d'Océanologie et de Géosciences, Université de Lille, CNRS, Université Littoral Côte d'Opale, UMR 8187, 62930 Wimereux, France

## ABSTRACT

In this study we examined the morphological anomalies of 2023 ovigerous females of the copepod *Eurytemora affinis* from laboratory and field samples. Field copepod collections were collected from 10 sites in Europe and North America. Laboratory cultured copepods were reared at 6 experimental conditions during several generations. The observation under the microscope of the ovigerous females' bodies revealed the presence of a 'Tumor Like Anomaly' (TLA) on a few individuals, which were manifested on the urosome. The occurrence of TLA was only noticed on copepods from the field samples. This result shows that under controlled laboratory conditions, *E. affinis* do not manifest any abnormal morphology, even when temperature and salinity conditions are stressful, which also demonstrates the high plasticity of this copepod species and its capacity to live at a wide range of environmental parameters. In this study, we also confirmed that *E. affinis* is capable of producing eggs with a wide range of sizes, ranging from very small and abnormal eggs of 75.79 $\mu$m only, to very large eggs, reaching a maximal diameter of 128.71 $\mu$m. The observed large eggs are abnormal compared to the average size of the other eggs produced within the same clutch. Other abnormalities on eggs, such as deformed eggs or eggs infested by fungi, are only observed from field samples at a few sampling sites. In the Seine estuary, several abnormalities in eggs and adult females were observed during February 2010. The critical period for *E. affinis*' development is sampling site-specific and needs to be accurately studied in the future to better explain the causes of the abnormalities described in this work, and their effects at population level.

Key words. — Large eggs, deformed eggs, fungi infested eggs, morphological anomalies, small eggs

[3]) Corresponding author; e-mail: anissa.souissi@univ-littoral.fr

 | DOI:10.1163/9789004465565_010

RÉSUMÉ

Dans cette étude, nous avons examiné les anomalies morphologiques de 2023 femelles ovigères du copépode *Eurytemora affinis* provenant de collectes sur le terrain et en conditions de laboratoire. Nous avons échantillonné des copépodes sur le terrain au niveau de 10 sites en Europe et en Amérique du Nord. Les copépodes cultivés en laboratoire ont été élevés dans 6 conditions expérimentales pendant plusieurs générations. L'observation au microscope des corps de femelles ovigères a révélé la présence d'une «anomalie ressemblant à une tumeur» (TLA en anglais) sur quelques individus qui se manifestait sur l'urosome. L'occurrence de TLA n'a été observée que sur les copépodes prélevés sur le terrain. Ces résultats montrent que dans des conditions de laboratoire contrôlées, *E. affinis* ne présente aucune morphologie anormale, même lorsque les conditions de température et de salinité sont stressantes, ce qui démontre également la plasticité élevée de cette espèce de copépode et sa capacité de vivre dans un large spectre de paramètres environnementaux. Dans cette étude, nous avons également confirmé que *E. affinis* est capable de produire des œufs d'une grande variété de tailles, allant de très petits œufs anormaux de diamètre égal à 75,79 $\mu$m à de très gros œufs atteignant un diamètre maximal de 128,71 $\mu$m. Les gros œufs observés sont anormaux par rapport à la taille moyenne des autres œufs produits dans la même couvée. D'autres anomalies sur les œufs, comme les œufs déformés ou les œufs infestés par des champignons, ne sont observées qu'à quelques sites d'échantillonnage sur le terrain. Dans l'estuaire de la Seine, en février 2010, plusieurs anomalies ont été observées au niveau des œufs et chez les femelles adultes. La période critique pour le développement d'*E. affinis* est propre au site d'échantillonnage et doit être étudiée avec précision à l'avenir afin de mieux expliquer les causes des anomalies décrites dans ce travail et leurs effets au niveau de la population.

## INTRODUCTION

Aquatic organisms, particularly invertebrate ectotherms (e.g., crustaceans) are adapted to live in a variable environment and encounter often several sources of stressors from natural and anthropogenic origins. Particularly planktonic copepods are present in all aquatic ecosystems and they are numerically the most dominant multicellular organisms in the oceans (Souissi et al., 2007). Despite the high diversity of copepod species, as well as their habitats, they are sensitive to stressors they may encounter during their short life cycle. For that reason, they are considered as good bioindicators and reference models in aquatic ecotoxicology (Kwok et al., 2015). Several mechanisms starting from gene expression (Jeong et al., 2017), behavioural patterns (Michalec et al., 2016), whole life-cycle approaches (Kadiene et al., 2019) and multi-generation responses (Souissi et al., 2016a) allowed to decipher the complex responses of copepods to their aquatic environment. For example, several factors such as temperature, salinity, food conditions, and the presence of pollutants can affect the growth rate of developmental stages of copepods as well as their reproduction. Consequently, the size of copepods, their clutch size or egg size, could be considered as key attributes to compare different conditions.

Although several studies have been dealing with the changes in copepod size and their reproductive output (mainly clutch size was used as endpoint)

(Devreker et al., 2009), detailed description of irregular or abnormal morphology in copepods received less attention. In fact, documenting the abnormalities in small organisms such as copepods requires a large number of samples, a regular methodology and systematic observation of all samples. However, focusing on abnormalities in copepods may enrich the indicative power of this group. It is consequently important to document these abnormalities by focusing on both field and experimental conditions.

Copepods can be subjected to different complications in their environment. They can, for example, deal with parasite or epibiont infestations as reported by Souissi et al. (2013) when the calanoid copepod *Eurytemora affinis* (Poppe, 1880) was infested by epibiotic ciliates on its entire body. Copepods can also exhibit different morphological anomalies. These anomalies are commonly called "Tumor-like Anomalies" (TLA), and they can differ in their size, shape, structure and location (Jagadeesan & Jyothibabu, 2016). Copepods with TLA were observed in different aquatic environments such as freshwater lakes, estuaries, coastal marine waters and deep ocean ridges (Crisafi & Crescenti, 1977; Omair et al., 1999; Messick et al., 2004; Skovgaard, 2004; Bhandare & Ingole, 2008; Mantha et al., 2013; Jagadeesan & Jyothibabu, 2016). The incidence of TLA is a rare event in copepods but Jagadeesan & Jyothibabu (2016) suggested that recording it properly is usually overlooked during zooplankton sampling. In their study, they monitored different areas in Indian waters and found 1 to 15% of the TLA incidence in 24 copepod species within the copepod community in different habitats. These anomalies occur in copepods mainly due to stressful environmental conditions. Other factors that together promote the occurrence of these phenomena might also be responsible, such as fluctuations in phytoplankton and bacteria populations, influx of fresh water into the system, pollution, and, the abundance of copepod hosts in case of parasite infestation (Manca et al., 1996; Souissi et al., 2013). For example, Jagadeesan & Jyothibabu (2016) suggested that TLA frequency in copepods was related to pre-southern monsoon periods in Indian waters. Wounding, puncturing, or damage of the copepod carapace could also be a major cause of TLA formation in copepods (Omair et al., 2001; Messik et al., 2004; Jagadeesan & Jyothibabu, 2016). These abnormalities may reduce or even compromise the fitness of individuals and decrease the life expectancy of copepods (Bridgeman et al., 2000; Souissi et al., 2013; Jagadeesan & Jyothibabu, 2016).

Liang & Uye (1997) pointed out the importance of copepod population dynamics, investigating the effect of environmental factors on egg production rate, recruitment rate from eggs to the plankton population, and mortality in the early naupliar stages. Reproductive parameters are often closely related to environmental conditions (Niehoff, 2007). Since it is difficult to study the mortality of copepods in the field, estimation of egg production was commonly used in the literature as a proxy of secondary production (Devreker et al., 2009). In fact, reproductive parameters are of great importance for the modulation of copepod population recruitment. The production of viable eggs that ensure the survival of individual populations is vital for the persistence of copepod populations (Acheampong et al., 2011) but when eggs fail to hatch, recruitment can be impacted. Egg hatching failure in copepods can be the result of the production of resting eggs, maternal assimilation of toxins, the production of unfertilized eggs (Lasley-Rasher et al., 2014) or of abnormal eggs. Since copepods function as a whole system that is affected by external factors, the functioning of the ovary can be impacted and therefore modulate the egg production rate, egg size, and egg energy content (Eckelbarger, 1994). Reproductive parameters are often closely related to environmental conditions (Niehoff, 2007). Inadequate nutrition or dietary toxins are known to negatively affect the hatching success of copepod eggs (Lasley-Rasher et al., 2014 and references therein) as also other environmental factors, such as temperature or food concentration (Lasley-Rasher et al., 2014 and references therein). Environmental factors can lead to the occurrence of copepod egg abnormalities such as abnormal-sized eggs within the same clutch (Hansen et al., 2012; Hammervold et al., 2015) or egg sac (Koski et al., 1999), and also to specific fungal attack on eggs or egg sac (Redfield & Vincent, 1979). However, studies documenting the variability of egg size as well as the occurrence of abnormalities on eggs, remain very limited.

In this paper, we aimed to list and describe the abnormalities that were observed on female copepods and on eggs, either in the laboratory under different controlled conditions, or in the field. The egg-bearing copepod used in this study is the estuarine calanoid copepod *Eurytemora affinis* that dominates the zooplankton community in several estuaries and other aquatic ecosystems in the Northern Hemisphere (Andersen & Nielson, 1997; Winkler et al., 2008, 2011). In macrotidal estuaries, *E. affinis* evolved several adaptations to keep the bulk of its population in the low salinity zone and is exposed to high variability of both salinity and temperature (Devreker et al., 2008, 2010). Several field and laboratory studies showed that the clutch size of *E. affinis* is highly variable and could be affected by food, temperature, and salinity (Ban, 1994; Devreker et al., 2009). To achieve the objective of this study, we combined several samples collected in the field and in the laboratory, with special attention to the observation of ovigerous females. Then, we combined the data to try and understand the possible sources inducing these abnormalities. On the one hand, the examination of samples from controlled laboratory conditions will allow us to understand whether the change of one or two factors could induce the appearance of some abnormalities caused by a stressor. On the other hand, the checking of different samples from different distribution areas of *E. affinis* at different sampling dates will allow recognition of the factors that could induce the appearance of abnormalities in copepods in their natural environment.

## MATERIAL AND METHODS

From previous experimental studies focusing on reproductive strategy and multigenerational effects (Souissi et al., 2010, 2016a, b), we obtained a high number of batches (i.e., generations) of *Eurytemora affinis*. The description of all samples obtained under controlled conditions is listed in table I. Moreover, a significant number of samples obtained from different sites and their geographical positions are given in fig. 1 and table II, the number of the field samples is indicated in table I.

The laboratory samples that were analysed in this study contain mainly ovigerous females of *E. affinis*. They constituted a collection of preserved copepods that were reared at different culture conditions during several generations (a sample of each of those generations was checked for this study) (table I). These laboratory copepod cultures were started from a sampling of an *E. affinis* population in the Seine estuary (France) during late autumn 2006 in the low salinity zone under the

TABLE I

Description of all experimental conditions from the laboratory and the sampling sites used in this study on *Eurytemora affinis* (Poppe, 1880)

| Samples | Condition | Salinity | Temperature (°C) | *n* |
|---|---|---|---|---|
| Laboratory samples | Condition 1 | 15 | 7 | 123 |
| | Condition 2 | 15 | 24 (from 7) | 257 |
| | Condition 3 | 15 | 20 | 206 |
| | Condition 4 | 15 | 24 (from 20) | 200 |
| | Condition 5 | 25 | 20 | 190 |
| | Condition 6 | 25 | 24 (from 20) | 147 |
| Total laboratory samples | | | | 1123 |
| In situ samples | Seine estuary (France) | | | 380 |
| | St Lawrence estuary (Canada) | | | 223 |
| | Chesapeake Bay (U.S.A.) | | | 89 |
| | Gironde estuary (France) | | | 55 |
| | Loire estuary (France) | | | 86 |
| | Scheldt estuary (Belgium) | | | 47 |
| | Other European estuaries | | | 24 |
| Total in situ samples | | | | 900 |
| Total samples | | | | 2023 |

The number of individuals (*n*) used in each condition is shown in the last column.

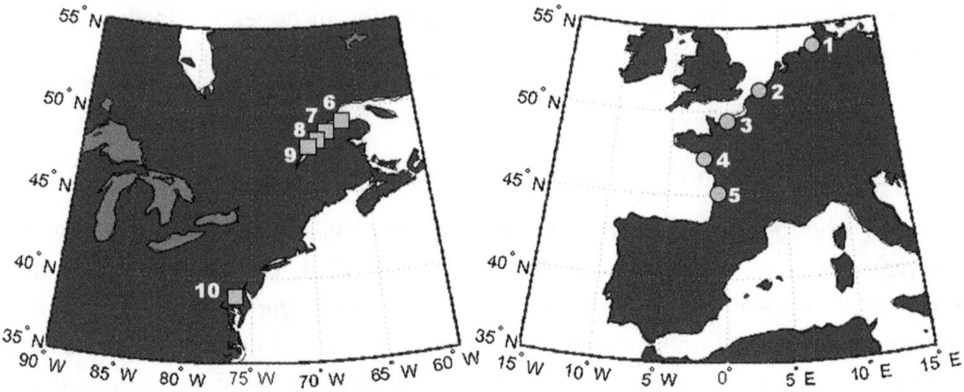


Fig. 1. Maps of the sampling locations of *Eurytemora affinis* (Poppe, 1880). 1-5, in Europe: 1, Elbe; 2, Scheldt; 3, Seine; 4, Loire; 5, Gironde. 6-10, in North America. 6-9, St Lawrence estuary sites: 6, Isle Verte; 7, St Jean Port Joli; 8, Montmagny; 9, Berthier sur Mer. 10, Chesapeake Bay.

Tancarville bridge (49°26′N 00°16′W) at a water temperature of 12°C and a salinity of 5. Copepods were then maintained in the laboratory under different culture conditions and fed ad libitum with the microalga *Rhodomonas marina* (P. A. Dangeard) Lemmermann, 1899 at its exponential phase of growth. The same procedure was applied to launch all the copepod cultures in beakers using each 40 ovigerous females sorted from the starter batch culture. After an acclimation phase, the first beaker culture was raised at 7°C with salinity 15 (condition T7S15) for 4 successive generations, the second culture was conducted at 20°C with salinity 15 (condition T20S15) for 7 generations and the last one at 20°C with salinity 25 (condition T20S25) for 6 generations. Then, all three copepod populations that were acclimated to those initial conditions (T7S15, T20S15 and T20S25, respec-

TABLE II

List of the coordinates of the sampling sites in the various estuaries studied in our research on *Eurytemora affinis* (Poppe, 1880) in Europe and in North America

| Continent | Estuary | Latitude | Longitude |
|---|---|---|---|
| Europe | Elbe | 53°32′06″N | 09°47′31″E |
| | Scheldt | 51°21′06″N | 04°14′58″E |
| | Seine | 49°28′33″N | 00°27′54″W |
| | Loire | 47°17′23″N | 02°01′52″W |
| | Gironde | 45°14′80″N | 00°43′50″W |
| North America | St Lawrence, Isle Verte | 48°00′20″N | 69°25′50″W |
| | St Lawrence, St Jean Port Joli | 47°12′59″N | 70°16′22″W |
| | St Lawrence, Montmagny | 46°59′26″N | 70°33′13″W |
| | St Lawrence, Berthier sur Mer | 46°56′07″N | 70°44′07″W |
| | Chesapeake Bay | 39°23.81′N | 76°03.32′W |

tively) experienced a temperature increase to 24°C and were followed during 5 generations in the three final conditions (T24 (from T7S15), T24 (from T20S15) and T24 (from T20S25)). The protocol of the generation follow-up is detailed in Souissi et al. (2010, 2016a, b).

A second set of copepod samples came from different estuarine sampling sites in the Northern Hemisphere. These sampling sites were located in a selection of estuaries in Europe and North America. They are representative of the transatlantic distribution of *E. affinis* as well as the diversity of its habitats (fig. 1, table II). These field samples were collected in the framework of a comparison study of transatlantic populations of *E. affinis*.

Regarding the temporal scale, the combined samples cover all seasons on both continents in different years from 2006 to 2009. We mention here that sampling in the St Lawrence estuary was not possible during winter and early spring, and that the highest number of samples was taken from the low salinity zone in the Seine estuary. In fact, the Seine estuary is considered as a reference site because of its intermediate latitude and the existence of several experimental studies on the reproductive strategy of its *E. affinis* population (Beyrend-Dur et al., 2009; Devreker et al., 2009; Dur et al., 2009; Souissi et al., 2016a, b).

From each sample, between 20 and 40 ovigerous females were selected to measure all morphological and reproductive traits taken into consideration. Each sorted female was observed under an inverted microscope. Afterwards, the clutch size was counted by tearing the egg sac carefully apart and individual measurements of 5 to 15 eggs per sac were done to estimate the egg size. All size measurements were conducted under an inverted microscope (Olympus IX71) using ImageJ software. The presence of abnormal eggs was verified in the egg sac directly, when counting and measuring the eggs. Then a second check was made on the photo database of the ovigerous females and their eggs.

## Statistical analysis

Egg size was calculated using the Estimated Spherical Diameter (ESD), with the area of the egg in the microscope image as a proxy of its size, and obtained using the software ImageJ. The following formula was used to calculate ESD:

$$\text{ESD} = 2((\text{Egg Area})/\pi)^{0.5}$$

Mean values, as well as standard deviations, were estimated for the ESD calculated per generation and per condition.

Thereafter, for each condition a confidence interval was calculated using only normal sized eggs to accurately determine outlier egg sizes by using the following limits:

$$ESD_{lower\_limit} = ESD_{mean} - 1.96(SD/n^{0.5}) \quad \text{and}$$
$$ESD_{upper\_limit} = ESD_{mean} + 1.96(SD/n^{0.5})$$

Where $n$ is the sample size of the eggs and SD is the standard deviation.

## RESULTS

It is worth mentioning that mainly ovigerous females of *Eurytemora affinis* were observed in the framework of this study. Ovigerous females are important in shaping the population dynamics of egg-bearing copepods, and they can be considered as a proxy for the status of the population as a whole. This is because it provides an insight into the eventual recruitment, i.e., the relative numbers of the new generation of offspring, and thereby provides an estimate of the general health of the population of the copepod species at issue.

The observation of copepod samples revealed different anomalies, first related to egg size with the presence of generally a single larger (fig. 2A) or a few smaller eggs (fig. 2B) in the same egg sac, then subsequently concerning the shape of the eggs (figs. 3, 4). The presence of infested eggs (fig. 5) or a whole infested egg sac (fig. 6) was also observed, and finally several ovigerous females exhibited anomalies (protuberances) on their urosome (fig. 7).

### Egg size abnormalities

The presence of a single (sometimes two or more) large eggs in the sac compared to the others of the same clutch occurred 17 times in 12 ovigerous

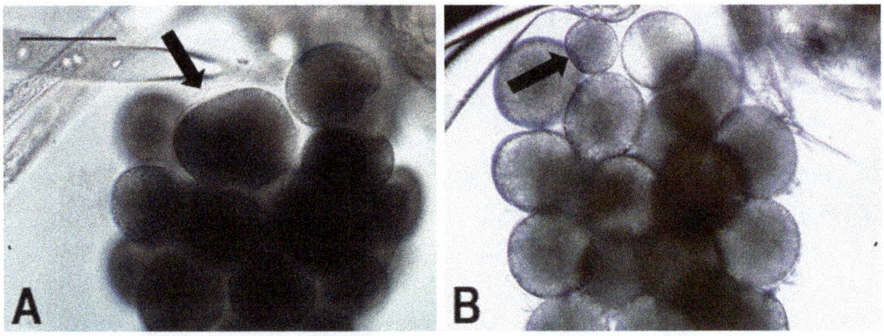


Fig. 2. Two photos of opposed conditions with regard to abnormal egg size occurrence in a single egg sac among otherwise normal eggs of *Eurytemora affinis* (Poppe, 1880): A, with a very large egg (originating from an ovigerous female from F17 at culture condition T24S25); and, B, a very small egg (originating from an ovigerous female from F15 at culture condition T20S15). Arrows designate these abnormal eggs. Horizontal scale bar (top left) = 100 $\mu$m.

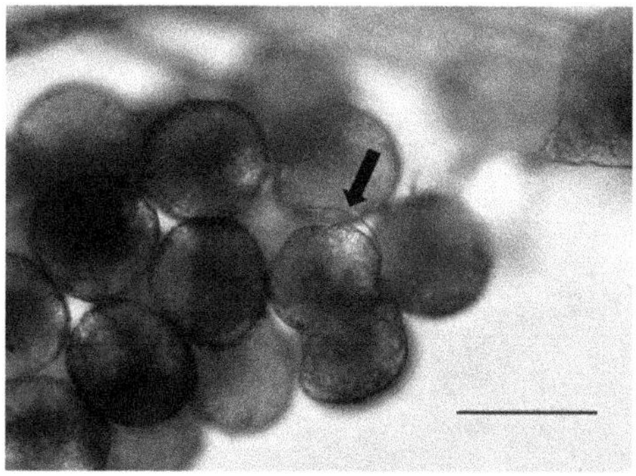

Fig. 3. Presence of a differently shaped egg in an egg sac of *Eurytemora affinis* (Poppe, 1880) (designated by arrow). Scale bar = 100 $\mu$m.

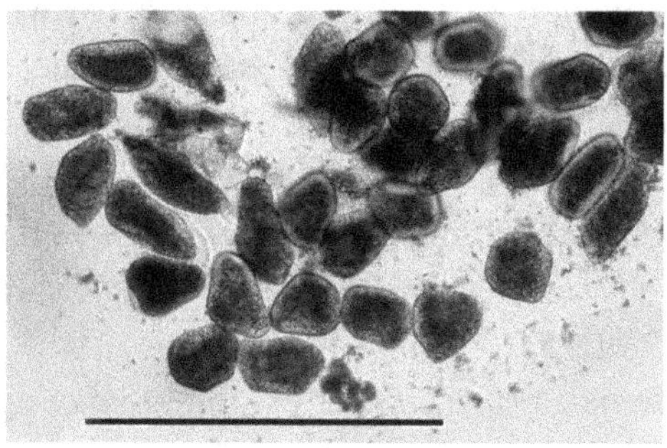

Fig. 4. Illustration of an example of deformed eggs of *Eurytemora affinis* (Poppe, 1880). Scale bar = 500 $\mu$m.

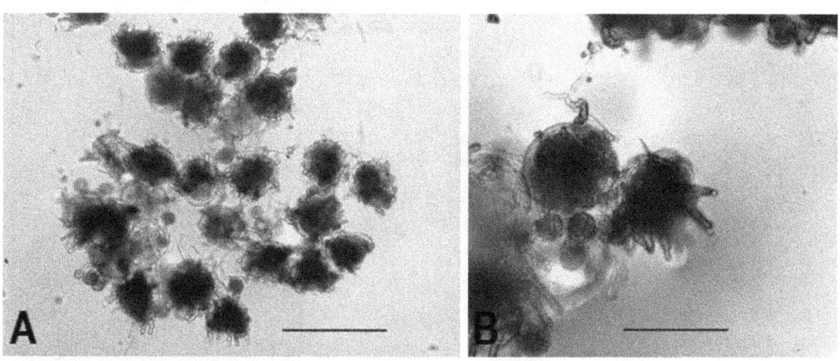


Fig. 5. Eggs from a single clutch of *Eurytemora affinis* (Poppe, 1880), infested by fungi: A, overview of a number of those eggs; B, detail of two of the infested eggs. Scale bars: A = 100 $\mu$m; B = 200 $\mu$m.

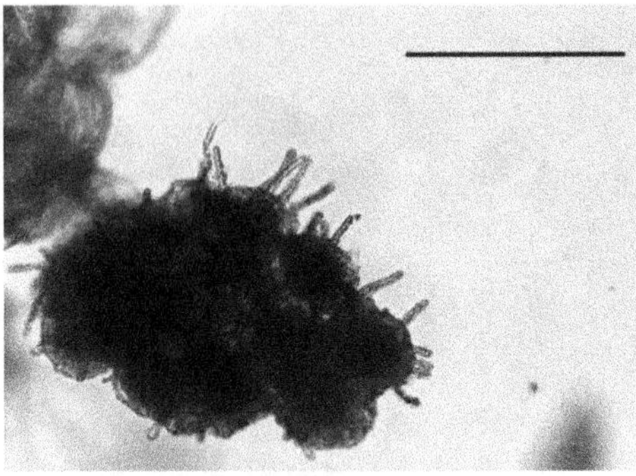

Fig. 6. Photo of an egg sac of *Eurytemora affinis* (Poppe, 1880) infested by fungi. Scale bar = 200 $\mu$m.

females. One large egg was observed in a single copepod from the Seine estuary field sampling during winter 2010 (10 February 2010) which represents 5.56% (1/18 sampled ovigerous females) of that sampling date, 0.26% (1/380) of all observed ovigerous females from the Seine estuary and 0.11% (1/900) of all field samples combined (table I). This egg presented an ESD of 111.73 $\mu$m whereas the other eggs from the same clutch measured on average 98.03 $\pm$ 2.52 $\mu$m ($n$ = 3). The remaining ovigerous females from that same sample showed a smaller average ESD of their eggs, i.e., of 95.01 $\pm$ 3.7 $\mu$m ($n$ = 8).

Eleven ovigerous females of *E. affinis* exhibited large eggs in their sac in the laboratory samples for all experimental conditions combined, which represent 0.89% (11/1123) of the total observed ovigerous *E. affinis* females (tables I, III). The detailed occurrence of these large eggs is as follows:

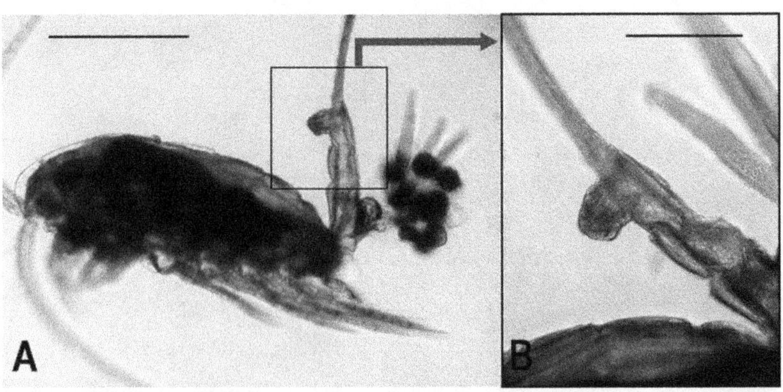


Fig. 7. Ovigerous female of *Eurytemora affinis* (Poppe, 1880) exhibiting a deformation (a "Tumor-like Anomaly") on the urosome: A, overview; B, detail of the box in A. Scale bars: A = 500 $\mu$m; B = 200 $\mu$m.

TABLE III
List of abnormal eggs (large above and small below) of *Eurytemora affinis* (Poppe, 1880) observed in the various experimental conditions

| | Laboratory condition | | | ESD | | | Confidence interval | |
|---|---|---|---|---|---|---|---|---|
| | Temperature | Salinity | Generation | Egg1 | Egg2 | Egg3 | Low limit | High limit |
| Large eggs | 7 | 15 | 4 | 113.49 | 112.14 | | 99.23 | 103.82 |
| | 7 | 15 | 5 | 128.71 | 126.83 | | 104.44 | 108.76 |
| | 24 (7) | 15 | 5 | 119.50 | 117.50 | | 110.97 | 116.72 |
| | 20 | 15 | 10 | 119.67 | 108.97 | 107.15 | 102.23 | 105.79 |
| | 24 (20) | 15 | 18 | 116.78 | 104.19 | | 96.79 | 101.01 |
| | 24 (20) | 15 | 20 | 104.19 | | | 89.71 | 92.01 |
| | 20 | 25 | 16 | 114.84 | | | 89.30 | 93.54 |
| | 24 | 25 | 1 | 125.71 | | | 90.81 | 94.73 |
| | 24 | 25 | 2 | 110.81 | | | 86.94 | 90.55 |
| | 24 | 25 | 4 | 117.60 | | | 91.87 | 96.30 |
| | 24 | 25 | 5 | 111.16 | | | 103.23 | 109.45 |
| Small eggs | 20 | 15 | 15 | 82.91 | | | 92.52 | 97.85 |
| | 24 | 15 | 19 | 75.79 | | | 99.76 | 104.81 |
| | 24 | 15 | 22 | 85.96 | | | 94.60 | 98.37 |

Each line corresponds to a single ovigerous female that can show 1, 2, or 3 abnormal eggs. The confidence interval of the egg size in the same clutch is indicated. ESD, Estimated Spherical Diameter in $\mu$m; temperature in °C.

In condition T7S15, two ovigerous females (representing 1.63% (2/123) of total observations), exhibited 2 larger eggs each in their sacs. One of the females exhibited the largest eggs observed among all measured eggs combined, with an ESD of 128.71 $\mu$m and 126.83 $\mu$m, respectively. The $ESD_{upper\_limit}$ of egg size for that female was 108.76 $\mu$m, whereas the mean values of ESD of eggs from the other females from the same condition ranged between 95 and 100 $\mu$m.

After transfer to 24°C (T24 (from T7S15)), one ovigerous female with 2 large eggs was found (representing 0.39% (1/257) of all observed females in the same condition). Also, another single observation was noted in condition T20S15 (0.49% (1/206)), but that ovigerous female produced 3 large eggs, thus being the sole female showing the highest number of large eggs in the same clutch.

In condition T24 (from T20) S15, two females presented large eggs. One had a single large egg and the other had 2 such eggs, which constituted 1% (2/200) of the total observed females at this condition.

In condition T20S25, which is a stressful haline condition, only one female showed a single large egg (which constitutes 0.53% (1/190) of the total observed ovigerous females in these condition samples). This contrasted with the double (haline and thermal) stressful condition T24 (from T20) S25, where 4 ovigerous females from different generations showed a single large egg in their sac, representing 2.72% (4/147) of the total condition sample; one of these females had the third largest egg observed, of 125.71 $\mu$m, the mean size of the other large eggs at the same condition was 113.19 $\mu$m.

The details of egg sizes and their comparison with other eggs in the female's egg sac, and other eggs in ovigerous females from the same generation, culture, and conditions, are presented in tables III and IV.

The presence of very small eggs in an egg sac compared to others in the same clutch of an ovigerous female (fig. 2B) was an exceptional event, happening only 3 times among all laboratory and field samples. These females originated from laboratory samples; one came from condition T20S15 (representing a proportion of 0.49% (1/206)) and two from condition T24 (from T20) S15 (with a proportion of 1% (2/200) of the total condition samples). The observed frequency of small eggs was of 0.27% (3/1123) among all ovigerous females from laboratory-controlled conditions, and 0.15% (3/2023) from all samples observed, including all laboratory and field samples. The smallest egg had an ESD of 75.79 $\mu$m, whereas the general egg size in the same sac was 102.28 $\pm$ 3.41 $\mu$m ($n$ = 7) (see tables III and IV). This ovigerous female paradoxically produced other eggs that were larger than the average size exhibited by the other ovigerous females from the same generation and condition (with simultaneously: 96.51 $\pm$ 2.97 $\mu$m ($n$ = 40) and 95.09 $\pm$ 3.52 $\mu$m ($n$ = 200)). The other small eggs had a size of 82.91 $\mu$m and 85.96 $\mu$m, the first one was found in an ovigerous female from condition T20S15 and the second from the same condition as the smallest eggs (i.e., T24 (from T20) S15) (see table III).

## Abnormalities in egg shape

In the egg sac of one ovigerous female raised in a laboratory culture at the condition T20S25, an unusually shaped egg was found that looked like a fusion between two eggs or a merged content of two eggs that cracked under high pressure in the egg sac (fig. 3). It was the single record among 2023 total observations of ovigerous female samples, representing thus a proportion of 0.05%.

Deformed eggs (fig. 4) were observed 11 times only in eggs of ovigerous females from the field, and mainly from one sampling date in the Seine estuary, where such an event was observed 10 times among 18 ovigerous females (55.55%) on 10 February 2010 (at that sampling date, salinity was 18 and water temperature 4.7°C) from a total of 380 field samples (representing 2.63%).

The same event was observed one time among 7 ovigerous females (constituting a proportion of 14.29% of the sample) analysed from the Montmagny sampling site in the St Lawrence estuary during May 2009 (the salinity was 0.04 and temperature 13.5°C) and among 27 copepod females observed at the same site from 2007 and

TABLE IV

Mean egg size of each female of *Eurytemora affinis* (Poppe, 1880) showing abnormal eggs (see details in table III) and its comparison with the mean egg size for all females in the same generation as well as the mean egg size under each experimental condition

| | Conditions | | | Mean ESD/egg sac female | | | Mean ESD/generation | | | Mean ESD/condition | | |
|---|---|---|---|---|---|---|---|---|---|---|---|---|
| | Temperature | Salinity | Generation | Mean | SD | *n* | Mean | SD | *n* | Mean | SD | *n* |
| Large eggs | 7 | 15 | 4 | 101.57 | 3.31 | 8 | 101.71 | 3.39 | 35 | 100.31 | 3.15 | 123 |
| | 7 | 15 | 5 | 106.64 | 3.12 | 8 | 100.85 | 2.73 | 40 | 100.31 | 3.15 | 123 |
| | 24 (7) | 15 | 5 | 113.89 | 3.59 | 6 | 104.23 | 3.29 | 22 | 98.91 | 4.20 | 116 |
| | 20 | 15 | 10 | 104.04 | 2.87 | 10 | 98.51 | 4.50 | 10 | 94.26 | 3.29 | 206 |
| | 24 (20) | 15 | 18 | 98.94 | 2.85 | 7 | 96.65 | 3.86 | 40 | 95.09 | 3.52 | 200 |
| | 24 (20) | 15 | 20 | 90.87 | 1.76 | 9 | 93.33 | 3.25 | 40 | 95.09 | 3.52 | 200 |
| | 20 | 25 | 16 | 91.46 | 2.87 | 7 | 93.86 | 3.22 | 19 | 94.07 | 3.61 | 186 |
| | 24 | 25 | 1 | 92.81 | 2.83 | 8 | 93.38 | 3.51 | 40 | 93.17 | 3.69 | 145 |
| | 24 | 25 | 2 | 88.77 | 2.43 | 7 | 91.99 | 4.04 | 40 | 93.17 | 3.69 | 145 |
| | 24 | 25 | 4 | 94.12 | 2.77 | 6 | 94.57 | 3.58 | 22 | 93.17 | 3.69 | 145 |
| | 24 | 25 | 5 | 106.45 | 5.02 | 10 | 93.21 | 3.30 | 40 | 93.17 | 3.69 | 145 |
| Small eggs | 20 | 15 | 15 | 95.26 | 4.08 | 9 | 94.27 | 2.45 | 40 | 94.26 | 3.29 | 206 |
| | 24 | 15 | 19 | 102.33 | 3.41 | 7 | 96.51 | 2.97 | 40 | 95.09 | 3.52 | 200 |
| | 24 | 15 | 22 | 96.54 | 3.47 | 13 | 95.76 | 2.65 | 40 | 95.09 | 3.52 | 200 |

ESD, Estimated Spherical Diameter in $\mu$m; temperature conditions: 24 (7), copepods raised first at 7°C and then transferred to 24°C; 24 (20), copepods raised first at 20°C and then transferred to 24°C.

2009 (3.7%), as well as among 223 copepod females from different sites in the St Lawrence estuary in the years from 2003 and 2009 (0.45% of the total St Lawrence estuary samples).

The total number of observations of deformed eggs represented a small proportion among the total observed field samples, with 1.22% (11/900) and 0.54% (11/2023) among all studied samples.

### Abnormalities in the general state of the eggs and the egg sac

The occurrence of eggs with fungi (fig. 5) was observed among eggs of ovigerous copepod females collected from St Jean Port Joli in the St Lawrence estuary. It was noted 8 times among 20 observations of ovigerous females (which constituted a proportion of 40% (8/20)) during July 2008 (salinity was 3.5 and water temperature 19.8°C) among 196 samples collected during 2003 and 2009, hence representing 4.08% of the ovigerous females. This proportion, hence, represented only 3.59% of the total observed ovigerous females in 223 samples from the St Lawrence estuary.

Ovigerous females with an egg sac infected by fungi (fig. 6) were registered 5 times among 11 observed individuals (45.45%) in a sample from Chesapeake Bay collected during April 2008. This represented 5.62% (5/89) of all ovigerous copepod females observed in Chesapeake Bay samples from different years. Fungal egg sac infection was geographically and temporally a rare phenomenon in ovigerous females with a proportion of 0.56% of the total studied field samples and 0.25% of all 2023 observed ovigerous females.

### Malformation of the body of *E. affinis* ovigerous females

Some observed ovigerous females from the field showed a protuberance on the last somite of their urosome (fig. 7), and we can refer this deformation to a Tumor Like Anomaly (TLA). These deformations were observed 5 times in all combined samples (0.25% (5/2023)). Four ovigerous females from the Seine estuary sampled during February 2010 had a TLA on their urosomes, representing a proportion of 22.22% (4/18) of the sampling date and 1.05% (4/380) of all Seine estuary samples. Only a single ovigerous female sampled in the salt marshes of Isle Verte from the St Lawrence estuary, during June 2007, had this TLA, representing thus 5% (1/20) of the sampling date and 0.45% (1/223) of the St Lawrence sampling.

## DISCUSSION

### Copepod abnormalities in their general morphology

Like all living organisms, copepods can show some abnormal development in their body structure at any stage of their life cycle. Because of their small sizes, it is logically difficult to detect with accuracy all these abnormalities without careful microscopic observations. Consequently, the morphological anomalies in copepods and zooplankton in general, are poorly documented. Few of these studies reported some anomalies in copepods that were qualified as "Tumor-like Anomalies" (TLA). A recent study realized by Jagadeesan & Jyothibabu (2016) sampled copepods from 15 different estuarine and coastal habitats and targeted the study of the occurrence of TLA on copepod communities. TLA was recorded in 24 species of copepods, which constituted approx. 1-15% of the community in different environments. According to their study, the occurrence of TLA was significantly higher on the dorsal surface of the prosome (70-90%) than on the urosome (10-30%). In our study, the TLA occurred only on the urosome part and no TLA was observed on the prosome. Looking for potential causes of TLA appearance, some studies noticed that shallower habitats showed higher percentages of TLA compared to deeper waters (Silina & Khudolei, 1994; Omair et al., 1999; Manca et al., 2004; Jagadeesan & Jyothibabu, 2016). This may be related to water quality as well as to the intensity of combined stressors that may alter the copepod's development. In fact, Jagadeesan & Jyothibabu (2016) observed a higher occurrence of TLA in copepods during the pre-southwest monsoon period, but the actual causality of such a correlation is not clear.

In our study, the TLA was observed on copepods sampled from the low salinity zone of the Seine estuary during a cold day in February 2010. Such cold winter conditions due to climatic anomalies were also observed in the late winter of 2005 and negatively affected the population dynamics of *E. affinis* in the Seine estuary (Devreker et al., 2010). We can suppose that most overwintering ovigerous females sampled in the Seine estuary in 2010 showed a poor status of health. The other observation of TLA comes from the sampling of the salt marches of Isle-Verte along the southern shore of the St. Lawrence estuary in June 2007. However, it is difficult to compare both sites, as the critical period of *E. affinis* development may be totally different due to differences in climatic regimes between both sites.

Because most TLA observed on copepods showed an internal connection with body muscle tissues (Jagadeesan & Jyothibabu, 2016), these manifestations seem to be caused by herniated body tissue. Consequently, there is no strong scientific evidence that the occurrence of TLA in copepods could be directly related to the presence of carcinogenic agents as speculated in several studies (Silina & Khudolei, 1994; Bhandare & Ingole, 2008; Mantha et al., 2013).

In contrast, other factors leading to small damages or punctures of the exoskeleton of copepods could be the main reason for TLA formation (Omair et al., 2001; Messick et al., 2004; Jagadeesan & Jyothibabu, 2016). These damages could be caused by external predators or parasites, leading to an initiation of herniation (Bridgeman et al., 2000, Omair et al., 2001; Messick et al., 2004).

In our study, we showed that a large number of individuals examined from different experimental conditions, did not show any TLA formation. This may confirm that all experimental conditions, even those realized at 24°C, which is very stressful, did not lead to the appearance of TLA formation. This consequently means that none of the observed individuals did show any herniated body tissue or any damage to the exoskeleton. Because *Eurytemora affinis* showed high plasticity regarding the experimentally tested global warming scenario of +4°C (Souissi et al., 2016a), a long-term exposure of copepods to different temperature and salinity conditions did not engender any change on the external morphology of the observed individuals. By contrast, damage of the integument was observed after short term exposure of adults of the calanoid *Pseudodiaptomus marinus* Sato, 1913 to very extreme osmotic and thermal shocks (Ibrahim et al., 2016). Because these experimental manipulations cannot occur in the field, more accurate studies are still needed in the field and in the laboratory to better elucidate the mechanisms leading to the occurrence of TLA in copepods.

## Abnormalities in copepod eggs: occurrence and ecological implications

The studies of morphological anomalies in copepods mainly focused on adult stages and only a few observations have focused on other developmental stages such as early nauplii (Ianora et al., 2004). Since *E. affinis* is an egg-bearing copepod, it was possible in our study to focus simultaneously on the morphology of ovigerous females as well as their eggs. Moreover, it is commonly admitted that egg production of copepods is a good indicator of the environmental conditions encountered by copepods in the past few days. In fact, copepods have to feed first to meet the basic metabolic requirement and eventually compensate additional metabolic costs related to any external stressor (e.g., osmoregulation, detoxification, etc.). Consequently, egg production, if proper study of the phenomenon is possible, can be directly linked to the physiological status of copepod females as well as their environmental conditions. Under unfavourable circumstances, copepods may decrease their egg production in order to ensure minimal investment leading to a successful development of the eggs (Anderson & Pond, 2000). Furthermore, copepods like *E. affinis* that carry their eggs in a sac may encounter additional constraints in their egg production by trying to optimize the number of offspring, the size of eggs, as well as the packing constraints (Beck & Beck, 2005). According to the importance of embryonic development for the success of recruitment in copepods (Ianora et al., 2004), we examined in this study systematically possible anomalies in eggs of *E. affinis* retrieved from the laboratory and the field.

The first anomaly that can be encountered in copepod egg development is related to its size. Uye & Sano (1995) suggested an empirical relationship between egg diameter and egg carbon content of copepods. In their relationship (see their fig. 9) the average egg diameter varied from 50 μm for the very small cyclopoid *Oithona davisae* F. D. Ferrari & Orsi, 1984 to approximately 200 μm (outer diameter) for the large Arctic copepod *Calanus glacialis* Jaschnov, 1955 (cf. Hirche & Bohrer, 1987). However, a single average value of egg diameter per copepod species was used (Uye & Sano, 1995) although a high intraspecific variability of egg size can easily be observed by comparing seasons (see this study). For stable conditions, the variability of egg size is supposed to be very limited and only a few studies already confirmed that larger nauplii hatched from larger eggs and exhibited higher fitness in terms of survival (Jamieson & Santer, 2003). This means that producing larger eggs may have an ecological advantage at the population level. But these few studies that reported this ecological advantage of producing larger eggs did not focus on the egg-size variability within the same clutch (i.e., the same female) or within the same cohort. In our study, we specifically targeted statistically significant outliers to identify abnormal eggs in terms of size.

We found that larger eggs occurred more frequently than smaller eggs. The largest egg observed in our study reached a size of 128.71 μm, which was 20.74% larger than the average of eggs produced under the same conditions (106.6 μm). The general morphology of these large eggs is similar to that of the other eggs produced by the same female and thus we can suppose that they are viable and could produce larger nauplii within the same offspring. Unfortunately, our experimental protocol did not allow us to test such hypothesis as we only focused on ovigerous females.

A large egg type of the free-spawning copepod *Acartia tonsa* Dana, 1849 was also observed (Hammervold et al., 2015). The rate of occurrence of this large type of egg, obtained in mass culture of *A. tonsa* used for aquaculture applications, was very low, about 1 large egg measuring on average 106.5 μm in diameter out of 100 000 normal-sized eggs (diameter 82.3 μm on average). In that study, the increase of egg diameter was significant and approached 30%. This study is interesting, because under constant conditions we may consider that the low rate of large egg occurrence (1/100 000) is similar to the natural rate of egg abnormalities in this copepod culture, grown in the laboratory for several generations under stable conditions. This provides evidence that copepods can produce eggs with different densities and diameters (Hammervold et al., 2015). In our study, we observed much higher rates of large egg occurrence and compared them to a single clutch or a set of eggs produced by the same cohort in the laboratory or collected in the same sample in the field. The higher rates of large egg occurrence observed in our study compared to *A. tonsa*, could be explained by interspecific differences between both species and/or by the high variability of experimental conditions used in our study compared to the mass culture of *A. tonsa* under stable conditions. Moreover,

Hansen et al. (2012) showed that adult euryhaline copepod species including *A. tonsa*, can osmoregulate, so the size of the egg changes with the experienced salinity. For *E. affinis*, such observation was not reported and should be tested in the future, although in our study abnormal eggs were obtained at the same experimental conditions of temperature and salinity.

In our study, we also observed very small eggs of *E. affinis* compared to average values. The mechanism leading to the production of such small eggs could be related to the lack of necessary resources to have equal-sized eggs.

Hammervold et al. (2015) hypothesized that the appearance of a large egg type is related to a relatively rare mutation that might be beneficial for the organisms without necessarily being inherited. Unfortunately, the protocol used in that study did not allow to test the heritability of egg size. But another study aiming to select the most productive population of the small calanoid copepod *Parvocalanus crassirostris* (F. Dahl, 1894) showed that the selected strain over 5 generations was composed of larger females that produced larger eggs (Alajmi et al., 2014). However, the demonstrated correlation between female size and egg size does not allow the conclusion that egg size is heritable in copepods, and further studies are needed to test this hypothesis.

In our study, we also reported two other types of egg abnormalities. First, we reported deformed eggs of *E. affinis* (see fig. 4) similar to those presented by Hansen et al. (2012) and consisting of subitaneous eggs of *A. tonsa* that experienced an abrupt and substantial change in salinity from 32 to 72. In our study, deformed eggs were observed from the field in the Seine estuary during the winter, and all morphological indicators seem to confirm that ovigerous females were under strong stress and the produced eggs may not be viable or of poor quality.

Finally, fungal infection of *E. affinis* eggs was reported here only from field observations. An earlier study by Redfield & Vincent (1979) described the various steps in, and the morphology of infested eggs in case of fungal disease. Their study focused on females of freshwater copepods of the genus *Diaptomus* sampled in Castle Lake in California. In our case study, the infested eggs come from ovigerous females collected from the St Lawrence estuary, from the sampling site Saint Jean Port Joli, with salinities around 3.5 during summer (July 2008) at a water temperature of 19.8°C (the temperature and salinity parameters are calculated as an average from the data of the other years at the same site for the same period of the year). So, we speculate that these environmental conditions were favourable for the development of fungi on copepod eggs. As described earlier by Redfield & Vincent (1979), only the eggs were infested by the fungi and not the ovigerous females themselves. A similar fungal infection was observed in the egg sac of an ovigerous female originating from Chesapeake Bay, sampled in April 2008 (temperature and salinity data were not available but for information, the following year in the same month (April 2009) the water temperature was 21°C and the salinity 0.1, so conditions were similar to fresh water). In that female, the body was not infected, and since opening the egg sac was not easy, we assume here that all the eggs (or their majority) were infected inside the egg sac, and that the fungi visible in fig. 6 are those on the surface of the eggs only.

The ecological consequences of fungal infections in copepods can be severe, by reducing naupliar recruitment as well as population density (Redfield & Vincent, 1979). An earlier study carried out in the laboratory considered that parasitism may substantially reduce the reproductive capability of cladocerans (Green, 1974). In copepods, Souissi et al. (2013) showed that a heavy infestation of *E. affinis* from the Seine estuary by epibiotic ciliates, significantly reduced the mating success and increased the mortality of individuals.

### Copepod egg abnormalities: a common mechanism and/or an indicator of stress?

Egg production of copepods has been intensively studied and often related to the external factors encountered by females. Like most copepod species, the egg production of *E. affinis* can be mainly affected by temperature, salinity, food quality, as well as by female age (Devreker et al., 2009). Most studies dealing with copepod reproduction focused on the effects of different factors on the number of eggs produced by the female, and their quality has often been assessed by the rate of hatching success. Because unhatched eggs could be unfertilized eggs, or non-viable eggs (i.e., of low quality), or resting eggs (i.e., diapausing eggs or delayed-hatching eggs), it is necessary to profoundly examine egg morphology in order to seek for additional descriptors of their quality.

The main factors commonly considered in estuarine copepod studies are temperature and salinity (Souissi et al., 2016a). Both factors can directly affect size and density of copepod eggs (Miller & Marcus, 1994). For example, due to osmoregulation, egg density is positively correlated with increased salinity tolerance (Hansen et al., 2012), whereas egg size was negatively correlated to temperature (Uye & Fleminger, 1976; Hansen et al., 2010) as warm-water copepod species more often produce eggs, through lacking the need of energy storage for overwintering (Guisande & Harris, 1995). In our case, however, the ovigerous females of *E. affinis* did not experience such a saline shock, since the salinity was 18 during sampling in the Seine estuary and the water temperature was 4.7°C during that period from February 2010. Usually at low temperature, the female body size is larger under laboratory conditions (optimal conditions) and the eggs were also larger, but no registration or previous records were found on deformation of eggs due to low temperature in either the laboratory, or in nature. Similarly, in the St

Lawrence estuary at the sampling site of Montmagny, one ovigerous female was found with deformed eggs, but the environmental conditions were opposite to the ones described for the Seine estuary. The salinity was very low, almost at a freshwater value (0.04), and the temperature was higher, with 13.5°C. In both cases, the stressful haline conditions that applied to *A. tonsa* according to Hansen et al. (2012), did not show the same effect on *E. affinis* eggs.

At extreme hypo-salinity (2.0), *A. tonsa* eggs did swell, but never hatched. The size of the perivitelline space of the egg increased significantly, but did not implode as in the case of hyper-saline exposure (Hansen et al., 2012). So, we conclude that the deformed eggs in the case of *E. affinis* in this study were due to other factors, such as insufficient, or poor quality food for the females. In fact, Koski et al. (1999) demonstrated that an inappropriate diet of the female could have detrimental effects on the hatching success of *E. affinis* eggs.

## CONCLUSION

Systematic observations of copepod abnormalities, and in particular regarding eggs, are rare. In this study, we confirmed that the estuarine copepod *Eurytemora affinis* can also show a "Tumour-like Anomaly" (TLA) but only on the urosome, whereas other studies have shown that a TLA is often observed dorsally on the prosome.

Detailed observations of a high number of ovigerous females allowed to detect that the occurrence of larger eggs within the same clutch was higher than the occurrence of smaller eggs, in each case as compared to the average size within a clutch. Among all experimental conditions, it seems that the double-stressed condition at a temperature of 24°C and a salinity of 25, generated the highest percentage of abnormal eggs. Nonetheless, under those conditions this percentage remains very low, and certainly below 5%. However, in field conditions and at certain sites and sampling dates, the percentage of some abnormal eggs, such as deformed eggs, can be higher than 50%. Even if these events can be localized in space and/or time, such high percentage of abnormalities and/or fungal infestation could negatively affect the whole population of the species in its development.

In this study, we confirmed that the egg-bearing copepod *E. affinis* has a high plasticity and can grow under several conditions. But we also confirmed, for the first time, that under stressful conditions copepod females were able to produce very large eggs compared to the average egg size. Knowing that larger eggs can produce larger nauplii, this situation could offer some ecological and evolutionary advantages to those individuals. Since detailed observations of egg size within the same clutch are rare, we can not identify the mechanisms behind such phenomenon as yet. We can state, however, that associating a single egg size to a given species is not ecologically relevant, and understanding the spatial and temporal variability of egg size is needed.

Under field conditions, *E. affinis* eggs can be completely deformed and most likely non-viable, or heavily infested by fungi. Our study confirmed that such abnormalities observed in the field can be an additional trait to be used as an indicator of the health status of copepods caused by stressful and poor environmental conditions, such as the winter conditions in 2010 in the Seine estuary.

## ACKNOWLEDGEMENTS

We are thankful to Natalia Sukhikh and Victor Alekseev for their kind invitation to contribute to this special issue targeting the species complex of *Eurytemora affinis*. We would like to thank Gesche Winkler and Dave Kimmel for providing the copepod samples from the St Lawrence estuary and Chesapeake Bay. We are very thankful to all past members of Sami Souissi's group who contributed to the field sampling in the Seine estuary as well as in the other European estuaries. A special thanks also goes to all persons who contributed to the realization of the long-term programmes of culturing *E. affinis* in the laboratory, and mainly the realization of the multi-generation experiments used in this study. All these research projects received funding from different sources, mainly from the GIP Seine-Aval, the FFCR (Fonds France-Canada pour la Recherche) and the Samuel-De Champlain bilateral research program between France and Quebec.

## REFERENCES

ACHEAMPONG, E., R. W. CAMPBELL, A. B. S. DIEKMANN & M. A. ST. JOHN, 2011. Food availability effects on reproductive strategy: the case of *Acartia tonsa* (Copepoda: Calanoida). Mar. Ecol. Prog. Ser., **428**: 151-159.

ALAJMI, F., C. ZENG & D. R. JERRY, 2014. Improvement in the reproductive productivity of the tropical calanoid copepod *Parvocalanus crassirostris* through selective breeding. Aquaculture, **420-421**: 18-23.

ANDERSON, C. M. & T. G. NIELSEN, 1997. Hatching rate of the egg-carrying estuarine copepod *Eurytemora affinis*. Mar. Ecol. Prog. Ser., **160**: 283-289.

ANDERSON, T. & D. W. POND, 2000. Stoichiometric theory extended to micronutrients: comparison of the roles of essential fatty acids, carbon, and nitrogen in the nutrition of marine copepods. Limnol. Oceanogr., **45**: 1162-1167.

BAN, S., 1994. Effect of temperature and food concentration on post-embryonic development, egg production and adult body size of calanoid copepod *Eurytemora affinis*. J. Plankton Res., **16**: 721-735.

BECK, C. W. & R. E. BECK, 2005. The effect of packing constraints on optimal investment in offspring. Evol. Ecol. Res., **7**: 1077-1088.

BEYREND-DUR, D., S. SOUISSI, D. DEVREKER, G. WINKLER & J. S. HWANG, 2009. Life cycle traits of two transatlantic populations of *Eurytemora affinis* (Copepoda: Calanoida): salinity effects. J. Plankton Res., **31**: 713-728.

BHANDARE, C. & B. S. INGOLE, 2008. First evidence of tumor-like anomaly infestation in copepods from the Central Indian Ridge. Indian J. Mar. Sci., **37**: 227-232.

BRIDGEMAN, T. B., G. MESSICK & H. A. VANDERPLOEG, 2000. Sudden appearance of cysts and ellobiopsid parasites on zooplankton in a Michigan lake: a potential explanation of tumor-like anomalies. Can. J. Fish. Aquat. Sci., **57**: 1539-1544.

CRISAFI, P. & M. CRESCENTI, 1977. Confirmation of certain correlation between polluted areas and tumor like conditions as well as tumor growths in pelagic copepods provening of numerous seas over the world. Rapp. Comm. Int. Mer Mediterranéen, **24**: 155.

DEVREKER, D., S. SOUISSI, J. C. MOLINERO, D. BEYREND-DUR, F. GOMEZ & J. FORGET-LERAY, 2010. Tidal and annual variability of the population structure of *Eurytemora affinis* in the middle part of the Seine estuary during 2005. Estuar. Coast. Shelf Sci., **89**: 245-255.

DEVREKER, D., S. SOUISSI, J. C. MOLINERO & F. NKIBUTO, 2008. Trade-offs of the copepod *Eurytemora affinis* in mega-tidal estuaries. insights from high frequency sampling in the Seine estuary. J. Plankton Res., **30**(12): 1329-1342.

DEVREKER, D., S. SOUISSI, G. WINKLER, J. FORGET-LERAY & F. LEBOULENGER, 2009. Effects of salinity and temperature on the reproduction of *Eurytemora affinis* (Copepoda; Calanoida) from the Seine estuary: a laboratory study. J. Exp. Mar. Biol. Ecol., **368**: 113-123.

DUR, G., S. SOUISSI, D. DEVREKER, V. GINOT, F. G. SCHMITT & J. S. HWANG, 2009. Effect of temperature and individual variability on the reproduction of the estuarine copepod *Eurytemora affinis*. Individual-based experimental and modelling approach. Ecol. Model., **220**: 1073-1089.

ECKELBARGER, K. J., 1994. Diversity of the metazoan ovaries and vitellogenic mechanisms: implications for life history theory. Proc. Biol. Soc. Wash., **107**: 193-218.

GREEN, J., 1974. Parasites and epibionts of Cladocera. Trans. Zool. Soc. Lond., **32**: 417-515.

GUISANDE, C. & R. HARRIS, 1995. Effect of total organic content of eggs on hatching success and naupliar survival in the copepod *Calanus helgolandicus*. Limnol. Oceanogr., **40**: 476-482.

HAMMERVOLD, S. H., R. N. GLUD, J. O. EVJEMO, A. HAGEMANN & B. W. HANSEN, 2015. A new large egg type from the marine live feed calanoid copepod *Acartia tonsa* (Dana) — perspectives for selective breeding of designer feed for hatcheries. Aquaculture, **436**: 114-120.

HANSEN, B. W., G. DRILLET, M. F. PEDERSEN, K. P. SJØGREEN & B. VISMANN, 2012. Do *Acartia tonsa* (Dana) eggs regulate their volume and osmolality as salinity changes? J. Comp. Physiol., (B) **182**(5): 613-623.

HIRCHE, H.-U. & R. N. BOHRER, 1987. Reproduction of the Arctic copepod *Calanus glacialis* in Fram Strait. Mar. Biol., **94**: 11-17.

IANORA, A., A. MIRALTO, S. A. POULET, Y. CAROTENUTO, I. BUTTINO, ET AL., 2004. Aldehyde suppression of copepod recruitment in blooms of a ubiquitous planktonic diatom. Nature, **429**: 403-407.

IBRAHIM, A., A. SOUISSI, A. LERAY, L. HÉLIOT, B. VANDENBUNDER & S. SOUISSI, 2016. Myofibril changes in the copepod *Pseudodiaptomus marinus* exposed to haline and thermal stresses. PLoS One, **11**(11): e0164770.

JAGADEESAN, L. & R. JYOTHIBABU, 2016. Tumour-like anomaly of copepods — an evaluation of the possible causes in Indian marine waters. Environ. Monit. Assess., **188**(4): 244.

JAMIESON, C. D. & B. SANTER, 2003. Maternal aging in the univoltine freshwater copepod *Cyclops kolensis*: variation in egg sizes, egg development times, and naupliar development times. Hydrobiologia, **510**: 75-81.

JEONG, C. B., H. M. KANG, M. C. LEE, D. H. KIM, J. HAN, D. S. HWANG, S. SOUISSI, S. J. LEE, K. H. SHIN, H. G. PARK & J. S. LEE, 2017. Adverse effects of microplastics and oxidative stress-induced MAPK/Nrf2 pathway-mediated defense mechanisms in the marine copepod *Paracyclopina nana*. Sci. Rep., **7**: 41323.

KADIENE, E. U., P. J. MENG, J. S. HWANG & S. SOUISSI, 2019. Acute and chronic toxicity of cadmium on the copepod *Pseudodiaptomus annandalei*: a life history traits approach. Chemosphere, **233**: 396-404.

KOSKI, M., J. ENGSTRÖM & M. VIITASALO, 1999. Reproduction and survival of the calanoid copepod *Eurytemora affinis* fed with toxic and non-toxic cyanobacteria. Mar. Ecol. Prog. Ser., **186**: 187-197.

KWOK, K. W. H., S. SOUISSI, G. DUR, E. J. WON & J. S. LEE, 2015. Copepods as references species in estuarine and marine waters. In: C. AMIARD-TRIQUET, J. C. AMAIRD & C. MOUNEYRAC (eds.), Aquatic ecotoxicology: advancing tools for dealing with emerging risks: 281-308. (Elsevier, Amsterdam).

LASLEY-RASHER, R. S., A. M. KRAMER, V. BURDETT-COUTTS & J. YEN, 2014. Assessing the in situ fertilization status of two marine copepod species, *Temora longicornis* and *Eurytemora herdmani*; how common are unfertilized eggs in nature? PLoS One, **9**(11): e112920. DOI:10. 1371/journal.pone.0112920.

LIANG, D. & S. UYE, 1997. Population dynamics and production of the planktonic copepods in a eutrophic inlet of the Inland Sea of Japan. IV. *Pseudodiaptomus marinus*, the egg-carrying calanoid. Mar. Biol., **128**: 415-421.

MANCA, M., M. BELTRAMI & D. SONVICO, 1996. On the appearance of epibionts on the crustacean zooplankton of a large subalpine lake undergoing oligotrophication (L. Maggiore, Italy). Memorie dell' Istituto Italiano di Idrobiologia, **54**: 161-171.

MANCA, M., A. CARNOVALE & P. ALEMANI, 2004. Exotopic protrusions and ellobiopsid infection in zooplanktonic copepods of a large, deep subalpine lake, Lago Maggiore, in northern Italy. J. Plankton Res., **26**: 1257-1263.

MANTHA, G., A. K. AWASTHI, A. M. AL-AIDAROOS & J. S. HWANG, 2013. Diversity and abnormalities of cyclopoid copepods around hydrothermal vent fluids, Kueishantao Island, northeastern Taiwan. J. Nat. Hist., **47**: 685-697.

MESSICK, G. A., H. A. VANDERPLOEG, J. F. CAVALETTO & S. S. TYLER, 2004. Histological characteristics of abnormal protrusions on copepods from Lake Michigan, USA. Zool. Stud., **43**: 314-322.

MICHALEC, F. G., M. HOLZNER, A. SOUISSI, S. STANCHEVA, A. BARRAS, R. BOUKHERROUB & S. SOUISSI, 2016. Lipid nanocapsules for behavioural testing in aquatic toxicology: time-response of *Eurytemora affinis* to environmental concentrations of PAHs and PCB. Aquat. Toxicol., **170**: 310-322.

MILLER, D. & N. H. MARCUS, 1994. The effects of salinity and temperature on the density and sinking velocity of eggs of the calanoid copepod *Acartia tonsa* Dana. J. Exp. Mar. Biol. Ecol., **179**: 235-252.

NIEHOFF, B., 2007. Life history strategies in zooplankton communities: the significance of female gonad morphology and maturation types for the reproductive biology of marine calanoid copepods. Prog. Oceanogr., **74**: 1-47.

OMAIR, M., B. NAYLOR, D. J. JUDE, J. QUDDUS, T. F. BEALS & H. A. VANDERPLOEG, 2001. Histology of herniations through the body wall and cuticle of zooplankton from the Laurentian Great Lakes. J. Invertebr. Pathol., **77**: 108-113.

OMAIR, M., H. A. VANDERPLOEG, D. J. JUDE & G. L. FAHNENSTIEL, 1999. First observations of tumor-like abnormalities (exophytic lesions) on Lake Michigan zooplankton. Can. J. Fish. Aquat. Sci., **56**: 1711-1715.

REDFIELD, G. W. & W. F. VINCENT, 1979. Stages of infection and ecological effects of a fungal epidemic on the eggs of a limnetic copepod. Freshw. Biol., **9**: 503-510.

SILINA, N. & V. KHUDOLEI, 1994. Tumor-like anomalies in planktonic copepods. Hydrobiol. J., **30**: 52-55.

SKOVGAARD, A., 2004. Tumour-like anomalies on copepods may be wounds from parasites. J. Plankton Res., **26**: 1129-1131.

SOUISSI, A., S. SOUISSI, D. DEVREKER & J. S. HWANG, 2010. Occurrence of intersexuality in a laboratory culture of the copepod *Eurytemora affinis* from the Seine estuary (France). Mar. Biol., **157**(4): 851-861.

SOUISSI, A., S. SOUISSI & B. W. HANSEN, 2016b. Physiological improvement in the copepod *Eurytemora affinis* through thermal and multigenerational selection. Aquat. Res., **47**: 2227-2242.

SOUISSI, A., S. SOUISSI & J. S. HWANG, 2013. The effect of epibiont ciliates on the behavior and mating success of the copepod *Eurytemora affinis*. J. Exp. Mar. Biol. Ecol., **445**: 38-43.

SOUISSI, A., S. SOUISSI & J. S. HWANG, 2016a. Evaluation of the copepod *Eurytemora affinis* life history response to temperature and salinity increases. Zool. Stud., **55**: 4. DOI:10.6620/ZS.2016.55-04.

SOUISSI, S., M. N. DALY YAHIA & J. S. HWANG, 2007. Foreword (Copepod special issue). J. Plankton Res., **29**: i1-i2.

UYE, S. & A. FLEMINGER, 1976. Effect of various environmental factors on egg development of several species of *Acartia* in southern California. Mar. Biol., **38**: 253-262.

UYE, S. & K. SANO, 1995. Seasonal reproductive biology of the small cyclopoid copepod *Oithona davisae* in a temperate eutrophic inlet. Mar. Ecol. Prog. Ser., **118**: 121-128.

WINKLER, G., J. J. DODSON & C. E. LEE, 2008. Heterogeneity within the native range. Population genetic analyses of sympatric invasive and noninvasive clades of the freshwater invading copepod *Eurytemora affinis*. Mol. Ecol., **17**: 415-430.

WINKLER, G., S. SOUISSI, C. POUX & V. CASTRIC, 2011. Genetic heterogeneity among several *Eurytemora affinis* populations in western Europe. Mar. Biol., **158**: 1841-1856.

First received 29 October 2019.
Final version accepted 3 December 2019.

[When citing this volume, please refer to *Crustaceana* 93 (2020), 241-547]

*Proceedings of the* Eurytemora *conference, St. Petersburg, 2019*

# DIFFERENTIAL ECO-PHYSIOLOGICAL PERFORMANCES OF TWO PSEUDOCRYPTIC SPECIES OF THE *EURYTEMORA AFFINIS* COMPLEX (COPEPODA, CALANOIDA) IN THE ST. LAWRENCE ESTUARINE TRANSITION ZONE: A RECIPROCAL TRANSPLANT EXPERIMENT

BY

JORY CABROL[1,2]), RÉJEAN TREMBLAY[1]) and GESCHE WINKLER[1,2,3])

[1]) Institut des Sciences de la Mer de Rimouski, Université du Québec à Rimouski, Rimouski, QC, Canada G5L 3A1

[2]) Québec-Océan, Rimouski, QC, Canada G5L 3A1

## ABSTRACT

The goal of this study was to evaluate and compare the short-term performances and the physiological plasticity of two cryptic species *Eurytemora carolleeae* and *Eurytemora affinis* (North-Atlantic clade) by simulating rapid advection from freshwater to brackish water conditions and reciprocally. To do so, two reciprocal transplant experiments without acclimation and under non-limited food condition were performed in the St. Lawrence estuarine transition zone during summer 2011. Results revealed that both species diverged in their short-term acclimation response when facing acute salinity changes that they might encounter when advected through the highly dynamic estuarine transition zone. We show that *E. carolleeae* could use the brackish environment without loss of performance and energy, while *E. affinis* needed to reallocate energy from other processes (i.e., reproduction) and required food intake to maintain itself in the freshwater environment. In addition, the transplant experiment highlighted that only 40% of the *E. affinis* showed short term capacity to acclimate to freshwater conditions, indicating that in situ advection by currents from brackish water to fresh water could be dramatic even for a short time period. Furthermore, the survivors of *E. affinis* in fresh water might not be able to reproduce, which limits establishment of a sustainable population of *E. affinis* (North-Atlantic clade) in the tidal freshwater part of the estuarine transition zone. Finally, we highlighted for the first time that both species of this pseudocryptic complex could use lipid remodelling to overcome temperature effects on membrane structure, but further studies are needed to determine the influence of membrane lipid remodelling on salinity tolerance.

## RÉSUMÉ

Le but de ce travail a été d'évaluer et de comparer les performances à court terme et la plasticité phénotypique de deux espèces cryptiques *Eurytemora carolleeae* et *Eurytemora affinis* (clade de l'Atlantique Nord) en simulant un transfert rapide d'eau douce en eau saumâtre et réciproquement.

[3]) Corresponding author; e-mail: gesche_winkler@uqar.ca

 | DOI:10.1163/9789004465565_011

Pour ce faire, deux expériences de transplantations réciproques sans acclimatation et avec une alimentation ad libitum ont été effectuées dans la zone de transition de l'estuaire du St Laurent durant l'été 2011. Les résultats ont montré que les deux espèces divergent dans leur réponse d'acclimatation à court terme face à un changement important de salinité, ce qu'elles sont susceptibles de rencontrer au cours d'un transfert à travers la zone de transition de l'estuaire. Nous montrons que *E. carolleeae* peut utiliser l'environnement saumâtre sans perte de performance et d'énergie, tandis que *E. affinis* a besoin de réattribuer de l'énergie à partir d'autres processus (i.e., reproduction) et a besoin de s'alimenter afin de se maintenir dans un environnement d'eau douce. De plus, l'expérience de transfert souligne que seulement 40% des *E. affinis* ont montré une capacité à court terme de s'acclimater à des conditions d'eau douce, indiquant que, in situ, le transport par les courants de l'eau saumâtre vers l'eau douce pourrait être dramatique même pour une courte période de temps. De plus, les survivants d'*E. affinis* en eau douce pourraient être incapable de se reproduire. Ce qui, limite l'établissement d'une population stable d'*E. affinis* (clade Nord-Atlatique) dans la partie douce de la marée dans la zone de transition de l'estuaire. Enfin, nous signalons pour la première fois que les deux espèces des ce complexe pseudo-cryptique pourraient utiliser le remodelage des lipides pour surmonter les effets de la température sur la structure des membranes, mais de nouvelles études sont nécessaires pour déterminer l'influence du remodelage des lipides membranaires sur la tolérance à la salinité.

## INTRODUCTION

The species complex of *Eurytemora affinis* (Poppe, 1880) (cf. Lee, 1999, 2000; Lee & Frost, 2002), consisting of at least 6 ancestral clades, plays a pivotal role in the functioning of coastal aquatic ecosystems throughout the Northern Hemisphere (Lee, 2000; Devreker et al., 2010, 2012; Winkler et al., 2016). They allow energy transfer from lower to higher trophic levels, supporting numerous ecological and economically relevant species (e.g., Winkler et al., 2003; Kimmel et al., 2006; Pierson et al., 2016). A recent taxonomic revision described the Atlantic clade within the *E. affinis* species complex as a new species, *Eurytemora carolleeae* Alekseev & Souissi, 2011 (cf. Alekseev & Souissi, 2011; Sukhikh et al., 2016). Morphological traits used to differentiate this and further species in this complex are nonadaptive (Alekseev & Souissi, 2011; Sukhikh et al., 2013), which might suggest *E. affinis* as a pseudocryptic complex (Lajus et al., 2015).

In the St. Lawrence estuarine transition zone (ETZ; Eastern Canada), *E. carolleeae* (hitherto *E. affinis* Atlantic clade) and the *E. affinis* North-Atlantic clade (hereafter *E. affinis*) co-occur and dominate mesozooplankton communities (Winkler et al., 2003; Favier & Winkler, 2014). However, both species show a marked spatial segregation in terms of dominance in particular zones of the ETZ. The form *E. carolleeae* inhabits predominantly the warm tidal freshwater habitat and hypersaline ponds, whereas the *E. affinis* dominates the colder brackish water stretch of the ETZ (Winkler et al., 2003, 2008; Favier & Winkler, 2014; fig. 1). This spatial pattern was hypothesized to result from differential 'species-specific' physiological tolerances to a combination of salinity × temperature conditions

(Winkler et al., 2008) and fostered by food availability (Favier & Winkler, 2014). To cope with the salinity variations and the gradient in the ETZ, the capacity of osmoregulation is of importance. Both species have been found to regulate their body ion concentrations, enabling them to perform rapid transition into various salinities (Lee et al., 2011, 2012; Johnson et al., 2013). However, in the laboratory they show differential osmoregulation capacities (Lee et al., 2011, 2012, 2013). *E. affinis* has rarely been observed in fresh water and seems mostly limited to mesohaline water (Favier & Winkler, 2014) as it can only compensate for the metabolic cost of osmoregulation in fresh water under high food condition (Lee et al., 2013). In contrast, *E. carolleeae* performs well (i.e., survival or hatching success) in both fresh- and brackish water conditions (Lee et al., 2011, 2013). Accordingly, these laboratory results suggest that *E. carolleeae* can use brackish water although proportional abundances of *E. carolleeae* found in this part of the ETZ were low, and always less than those of *E. affinis* (cf. Favier & Winkler, 2014; Cabrol et al., 2015). Furthermore, the ETZ of the St. Lawrence is characterized by strong tidal currents (St-Onge-Drouin et al., 2012). Under these conditions, advection should homogenize zooplankton distribution throughout the ETZ, suggesting that individuals of both species might experience sudden changes in environmental conditions due to transport through the marked salinity and temperature gradients occurring in the ETZ. However, short term physiological acclimation capacities (plasticity) of these two ecologically relevant species when facing an acute change of environmental conditions, are unknown.

Metabolic and physiological phenotypic plasticity are known to be the first 'mechanisms' of response when facing adverse conditions (e.g., Ghalambor et al., 2007). Plastic responses also come with a metabolic cost (i.e., reallocation of the available energy budget), in order to preserve positive life-history traits (e.g., Hoffmann, 1995; Dewitt et al., 1998; Calosi et al., 2013). In addition, food supply (e.g., Lee et al., 2013; Favier & Winkler, 2014; Hammock et al., 2016), temperature (e.g., Souissi et al., 2013; Rahlff et al., 2017), life stage (e.g., Lee & Petersen, 2003; Dur & Souissi, 2018) and life history (e.g., Beyrend-Dur et al., 2009; Kaarlson et al., 2018) are known to affect the metabolic cost of *Eurytemora* spp. and thus, their overall acclimation capacities. Short term acclimation capacities of the *E. affinis* pseudocryptic species complex have received little attention so far, except in the study by Lee & Petersen (2003). Instead, most studies focused on adaptation under natural selection over multiple generations (e.g., Lee & Petersen, 2003; Souissi et al., 2013; Souissi et al., 2016; Karlsson et al., 2018; Karlsson & Winder, 2018).

The goal of this study was to evaluate and compare the short-term performances and physiological plasticity of the two *E. affinis* pseudocryptic species by simulating rapid advection from fresh water to brackish water and reciprocally, during a period of high food supply. Based on previous laboratory experiments, we hypothesize that *E. carolleeae* shows similar performances in both habitats, whereas *E. affinis*, which seemed to be restricted to mesohaline waters, exhibits lower performances during transplantation into the freshwater habitat. To tackle this hypothesis, we performed two short term (24 hours) reciprocal transplant experiments without acclimation. Effects of short-term acclimation were determined by comparing eco-physiological responses of the two pseudocryptic species between the new (transplant) and their native (home) habitat. To quantify short term performance, survival rates and spawning success were measured. In addition, physiological changes in relation to the experienced environmental conditions were measured by comparing: (i) fatty acid content of neutral lipids as a measure of short-term energy utilization (Fraser et al., 1989; Antonio & Richoux, 2014; Cabrol et al., 2015) and (ii) fatty acid composition of polar lipids representing physiological acclimation in membranes (Hazel & Williams, 1990; Hazel, 1995; Stillwell & Wassall, 2003). These results will help to explain spatial segregation occurring in the St. Lawrence ETZ. Furthermore, they will enhance our understanding of acclimation capacity after an acute environmental change, not only in the estuarine transition zone, but also when advected or introduced into new environments.

## MATERIAL AND METHODS

### Experimental design

According to the geographic segregation found between *Eurytemora carolleeae* (fresh water) and *Eurytemora affinis* (brackish water) in the south channel of the estuarine transition zone (Winkler et al., 2003; Favier & Winkler, 2014; fig. 1), in situ transplant experiments were conducted in July (18 to 22) and August (15 to 19) 2011 to determine short term performance by simulating rapid advection of individuals from the home habitat to a new one. During each period, moorings composed of 6 polyethylene bags (25 l each). Three bags (replicates) containing *E. carolleeae* and three bags containing *E. affinis* were deployed in each habitat at 1 m depth for 24 hours. For logistical reasons, moorings were deployed with a 24 hours delay in the two habitats. To equip each mooring, zooplankton was sampled as presented in detail in Cabrol et al. (2015). Briefly, adults of the two species were caught by short ($<2$ min) and successive oblique tows using a 1 m round net with 303 $\mu$m mesh size. *E. carolleeae* was sampled near Berthier-sur-Mer, and the North Atlantic clade of *E. affinis* near Saint-Jean-Port-Joli (fig. 1). Half of the sample was preserved in ethanol for a posteriori species identification. The remaining individuals were kept alive in 63 $\mu$m pre-filtered water at the same ambient water temperature of the two habitats. Adults without injuries (e.g., lack

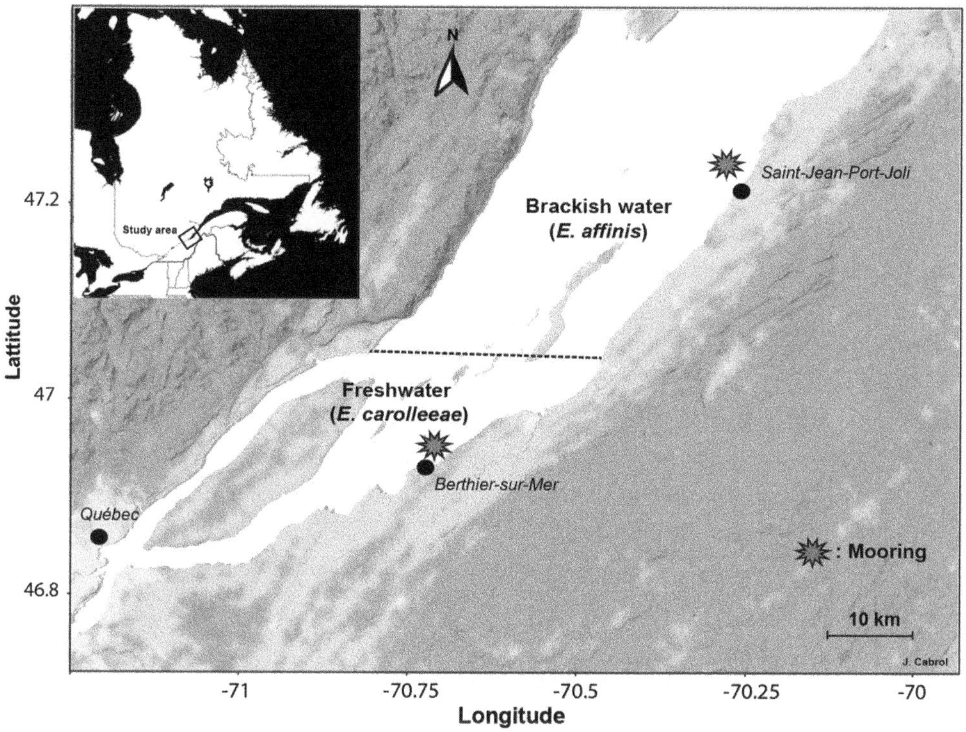


Fig. 1. Sampling area in July and August 2011. Dashed line represents a schematic limit of the tidal freshwater and the brackish water habitats, home of *Eurytemora carolleeae* Alekseev & Souissi, 2011 and *Eurytemora affinis* (Poppe, 1880) North-Atlantic clade, respectively.

of antennae or limbs), egg sacs, or attached spermatophores, were picked up one by one under a stereomicroscope (Leica MZ12.5). At the start of the experiments, 60 non-ovigerous females and 60 males were added to each 25 l polyethylene bag, which bags were pre-filled with 63 $\mu$m filtered-water from the habitat where the mooring was deployed. The 63 $\mu$m mesh sieve was used as a trade-off to remove all macro- and mesozooplankton without altering the composition of potential food supply (Cabrol et al., 2015). In addition, *E. carolleeae* and *E. affinis* were flash frozen separately and kept under nitrogen atmosphere at $-80$°C to determine the state of the physiological condition at the beginning of each experiment. Salinity was measured at the beginning and the end of the experiment in every replicate bag using a YSI-professional probe. In situ temperature variations were followed during all experiments using two SBE37-SMP MicroCAT situated at the extremities of each mooring.

At the end of each experiment, copepods were carefully and quickly retrieved from each bag using a 20 $\mu$m mesh size sieve. Special attention was given to the emptying and rinsing of the bags to avoid the loss of individuals due to manipulation. The individuals were flash-frozen and kept at −80°C until laboratory analyses (<6 months) to determine survival rates, spawning success and physiological conditions.

## Species identification

To confirm proportions of *E. carolleeae* and *E. affinis* in the experiments, the restriction fragment length polymorphism (RFLP) approach was used according to Favier & Winkler (2014). This method generated two pairs of different fragment lengths for the two species (*E. carolleeae* = 61 and 591 bp and *E. affinis* = 211 and 451 bp) allowing visual identification after standard electrophoresis migration. Methodological details are given in Cabrol et al. (2015). We analysed 288 individuals (144 per habitat). In the freshwater habitat, 100% of the individuals belonged to *E. carolleeae* and in the brackish water habitat 99.3% of the individuals belonged to the *E. affinis* NA clade and 0.7% to *E. carolleeae* (cf. Cabrol et al., 2015), confirming the spatial segregation previously found by Winkler et al. (2003) and Favier & Winkler (2014).

## Survival rate and spawning success

Survival rates for both males and females were calculated for each replicate (bag) by counting intact individuals under the stereomicroscope (Leica $MZ12_5$) and are expressed in % survival. Female spawning success, defined as the proportion of females with egg-sac spawned during the experiment, was monitored and expressed in %.

## Characterization of the fatty acid contents and compositions

Individuals were gently rinsed one by one to eliminate stuck particles (e.g., phytoplankton). As males and females exhibit differences in their morphology and physiology, analyses were separated by sex. In addition, all males and all females from one replicate (bag) were pooled to ensure sufficient material for fatty acid (FA) analysis.

FA compositions were analysed following methods presented in detail in Cabrol et al. (2015, 2018). Briefly, bulk lipids were extracted by grinding, using a modified Folch procedure (Chen et al., 1981). Neutral lipids (e.g., wax esters, triacylglycerol, free FAs), representative of energetic lipid reserves (Fraser et al., 1989; Cabrol et al., 2015), and polar lipid fractions (e.g., phospholipids), representative of the membrane composition, were separated on a silicate gel (Marty et al., 1992). Fatty acid methyl esters (FAME) of both fractions were obtained by direct transesterification following methods in Lepage & Roy (1992).

FAMEs were separated using an Omegawax 250 (Supelco) capillary column and determined using a Polaris Q ion trap coupled to a multichannel gas chromatograph "Trace GC ultra" (Thermo Scientific) and a mass detector model ITQ900 (Thermo Scientific). FAME were identified and quantified using known standards (Supelco 37 Component FAME Mix and menhaden oil; Supleco) or according to their mass spectra for peaks not related to a known standard with the Xcalibur v.3.1 software.

## Data analyses

To determine whether environmental conditions and individual performances varied significantly during the transplant experiments, permutational multivariate analysis of variance (PERMANOVA), designed as an ANOVA (see Anderson et al., 2008), was applied using PRIMER® version 7.0.13 (Clarke & Gorley, 2015) with PERMANOVA+ add-on (Anderson et al., 2008). The PERMANOVA approach was chosen as it allows analysis of variance in presence of an unbalanced design and does not assume normality of the data (Anderson & Walsh, 2013). To test differences in environmental conditions, two-way PERMANOVAs (Habitat × Month) were performed on salinity and temperature. Three-way PERMANOVAs (Habitat × Month × Species), were applied on males and females to explore potential differences among survival rates, spawning success, and content and composition of energy lipid reserves (neutral lipids) after transplantations. In the case of significant interactions, post-hoc tests (pairwise multiple comparisons routine) were carried out to identify differences among levels within factors. $P$-values were calculated using 9999 permutations, partial sums of squares (Type III) and Monte-Carlo random draw probability (hereafter $P_{(MC)}$) under a reduced model to avoid issues of too few available permutations with small sample sizes (Anderson & Robinson, 2003). In addition, similar percent analyses routine (SIMPER; Clarke & Gorley, 2015) was performed to identify FAs, which contribute to dissimilarities between groups.

To explore potential lipid remodelling during transplant experiments, canonical analyses of principal coordinates (CAP, Anderson & Robinson, 2003; Anderson & Willis, 2003) and "trace" tests (equivalent to the Pillai's trace test in a traditional MANOVA; Anderson et al., 2008) were performed on polar FAs which contributed to $>1\%$ of total polar FAs. FAs responsible for among group differences were identified by examining Pearson correlations and Student's $t$-tests were performed on correlated FA to determine if they differ significantly between habitats for each species and sex (table I) using JMP software (V13.1). To detect if the responses to transplantation of both species were similar in terms of the extent of FA variation in the new habitat compared to their native habitats, multidimensional dispersion routines (PERMDISP) were performed using PRIMER. In addition, to detect if

TABLE I

Proportion of neutral FAs (in % of total FAs) which contributes to the differences between habitats found at the end of the reciprocal transplant experiments (after 24 hours) for males and females of *Eurytemora carolleeae* Alekseev & Souissi, 2011 and *Eurytemora affinis* (Poppe, 1880)

| | Male | | | | Female | | | |
|---|---|---|---|---|---|---|---|---|
| | *E. carolleeae* | | *E. affinis* | | *E. carolleeae* | | *E. affinis* | |
| | Freshwater habitat | Brackish water habitat | Freshwater habitat | Brackish water habitat | Freshwater habitat | Brackish water habitat | Freshwater habitat | Brackish water habitat |
| 14:0 | 3.1 ± 1.2 | 3.6 ± 0.6 | 3.1 ± 0.7 | 2.8 ± 0.4 | 4.0 ± 1 | 3.6 ± 0.6 | 4.4 ± 0.5 | 3 ± 0.6 |
| 15:0i | 0.1 ± 0.01 | 0.2 ± 0.1 | 0.3 ± 0.1 | 1.1 ± 1.1 | 0.2 ± 0.1 | 0.3 ± 0.01 | 0.3 ± 0.1 | 0.2 ± 0.1 |
| 15:0 | 0.6 ± 0.1 | 1.6 ± 0.5 | 1.3 ± 0.3 | 1.0 ± 0.2 | 0.7 ± 0.01 | 1.4 ± 0.1 | 1.4 ± 0.1 | 0.8 ± 0.1 |
| 16:0 | 24.6 ± 1.7 | 21.6 ± 3.6 | 19 ± 1.9 | 23.0 ± 1.1 | 22.5 ± 1.5* | 15.3 ± 0.5* | 13.6 ± 0.8* | 18.6 ± 1.8* |
| 17:0i | 0.4 ± 0.1 | 0.4 ± 0.2 | 0.9 ± 0.4 | 0.2 ± 0.1 | 0.3 ± 0.1 | 0.4 ± 0.01 | 0.4 ± 0.1 | 0.4 ± 0.1 |
| 17:0ai | 0.4 ± 0.2 | 0.6 ± 0.2 | 0.9 ± 0.4 | 0.3 ± 0.2 | 0.2 ± 0.1 | 0.2 ± 0.01 | 0.3 ± 0.1 | 0.3 ± 0.1 |
| 16:2n4 | 3 ± 0.5 | 3.2 ± 0.6 | 2.4 ± 0.4 | 3.3 ± 0.8 | 4 ± 0.5 | 4.9 ± 0.5 | 5.9 ± 0.5* | 2.9 ± 0.6* |
| 16:3n3 | 1.9 ± 0.3 | 2.1 ± 0.4 | 1.6 ± 0.5 | 1.9 ± 0.5 | 2.2 ± 0.4 | 2.9 ± 0.2 | 4.4 ± 0.5 | 1.7 ± 0.3 |
| 16:1n7 | 12.9 ± 2.6 | 10.3 ± 3.4 | 5.6 ± 0.6 | 14.2 ± 4.1 | 14.4 ± 3.2 | 9.7 ± 1 | 11.6 ± 0.9 | 12.2 ± 1.8 |
| 17:1n7 | 0.5 ± 0.2 | 0.6 ± 0.3 | 0.6 ± 0.4 | 0.9 ± 0.3 | 0.4 ± 0.2 | 0.2 ± 0.01 | 0.2 ± 0.1 | 0.3 ± 0.1 |
| 18:0 | 9.7 ± 2.6 | 6.5 ± 1.2 | 13.3 ± 1.4 | 11.4 ± 4.4 | 6.45 ± 1.8 | 5.4 ± 0.8 | 3.8 ± 0.6* | 7.7 ± 2.2* |
| 18:1n9 | 6.5 ± 0.8 | 9.6 ± 3.6 | 5.5 ± 1.3 | 4.7 ± 1.5 | 4.3 ± 0.9 | 3.6 ± 1.2 | 1.7 ± 0.3* | 9.1 ± 4.3* |
| 18:1n7 | 2.7 ± 0.1 | 3.7 ± 0.5 | 4.1 ± 0.4 | 3.1 ± 0.3 | 2.6 ± 0.2 | 4.6 ± 0.3 | 4.2 ± 0.2 | 4.2 ± 1 |
| 18:2n6 | 2.2 ± 0.5 | 3 ± 1 | 1.7 ± 0.5 | 1.8 ± 0.3 | 2 ± 0.3 | 1.3 ± 0.2 | 1.1 ± 0.1 | 2.8 ± 0.9 |
| 18:3n3 | 5.1 ± 1.4 | 1.4 ± 0.4 | 0.9 ± 0.3 | 1.8 ± 0.3 | 4.4 ± 1 | 0.6 ± 0.1 | 0.6 ± 0.1 | 1.3 ± 0.2 |
| 18:4n3 | 4.1 ± 0.7 | 1.7 ± 0.5 | 1.3 ± 0.2 | 2.4 ± 0.6 | 3.3 ± 0.4 | 1.4 ± 0.1 | 1.9 ± 0.1 | 1.9 ± 0.3 |

TABLE I
(Continued)

| | Male | | | | Female | | | |
|---|---|---|---|---|---|---|---|---|
| | *E. carolleeae* | | *E. affinis* | | *E. carolleeae* | | *E. affinis* | |
| | Freshwater habitat | Brackish water habitat | Freshwater habitat | Brackish water habitat | Freshwater habitat | Brackish water habitat | Freshwater habitat | Brackish water habitat |
| 20:0 | 0.6 ± 0.1 | 0.9 ± 0.2 | 1.1 ± 0.4 | 0.6 ± 0.1 | 0.6 ± 0.1 | 0.5 ± 0.01 | 0.5 ± 0.01 | 0.6 ± 0.1 |
| 20:5n3 (EPA) | 8.1 ± 0.3 | 10.4 ± 2 | 13.8 ± 2.4 | 10.2 ± 0.7 | 11.6 ± 0.6* | 23 ± 1* | 22.3 ± 1.0* | 13.7 ± 2.7* |
| 20:6n3 (DHA) | 5 ± 1.5 | 7.4 ± 1.6 | 12.9 ± 3.1 | 6.4 ± 1.3 | 6.8 ± 2* | 10.6 ± 0.5* | 10.8 ± 0.6* | 7.6 ± 1.3* |
| ΣSFA | 41.8 ± 3.1 | 38.7 ± 4.2 | 42.7 ± 4.4 | 42.8 ± 6.6 | 36.8 ± 1.1 | 28.8 ± 1.4 | 26.9 ± 2.1 | 33.7 ± 4.8 |
| ΣMUFA | 23.9 ± 2.2 | 25.9 ± 4 | 17 ± 1.5 | 24.1 ± 3.2 | 23 ± 2.7 | 19.3 ± 0.9 | 18.8 ± 0.6 | 27.9 ± 4.3 |
| ΣPUFA | 34.3 ± 3.7 | 35.4 ± 4 | 40.3 ± 5.6 | 33.0 ± 4.4 | 40.3 ± 2.9 | 51.9 ± 1.7 | 54.2 ± 2.2 | 38.5 ± 5 |
| TFA (μg/mg) | 36.3 ± 11.3 | 29.6 ± 9 | 19.7 ± 6.1 | 22.0 ± 3.9 | 39.2 ± 13.8 | 52.4 ± 13.3 | 32.8 ± 6.3 | 26.8 ± 4.8 |

The data from the two months are combined. Values followed by an asterisk (*) are significantly different between habitats (Student's $t$-test; $p<0.05$).
Fatty acids are expressed as % of total fatty acids. Total fatty acids are expressed in μg/mg wet weight of each copepod species. Only fatty acids contributing >1% in at least one component are reported. Values are mean ± SE.
Abbreviations: i, iso; ai, ante-iso; EPA, eicosapentaenoic acid; DHA, docosahexaenoic acid; SFA, saturated fatty acids; MUFA, monounsaturated fatty acids: PUFA, polyunsaturated fatty acids; TFA, total fatty acids.

the variation of the response was different for transplanted and native individuals, multidimensional dispersion routines (PERMDISP), based on the deviation from centroids (Anderson et al., 2008), were performed separately for males and females of each species.

## RESULTS

### Characterization of environmental conditions

No significant variation of the in situ temperature (variations <0.2°C) was observed between the start and the end of all the experiments. However, the average in situ temperature found during both experiments was higher in the freshwater habitat (home of *Eurytemora carolleeae*) situated in the freshwater part of the estuarine transition zone, compared to the brackish water habitat, home of *Eurytemora affinis* (23.2 ± 0.13 and 20.9 ± 0.6°C, respectively, fig. 2A). The freshwater habitat showed a similar temperature in July and August, whereas in the brackish water habitat the temperature was 1.5°C higher in July than in August (fig. 2A). Mean salinity was very low, 0.1 ± 0.02 PSU in the freshwater habitat, whereas mean salinity was 3.7 ± 0.4 PSU in the brackish water habitat, which is more influenced by the intrusion of salt water (fig. 2B). Salinity patterns were similar in July and August.

### Survival rates

Survival rates of *E. carolleeae* after 24 hours of the experiment did not differ between habitat or sex for an average of 88.6 ± 3% (fig. 3A, B). Similarly, the survival rates of *E. affinis* males were similar in their home and transplant habitats for an average of 44.4 ± 3.2% (fig. 3A). In contrast, the survival rates of *E. affinis* females were 33% lower in the transplant (fresh water) than in their home habitat (brackish water; fig. 3B). In addition, we found that the survival rate of *E. affinis* males was lower than that found in females in their home habitat (brackish water; fig. 3A, B). In general, the survival rate varied in time independently of the species or sex, being significantly higher in the brackish water habitat in August compared to July (fig. 3C, D).

### Spawning success

No variation in the proportion of egg-carrying females was found between months (PERMANOVA; Pseudo-$F_{1;23}$ = 0.48; $P_{(MC)}$ = 0.5). Similarly, in *E. carolleeae* no difference was found after the 24 hours of transplantation into the brackish water habitat compared to the freshwater habitat for an average of

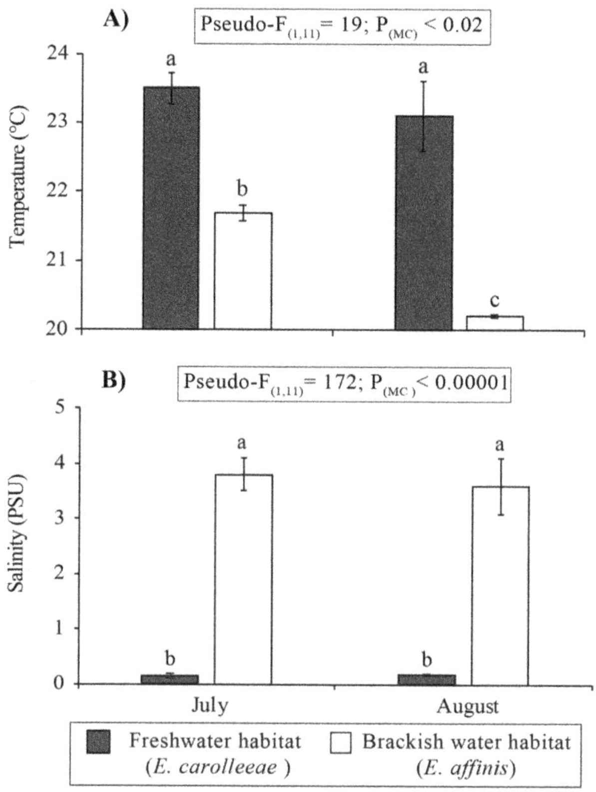


Fig. 2. A, Temperature; and B, salinity found in the tidal freshwater habitat (dark grey) and the brackish water habitat (white) in July and August 2011. Letters indicate significant differences (pairwise comparison; $P>0.05$). Mean ± SE.

54.8 ± 2.3% of egg-carrying females (fig. 4). However, the spawning success was significantly lower for the *E. affinis* in the freshwater habitats compared to females of the home habitat (fig. 4), as only 2.1 ± 1.5% of *E. affinis* females had spawned in the fresh water (fig. 4). In addition, *E. affinis* showed a lower proportion of egg-carrying females than *E. carolleeae* in their respective home habitats.

## Energy lipid reserves: content and composition

The quantity of neutral FAs, representative of the energy lipid content, was similar between species and sexes (PERMANOVA; Pseudo-$F_{2;43} = 0.06$; $P_{(MC)} = 0.94$) for an average neutral FA content of approx. 29.2 $\mu$g/mg DW. On the other hand, lipid contents after transplantation were highly variable (table I), with values ranging from 4.63 to 78.4 (all factors confounded), indicating the presence of a high heterogeneity in response when facing new environmental conditions, which could be directly related to the phenotypic variance among survivors expressed

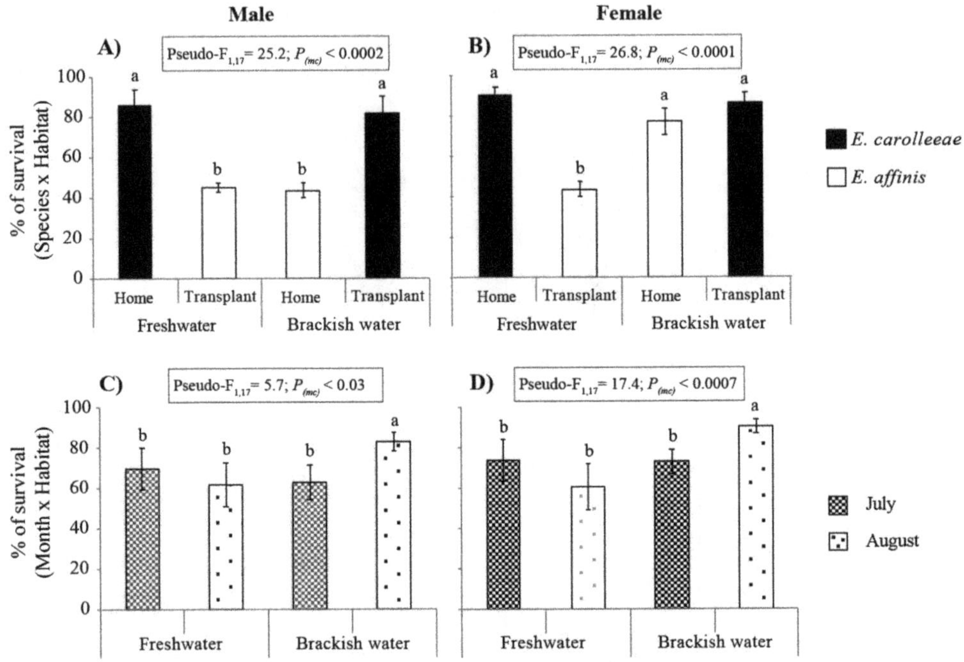


Fig. 3. Survival rate (%) found at the end of the experiment (after 24 hours) of: A, C, males; and, B, D, females; of *Eurytemora carolleeae* Alekseev & Souissi, 2011 (black) and *Eurytemora affinis* (Poppe, 1880) (white), for all months combined and for both species pooled in July and August. Levels not connected by the same letter are significantly different (pairwise comparison; $P>0.05$). Mean ± SE.

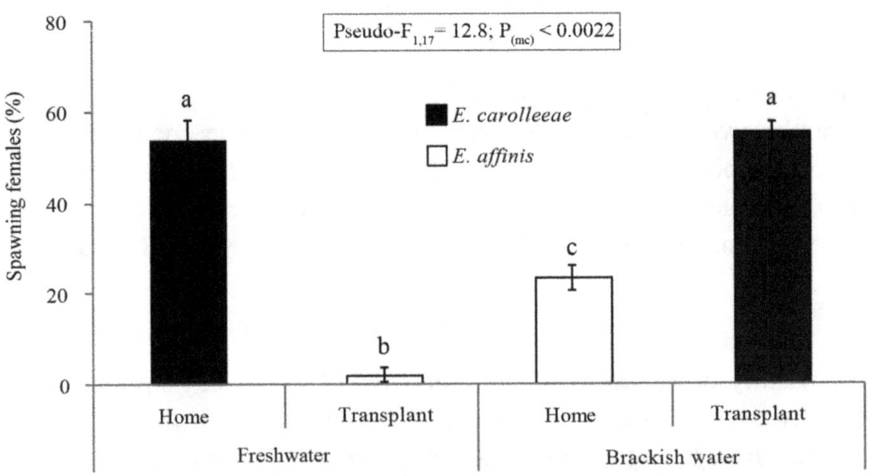


Fig. 4. Spawning success, defined as the number of females with egg sac found at the end of the experiment, of *Eurytemora carolleeae* Alekseev & Souissi, 2011 (black) and *Eurytemora affinis* (Poppe, 1880) (white) for all months combined. Levels not connected by the same letter are significantly different (pairwise comparison; $P>0.05$). Mean ± SE.

during transplantations. In contrast, FA compositions of neutral lipids, which are classically related to the diet, differed between females in the transplant and the home habitats (PERMANOVA; Pseudo-$F_{1;15} = 9.23$; $P_{(MC)} = 0.003$), but did not in males (PERMANOVA; Pseudo-$F_{1;15} = 2.14$; $P_{(MC)} = 0.06$; table I). The difference in the FA composition of females was mostly due to the variations of five FAs (in order of importance: 20 : 5n3, 22 : 6n3, 18 : 1n9, 16 : 0 and 18 : 0), which contributed together almost 80% of the total dissimilarity of the FA composition between the two habitats (SIMPER analyses).

### Remodelling of membrane lipids

Polar lipid (PL) composition of both *E. carolleeae* and *E. affinis* differed significantly in the transplant habitats compared to their respective home habitats (fig. 5). After 24 hours, the FA composition of PL of *E. carolleeae* in the transplant habitat (brackish water) tended to be more similar to the FA composition of *E. affinis* in the same habitat (brackish water) as compared to *E. carolleeae* in its home (freshwater) habitat (see fig. 5A,B). *E. affinis* males showed a similar pattern. However, PL composition of *E. affinis* females in the freshwater habitat tended to differ in terms of composition to that of *E. carolleeae* in the same habitat (fig. 5B).

Although lipid remodelling for males and females differed for specific FAs (table II), they showed a similar trend when transplanted into the new habitat (fig. 5). Individuals of *E. carolleeae* decreased their proportion of saturated FAs (SFAs) to increase the proportion of highly polyunsaturated FAs (PUFAs), such as the eicosapentaenoic acid (EPA) and the docosahexaenoic acid (DHA) in the brackish habitat compared to the freshwater habitat (table II). *E. affinis* showed contrasting patterns with an increase of saturated and monounsaturated FAs and a marked decrease of PUFAs, at least for males, when transplanted into the freshwater habitat (fig. 5, table II). Finally, polar FA modifications have been found to be more variable for *E. affinis* when transplanted into the fresh water compared to *E. carolleeae*, when transplanted into the brackish water (PERMDISP; $T = 2.7$; $P = 0.007$)

## DISCUSSION

### Short term response of *E. carolleeae*

In general, performances of *Eurytemora carolleeae* were not impaired when they encountered new environmental conditions of higher salinity and lower temperatures compared to their freshwater habitat. After 24 hours, survival rate and spawning success (figs. 3, 4; respectively) were similar in the transplant relative

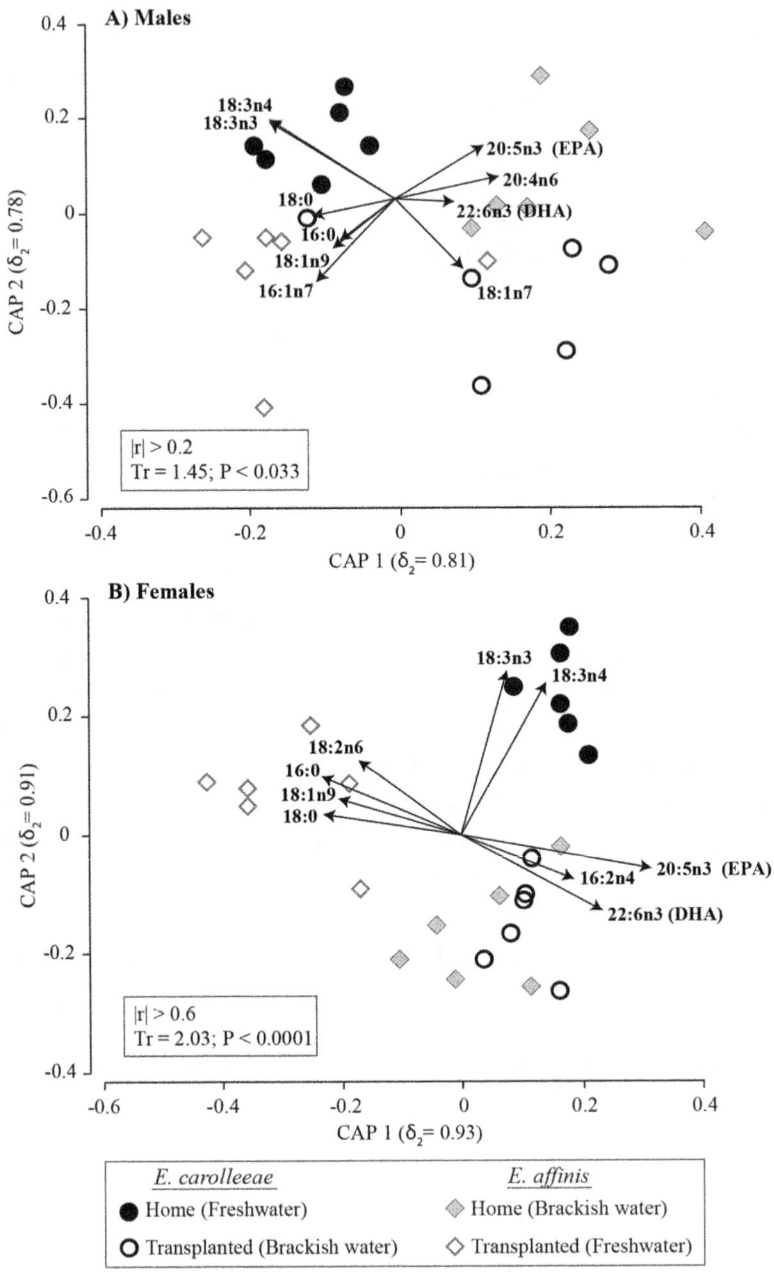


Fig. 5. Canonical analysis of principal coordinates (CAP) of polar fatty acid of A, males; and, B, females of *Eurytemora carolleeae* Alekseev & Souissi, 2011 (black) and *Eurytemora affinis* (Poppe, 1880) (light grey) in their native habitat (filled) and transplanted habitat (empty). Vector overlays are Pearson correlations.

TABLE II

Proportion of polar FAs (in % of total FAs) which contributes to the differences between habitats found at the end of the reciprocal transplant experiments (after 24 hours) for: A, males; and, B, females; of *Eurytemora carolleeae* Alekseev & Souissi, 2011 and *Eurytemora affinis* (Poppe, 1880)

| | Male | | | | Female | | | |
|---|---|---|---|---|---|---|---|---|
| | *E. carolleeae* | | *E. affinis* | | *E. carolleeae* | | *E. affinis* | |
| | Freshwater habitat | Brackish water habitat | Freshwater habitat | Brackish water habitat | Freshwater habitat | Brackish water habitat | Freshwater habitat | Brackish water habitat |
| 14:0 | 1.69 ± 0.19 | 1.29 ± 0.27 | 1.74 ± 0.66 | 1.78 ± 0.25 | 1.14 ± 0.28 | 1.17 ± 0.16 | 1.2 ± 0.31 | 1.82 ± 0.4 |
| 16:0 | 17.55 ± 0.75* | 14.27 ± 0.64* | 25.7 ± 3.69* | 17.18 ± 0.71* | 19.8 ± 2.06 | 19.7 ± 3.44 | 24.34 ± 1.61 | 21.43 ± 2.4 |
| 16:1n7 | 3.31 ± 0.45* | 2.08 ± 0.25* | 3.39 ± 0.88 | 3.18 ± 0.32 | 1.61 ± 0.25 | 1.95 ± 0.26 | 2.69 ± 0.54* | 1.49 ± 0.12* |
| 16:2n4 | 0.91 ± 0.1 | 0.89 ± 0.09 | 0.45 ± 0.08 | 1.26 ± 0.16 | 0.47 ± 0.06 | 0.56 ± 0.07 | 0.35 ± 0.11 | 0.61 ± 0.05 |
| 17:0 | 0.87 ± 0.07 | 0.89 ± 0.03 | 0.77 ± 0.11 | 0.91 ± 0.09 | 1.21 ± 0.13 | 0.87 ± 0.13 | 0.88 ± 0.13* | 1.27 ± 0.06* |
| 18:0 | 9.78 ± 0.63 | 10.1 ± 0.51 | 20.51 ± 4.57* | 11.16 ± 0.78* | 19.78 ± 2.91 | 16.32 ± 4.91 | 25.16 ± 2.58* | 20.48 ± 2.06* |
| 18:1n9 | 2.29 ± 0.33 | 2.83 ± 0.79 | 7.1 ± 2.37* | 1.61 ± 0.15* | 5.03 ± 1.43 | 5.61 ± 2.4 | 6.77 ± 1.44* | 3.67 ± 0.72* |
| 18:1n7 | 3.93 ± 0.3* | 5.76 ± 0.28* | 4.1 ± 1.23 | 5.36 ± 0.43 | 2.92 ± 0.25 | 4.26 ± 0.87 | 3.77 ± 0.41 | 3.65 ± 0.37 |
| 18:2n6 | 1.11 ± 0.07* | 0.81 ± 0.05* | 1.53 ± 0.26 | 0.9 ± 0.08 | 1.16 ± 0.11 | 1.33 ± 0.37 | 1.18 ± 0.24 | 1.22 ± 0.11 |
| 18:3n3 | 1.94 ± 0.28* | 0.31 ± 0.02* | 0.84 ± 0.24 | 0.3 ± 0.02 | 1.28 ± 0.22* | 0.41 ± 0.09* | 0.65 ± 0.1 | 0.49 ± 0.09 |
| 18:4n3 | 1.36 ± 0.05* | 0.44 ± 0.01* | 0.61 ± 0.15 | 0.52 ± 0.03 | 0.77 ± 0.09* | 0.32 ± 0.03* | 0.53 ± 0.12 | 0.46 ± 0.04 |

TABLE II
(Continued)

| | Male | | | | Female | | | |
|---|---|---|---|---|---|---|---|---|
| | *E. carolleeae* | | *E. affinis* | | *E. carolleeae* | | *E. affinis* | |
| | Freshwater habitat | Brackish water habitat | Freshwater habitat | Brackish water habitat | Freshwater habitat | Brackish water habitat | Freshwater habitat | Brackish water habitat |
| 20 : 0 | 0.31 ± 0.05 | 0.39 ± 0.05 | 0.65 ± 0.08* | 0.43 ± 0.05* | 0.7 ± 0.14 | 0.46 ± 0.15 | 0.66 ± 0.08 | 1.05 ± 0.07 |
| 20 : 4n6 | 2.04 ± 0.12* | 1.80 ± 0.05* | 0.95 ± 0.2* | 1.49 ± 0.08* | 1.48 ± 0.17 | 1.45 ± 0.22 | 0.89 ± 0.13* | 1.28 ± 0.15* |
| 20 : 5n3 (EPA) | 21.37 ± 0.21 | 21.73 ± 1.03 | 8.43 ± 2.09* | 19.81 ± 0.84* | 12.82 ± 1.37 | 13.15 ± 2.68 | 8.11 ± 1.37* | 12.62 ± 1.8* |
| 20 : 5n6 | 3.28 ± 0.2* | 2.22 ± 0.2* | 0.93 ± 0.26* | 1.54 ± 0.1* | 2.34 ± 0.29* | 1.41 ± 0.29* | 0.91 ± 0.15 | 1.38 ± 0.27 |
| 20 : 6n3 (DHA) | 21.94 ± 1.16* | 25.14 ± 1.31* | 12.97 ± 3.17* | 24.94 ± 1.24* | 19.39 ± 1.82* | 23.88 ± 2.67* | 15.33 ± 2.3 | 17.52 ± 3.26 |
| ΣSFA | 32.05 ± 1.4 | 30.34 ± 1.52 | 52.12 ± 8.22 | 34.28 ± 1.47 | 45.58 ± 4.73 | 41.15 ± 8.95 | 55.01 ± 0.49 | 50.88 ± 4.8 |
| ΣMUFA | 10.55 ± 0.5 | 12.50 ± 1.43 | 18.41 ± 5.08 | 11.07 ± 0.43 | 11.37 ± 1.62 | 13.56 ± 3.83 | 14.49 ± 1.91 | 10.34 ± 1.08 |
| ΣPUFA | 57.40 ± 1.5 | 57.16 ± 2.11 | 29.48 ± 6.68 | 54.65 ± 1.74 | 43.05 ± 4.78 | 45.29 ± 9.06 | 30.50 ± 4.26 | 38.78 ± 4.88 |
| TFA ($\mu$g/mg) | 27.71 ± 2.89 | 20.43 ± 1.91 | 32.83 ± 8.93 | 13.89 ± 1.91 | 24.66 ± 6.07 | 39.45 ± 11.45 | 24.20 ± 5.22 | 10.02 ± 1.56 |

The data from the two months are combined. Values followed by an asterisk (*) are significantly different between habitats (Student's *t*-test; $p<0.05$). Fatty acids are expressed as % of total fatty acids. Total fatty acids are expressed in $\mu$g/mg wet weight of each copepod species. Only fatty acids contributing >1% in at least one component are reported. Values are mean ± SE.

Abbreviations: i, iso; ai, ante-iso; EPA, eicosapentaenoic acid; DHA, docosahexaenoic acid; SFA, saturated fatty acids; MUFA, monounsaturated fatty acids: PUFA, polyunsaturated fatty acids; TFA, total fatty acids.

to their native habitats. Both traits indicate that low physiological stress was induced by the acute change of environmental conditions. Short term physiological plasticity of *E. carolleeae* to acute environmental change from warm fresh water to cooler brackish water (3.7 ± 0.4 PSU) was high in this in situ transplant experiment, confirming previous laboratory experiments (Lee et al., 2007, 2013). Due to the fact that the *Eurytemora affinis* complex derived from saline ancestors, the tidal freshwater habitat of *E. carolleeae* represents a challenge in terms of osmoregulation. Probably the tidal freshwater population of *E. carolleeae* is well adapted to its environment, as has been observed for the freshwater population of *E. carolleeae* of the Great Lakes (Lee et al., 2007), which most likely invaded from the source population of the tidal freshwater stretch of the St. Lawrence estuarine transition zone (Winkler et al., 2008). In fresh water *E. carolleeae* osmoregulates to maintain higher osmotic pressure inside the body compared to ambient conditions (Lee et al., 2011; Johnson et al., 2013). In addition, the adapted population of the Great Lakes has significantly increased its V-ATPase enzyme activity relative to its saline ancestor (Lee et al., 2011, 2012). V-ATPase generates an $H^+$ gradient to allow cation transportation into the cell (Beyenbach & Wieczorek, 2006). This freshwater adaptation is often accompanied by a loss of saltwater tolerance after invading fresh water (Lee et al., 2003, 2007). However, the transfer from tidal fresh water into brackish conditions did not show evidence of tolerance loss, as the salinity change experienced during our transplantation (approx. 4 PSU) might not be large enough for us to notice such loss of saltwater tolerance as found in previous laboratory experiments using higher salinities of 25 PSU (Lee et al., 2003, 2007). This is also supported by results from energy lipid contents of *E. carolleeae*, which were similar in both habitats, indicating that: (i) overall metabolic cost, including osmoregulation, was likely similar at environmental conditions of 0.1 PSU, 23°C and approx. 4 PSU, 21°C; and/or (ii) energetic costs due to plastic physiological changes were fuelled directly by food intake and/or reserve replacement. Some evidence might be retrieved from FA composition of neutral lipids that energy reserves were used and replenished immediately, so that no energy content loss was noticeable after 24 hours. The significant change of reserve lipid composition found at least in females during the transplantation suggests the consumption of energy reserves. Proportions of SFA decreased in the transplant habitat. Since energy is released more efficiently by SFAs than PUFAs b-oxidation (Langdon & Waldock, 1981), SFA might provide energy for metabolic processes or plastic physiological acclimation (i.e., osmoregulation, membrane lipids remodelling), when transplanted into new conditions. Therefore, these results confirm that the population of *E. carolleeae* of the St. Lawrence ETZ could acclimate to an acute increase of salinity without any significant loss of performance or energy, indicating that at least adult *E. carolleeae*, are not impaired when advected into the brackish water habitat of the ETZ. This suggests that *E. carolleeae*, occurring in low proportions compared to the *E. affinis* North Atlantic clade in the brackish water of the ETZ in some years (Favier & Winkler, 2014; Winkler et al., 2016), are physiologically competent in this part of the ETZ.

### Short-term response of *E. affinis*

These in situ reciprocal transplant experiments highlighted that performances of *Eurytemora affinis* strongly decreased when transplanted into the warmer freshwater habitat, confirming our hypotheses which suggested lower short-term performances and acclimation capacity compared to *E. carolleeae*. Survival rates of *E. affinis* females were almost half compared to that in the home habitat after 24 hours (fig. 3). Surprisingly, the survival rate of males was similar and low in home and transplant habitats. The low number of surviving males of *E. affinis* suggests that the ambient conditions might have been suboptimal for males in both habitats, and that *E. affinis* males would have very different requirements compared to females, which exhibited higher performances. Bradley (1978) showed also differences in tolerance and acclimation to temperature between females and males of an *E. affinis* population in Chesapeake Bay, being higher in females compared to males. Although we do not know the upper tolerance limits of temperature of the *E. affinis* North Atlantic clade in the ETZ, temperatures of >20°C might negatively affect the male survival rate. Accordingly, we found that the overall survival rate (both species and sexes combined) was higher when temperatures decreased, like in August compared to July (fig. 3B). A further indication that the acute change to warmer fresh water was stressful to *E. affinis* is the almost complete suppression of spawning activity. Spawning is an energy demanding process, which can be suppressed when energy is needed for other metabolic processes that cope with environmental change (e.g., Hoffmann, 1995; Calow & Forbes, 1998). The metabolic cost of osmoregulation in fresh water should be higher than in brackish water as *E. affinis* is hyperosmotic until 25 PSU (Lee et al., 2012, 2013). However, no evidence of a significant decrease of energy reserve content was found in *E. affinis* in the transplant habitat after 24 hours, suggesting that the energy reserve consumption was replenished by food intake in survivors, as already observed for *E. carolleeae*. On the other hand, survival when facing adverse conditions can be directly related to the capacity to allocate energy. Since the survival rates were significantly lower in the transplanted habitats (fresh water), it is possible that energy decrease occurred for *E. affinis*, but might be confounded with mortality. Food replenishment in the survivors is consistent with previous work showing that the salinity tolerance (i.e., survival) of *Eurytemora* was positively related to the quantity of available food (Kimmel & Bradley,

2001; Lee et al., 2013; Hammock et al., 2016). In both habitats food supply was similar in terms of quantity and quality (Cabrol et al., 2015), and assumed to be sufficient to support high secondary production (Winkler et al., 2003; Favier & Winkler, 2014). Consequently, differential performances were more likely due to the difference in salinity × temperature tolerances within the *E. affinis* population in the ETZ. According to Lee & Peterson (2013), selection of individual tolerance to fresh water during an acute transfer occurred mostly in the first hour. Thus, after 24 hours in the transplant experiment, survivors were expected to be competent in fresh water, likely expressing a trade-off in energy allocation to maintain physiological processes for survival. Energy might be reallocated from processes such as reproduction and received via food intake to fuel processes such as osmoregulation and membrane remodelling. However, to evaluate long-term tolerance to temperature × salinity conditions of *E. affinis* further investigations are needed, and including all life stages of *E. affinis*.

## Short-term acclimation to temperature: remodelling of membrane lipids

The two *Eurytemora* spp. were able to quickly modify (within 24 hours) the composition of their membrane lipids. Polar FAs composition changed in the transplant treatments for both species (fig. 5). This lipid remodelling is called "homeoviscous adaptation" (Hazel, 1995) indicating physiological acclimation to temperature variations in order to counteract the potential effect of temperature changes on membrane structure or viscosity. Previous studies on Arctic copepods showed that phospholipids, which are the first cell membrane constituents, consisted mostly of two kinds of FAs (SFAs and highly PUFAs), that are largely independent of dietary changes due to their specialized functions (Parrish, 2009). Accordingly, phospholipids of both *E. carolleeae* and *E. affinis* rely on the same FAs (see table II). We also found a high proportion of two monounsaturated FAs, namely 16: 1n7 and 18: 1n9, which seemed to contribute to cell membrane structuring. During the transplantation into the colder brackish habitat, *E. carolleeae* remodelled its polar lipid composition resembling that of *E. affinis* in the same habitat. The PUFA concentrations (mostly EPA and DHA) in phospholipids were higher in the colder brackish habitat compared to those in *E. carolleeae* from the warmer freshwater habitat. Specifically, DHA plays a central role in membrane fluidity due to its unique structure that facilitates rapid conformational change in membrane proteins (Turner et al., 2003). Several studies have demonstrated that poikilothermic animals, including copepods, subject to colder temperatures tend to increase the proportion of PUFA's such as DHA in membranes in order to increase membrane fluidity (e.g., Stillwell & Wassall, 2003; Schlechtriem et al., 2006; Pernet et al., 2007). Although membrane fluidity was not directly measured here, our results clearly point to the presence of counteracting mechanisms against a disordering effect of membrane lipids by temperature variations. *E. affinis*, in contrast, showed the opposite pattern when transplanted into the warmer fresh water compared to its native brackish water habitat, with a marked decrease of EPA and DHA proportions. To our knowledge this is the first time that this remodelling mechanism is observed in the pseudocryptic species complex *Eurytemora affinis*. Such a mechanism, which acts on a short time scale, could play an important role in the actual distribution of the *Eurytemora affinis* complex by promoting the utilization of temperature fluctuating environments and potentially enhancing the colonization of new habitats.

Furthermore, membrane lipid remodelling might also indirectly influence osmoregulation capacities. Membrane structure and composition are known to affect functioning of embedded enzymes in the cell membranes (Gupta & Berridge, 1966; Kimelberg & Papahadjopoulos, 1972; Cooke & Burden, 1990). For example, East et al. (1984) and Froud et al. (1986) showed that the ATPase activity strongly varies in relation to the FA composition of the membrane. Similarly, Turner et al. (2003) report a high positive correlation between the V-ATPase activity (ATP $min^{-1}$) and the content of DHA in the surrounding membrane bilayer. Since the V-ATPase plays a critical role in osmoregulation of the *Eurytemora* complex (Lee et al., 2011, 2012), by allowing proton transportation (Towle & Weihrauch, 2001; Beyenbach & Wieczorek, 2006), changes in membrane fluidity and composition could potentially affect enzyme functioning and thus, salinity tolerance. However, this effect of membrane fluidity and composition on V-ATPase functioning still remains understudied, especially for marine crustaceans. Further studies are needed to confirm if membrane remodelling influences the V-ATPase performance and thereby, the capacity of the *Eurytemora* complex to acclimate to new hyper- or hypotonic conditions relative to its native environments. This information could be important to better understand the interaction between temperature and salinity effects on eco-physiological processes and plasticity, in particular for the dispersal stages such as nauplii, that do not have a strong swimming capacity and are subject to advection.

## Spatial segregation

Results of the present in-situ transplant experiments provide partial confirmation to the geographic distribution patterns of the pseudocryptic species complex *E. affinis* in the St. Lawrence estuary. In general, strong spatial segregation between *E. carolleeae*, predominant in the tidal fresh water, and *E. affinis*, occupying mostly the mesohaline part, is found in the ETZ in summer (Winkler et al., 2008; Favier & Winkler, 2014; Cabrol et al., 2015; Winkler et al., 2016). This segregation was hypothesized to result from 'species-specific' physiological tolerances to salinity × temperature interactions (Winkler et al., 2008). According to the present results and previous laboratory studies (Lee et al., 2003, 2007), *E. carolleeae* could use the brackish water habitat, whereas *E. affinis* was strongly impaired in the tidal freshwater habitat and therefore might be "limited" to the mesohaline water at least in summer. Only 40% of the transplanted *E. affinis* survived, which showed signs of stress (suppression of spawning activity), so that even when surviving an advection into the fresh water, these individuals might not be competent to reproduce, as energy may be reallocated to osmoregulation. This may explain why *E. affinis* is generally absent from the freshwater habitat. However some intrusion of *E. affinis* was reported in early June, when water temperatures were still below 20°C and food quality and supply was high (Favier & Winkler, 2014). This suggests an increase in salinity tolerance, when food conditions are good, as has been observed in laboratory experiments (Lee et al., 2013). Furthermore, *E. carolleeae* was absent from the tidal freshwater habitat at that moment in time (Favier & Winkler, 2014), so that this species did not represent competition.

High plasticity and the salinity tolerance to brackish water of *E. carolleeae* revealed by several studies (see review of Lee, 2016), including the present study, are contradicting the distribution pattern found in the St. Lawrence ETZ. The absence or low proportion of *E. carolleeae* seen in the mesohaline brackish water clearly suggests the presence of more complex mechanisms causing segregation. Food preferences, predation and competition between both pseudocryptic species could be possible, as all three are known to shape community structure (Polis et al., 1989). Cabrol et al. (2015) showed that *E. carolleeae* and *E. affinis* both ingest phytoplankton and more specifically dinoflagellates, probably in order to accumulate DHA, which is an essential fatty acid, as its availability might be limited in the St. Lawrence ETZ. However, no study looked at fine scale differences in feeding of both species in the same environment, so that no conclusion can be drawn on the species-specific preferences of food. No controlling effect of potential predators was found so far on this pseudocryptic species complex (Sirois & Dodson, 2000; Winkler et al., 2003, 2007). Even if food supply in general (i.e., phytoplankton standing stock) is not limited, specific essential components could be (Cabrol et al., 2015). For example, essential FAs are critical compounds, which play a central role in crustacean physiology by regulating key processes such as membrane functioning, hormone production, or growth (Müller-Navarra et al., 1995; Brett & Müller-Navarra, 1997; Parrish, 2009). However, they show insufficient biosynthetic production in marine animals (Glencross, 2009) and must be obtained from food. Both species might compete for specific food sources in a given habitat, which may be limited and not representative of the overall food availability. Inferring this to *E. affinis* and *E. carolleeae* in the mesohaline habitat, *E. affinis* might be the more competent than *E. carolleeae*. According to the competitive exclusion theory, one of the two species might adopt a compromise that benefits it under specific conditions (here distribution in tidal fresh water) to avoid competition (Chesson, 2000). Then, based on our experiments and previous studies (Sirois & Dodson, 2000; Winkler et al., 2003, 2007; Favier & Winkler, 2014; Cabrol et al., 2015), we propose that the observed segregation might result from a spatial niche partitioning based on a trade-off between their differential salinity × temperature tolerance capacities and the competition avoidance. However, further studies on the effect of the competition need to be carried out to confirm our hypothesis.

## CONCLUSION

Based on in situ reciprocal transplant experiments, results revealed that *Eurytemora carolleeae* and *Eurytemora affinis* differed in their short-term acclimation response when facing brackish water and freshwater conditions, respectively, that they might encounter when advected through the highly hydrodynamic estuarine transition zone or introduced from ballast water. Short-term performance of *E. carolleeae* was not impacted by an acute change of environmental conditions, whereas *E. affinis* showed significantly lower performance after 24 hours under new environmental conditions. Therefore, physiological plasticity was higher in *E. carolleeae* towards colder brackish conditions than that of *E. affinis* towards the warmer fresh water when transplanted without acclimation. Furthermore, we showed that the two pseudocryptic species are able to quickly modify their polar lipid compositions of the membranes, probably to adjust to the new ambient temperature conditions, which might indirectly affect osmoregulation performance when introduced into new salinity × temperature habitats. Although further studies are needed, we highlight crucial information to better understand the actual geographic distribution and habitat utilization of the *Eurytemora* pseudocryptic species complex.

## ACKNOWLEDGEMENTS

We thank our captain S. Leblanc of the "Macoma" for help in the field. Special thanks are extended to J. B. Favier and L. DeChamplain for their invaluable assistance in the laboratory and the field. We also thank M. J. Anderson for support in statistical analysis. We extend special thanks to M. Babin for his help in fatty acids analysis and to Dr. C. Hammer for revising the English language and manuscript improvement. This research was supported by the discovery grant from the Natural Sciences and Engineering Research Council of Canada to G.W. This is a contribution to the program of research cluster "Quebec–Ocean".

## REFERENCES

ALEKSEEV, V. R. & A. SOUISSI, 2011. A new species within the *Eurytemora affinis* complex (Copepoda: Calanoida) from the Atlantic coast of USA, with observations on eight morphologically different European populations. Zootaxa, **2767**: 41-56.

ANDERSON, M. J., R. N. GORLEY & K. R. CLARKE, 2008. PERMANOVA for PRIMER: guide to software and statistical methods: 1-214. (PRIMER-E, Plymouth).

ANDERSON, M. J. & J. ROBINSON, 2003. Generalized discriminant analysis based on distances. Aust. N.Z. J. Stat., **45**: 301-318.

ANDERSON, M. J. & D. C. WALSH, 2013. PERMANOVA, ANOSIM, and the Mantel test in the face of heterogeneous dispersions: what null hypothesis are you testing? Ecol. Monographs, **83**: 557-574.

ANDERSON, M. J. & T. J. WILLIS, 2003. Canonical analysis of principal coordinates: a useful method of constrained ordination for ecology. Ecol., **84**: 511-525.

ANTONIO, E. S. & N. B. RICHOUX, 2014. Trophodynamics of three decapod crustaceans in a temperate estuary using stable isotope and fatty acid analyses. Mar. Ecol. Prog. Ser., **504**: 193-205.

BEYENBACH, K. W. & H. WIECZOREK, 2006. The V-type $H^+$ ATPase: molecular structure and function, physiological roles and regulation. J. Exp. Biol., **209**: 577-589.

BEYREND-DUR, D., S. SOUISSI, D. DEVREKER, G. WINKLER & J. S. HWANG, 2009. Life cycle traits of two transatlantic populations of *Eurytemora affinis* (Copepoda: Calanoida): salinity effects. J. Plankton Res., **31**: 713-728.

BRADLEY, B. P., 1978. Increase in range of temperature tolerance by acclimation in the copepod *Eurytemora affinis*. Biol. Bull., **154**: 177-187.

BRETT, M. & D. MÜLLER-NAVARRA, 1997. The role of highly unsaturated fatty acids in aquatic foodweb processes. Freshwater Biol., **38**: 483-499.

CABROL, J., R. TREMBLAY & G. WINKLER, 2015. Physiological condition and differential feeding behavior in the cryptic species complex *Eurytemora affinis* in the St Lawrence estuary. J. Plankton Res., **37**: 372-387.

CABROL, J., T. TROMBETTA, A. AMAUDRUT, F. AULANIER, R. SAGE, T. TREMBLAY, C. NOZAIS, M. STARR, S. PLOURDE & G. WINKLER, 2019. Trophic niche partitioning of dominant North-Atlantic krill species, *Meganyctiphanes norvegica*, *Thysanoessa inermis* and *T. raschii*. Limnol. Oceanogr., **64**: 165-181.

CALOSI, P., S. P. RASTRICK, C. LOMBARDI, H. J. DE GUZMAN, L. DAVIDSON, M. JAHNKE & M. C. GAMBI, 2013. Adaptation and acclimatization to ocean acidification in marine ectotherms: an in-situ transplant experiment with polychaetes at a shallow $CO_2$ vent system. Phil. Trans. R. Soc., (B), **368**: 20120444.

CALOW, P. & V. E. FORBES, 1998. How do physiological responses to stress translate into ecological and evolutionary processes? Comp. Biochem Physiol., **120**: 11-16.

CHEN, I. S., C. S. J. SHEN & A. J. SHEPPARD, 1981. Comparison of methylene chloride and chloroform for the extraction of fats from food products. J. Am. Oil Chem. Soc., **58**: 599-601.

CHESSON, P., 2000. General theory of competitive coexistence in spatially varying environments. Theor. Popul. Biol., **58**: 211-237.

CLARKE, K. R., R. N. GORLEY, P. J. SOMERFIELD & R. M. WARWICK, 2015. Change in marine communities: an approach to statistical analysis and interpretation (3rd ed.): 1-321. (PRIMER-E, Plymouth).

COOKE, D. T. & R. S. BURDEN, 1990. Lipid modulation of plasma membrane-bound ATPases. Physiol. Planta., **78**: 153-159.

DEVREKER, D., J. J. PIERSON, S. SOUISSI, D. G. KIMMEL & M. R. ROMAN, 2012. An experimental approach to estimate egg production and development rate of the calanoid copepod *Eurytemora affinis* in Chesapeake Bay, USA. J. Exp. Mar. Biol. Ecol., **416**: 72-83.

DEVREKER, D., S. SOUISSI, J. C. MOLINERO, D. BEYREND-DUR, F. GOMEZ & J. FORGET-LERAY, 2010. Tidal and annual variability of the population structure of *Eurytemora affinis* in the middle part of the Seine estuary during 2005. Estuar. Coast. Shelf Sci., **89**: 245-255.

DEWITT, T. J., A. SIH & D. S. WILSON, 1998. Costs and limits of phenotypic plasticity. Trends Ecol. Evol., **13**: 77-81.

DUR, G. & S. SOUISSI, 2018. Ontogenetic optimal temperature and salinity envelops of the copepod *Eurytemora affinis* in the Seine estuary (France). Estuar. Coast. Shelf Sci., **200**: 311-323.

EAST, J. M., O. T. JONES, A. C. SIMMONDS & A. G. LEE, 1984. Membrane fluidity is not an important physiological regulator of the ($Ca^{2+}$-$Mg^{2+}$)-dependent ATPase of sarcoplasmic reticulum. J. Biol. Chem., **259**: 8070-8071.

FAVIER, J.-B. & G. WINKLER, 2014. Coexistence, distribution patterns and habitat utilization of the pseudocryptic species complex *Eurytemora affinis* in the St Lawrence estuarine transition zone. J. Plankton Res., **36**: 1247-1261.

FRASER, A. J., J. R. SARGENT & J. C. GAMBLE, 1989. Lipid class and fatty acid composition of *Calanus finmarchicus* (Gunnerus), *Pseudocalanus* sp. and *Temora longicornis* (Müller) from a nutrient enriched seawater enclosure. J. Exp. Mar. Biol. Ecol., **130**: 81-92.

FROUD, R. J., J. M. EAST, O. T. JONES & A. G. LEE, 1986. Effects of lipids and long-chain alkyl derivatives on the activity of calcium-magnesium ATPase. Biochem., **25**: 7544-7552.

GHALAMBOR, C. K., J. K. MCKAY, S. P. CARROLL & D. N. REZNICK, 2007. Adaptive versus non-adaptive phenotypic plasticity and the potential for contemporary adaptation in new environments. Funct. Ecol., **21**: 394-407.

GLENCROSS, B. D., 2009. Exploring the nutritional demand for essential fatty acids by aquaculture species. Rev. Aquacult., **1**: 71-124.

GUPTA, B. L. & M. J. BERRIDGE, 1966. Fine structural organization of the rectum in the blowfly, *Calliphora erythrocephala* (Meig.) with special reference to connective tissue, tracheae and neurosecretory innervation in the rectal papillae. J. Morphol., **120**: 23-81.

HAMMOCK, B. G., S. LESMEISTER, I. FLORES, G. S. BRADBURD, F. H. HAMMOCK & S. J. TEH, 2016. Low food availability narrows the tolerance of the copepod *Eurytemora affinis* to salinity, but not to temperature. Estuar. Coast., **39**: 189-200.

HAZEL, J. R. & E. E. WILLIAMS, 1990. The role of alterations in membrane lipid composition in enabling physiological adaptation of organisms to their physical environment. Prog. Lip. Res., **29**: 167-227.

HOFFMANN, A. A., 1995. Acclimation: increasing survival at a cost. Trends Ecol. Evol., **10**: 1-2.

JOHNSON, K. E., L. PERREAU, G. CHARMANTIER, M. CHARMANTIER-DAURES & C. E. LEE, 2013. Without gills: localization of osmoregulatory function in the copepod *Eurytemora affinis*. Physiol. Biochem. Zool., **45**: 310-324.

KARLSSON, K., S. PUIAC & M. WINDER, 2018. Life-history responses to changing temperature and salinity of the Baltic Sea copepod *Eurytemora affinis*. Mar. Biol., **165**: 30.

KARLSSON, K. & M. WINDER, 2018. Ecosystem effects of morphological and life history traits in two divergent zooplankton populations. Front. Mar. Sci., **5**: 408.

KIMELBERG, H. K. & D. PAPAHADJOPOULOS, 1972. Phospholipid requirements for ($Na^{++}$ $K^{+}$)-ATPase activity: head-group specificity and fatty acid fluidity. Biochimica et Biophysica Acta (BBA) — Biomembranes, **282**: 277-292.

KIMMEL, D. G. & B. P. BRADLEY, 2001. Specific protein responses in the calanoid copepod *Eurytemora affinis* (Poppe, 1880) to salinity and temperature variation. J. Exp. Mar. Biol. Ecol., **266**: 135-149.

KIMMEL, D. G., W. D. MILLER & M. R. ROMAN, 2006. Regional scale climate forcing of mesozooplankton dynamics in Chesapeake Bay. Estuar. Coast., **29**: 375-387.

LAJUS, D., N. SUKHIKH & V. ALEKSEEV, 2015. Cryptic or pseudocryptic: can morphological methods inform copepod taxonomy? An analysis of publications and a case study of the *Eurytemora affinis* species complex. Ecol. Evol., **5**: 2374-2385.

LANGDON, C. & M. J. WALDOCK, 1981. The effect of algal and artificial diets on the growth and fatty acid composition of *Crassostrea gigas* spat. J. Mar. Biol. Assoc. U. K., **61**: 431-448.

LEE, C. E., 1999. Rapid and repeated invasion of fresh water by the copepod *Eurytemora affinis*. Evol., **53**: 1423-1434.

LEE, C. E., 2000. Global phylogeography of a cryptic copepod species complex and reproductive isolation between genetically proximate "populations". Evol., **54**: 2014-2027.

LEE, C. E., 2016. Evolutionary mechanisms of habitat invasions, using the copepod *Eurytemora affinis* as a model system. Evol. App., **9**: 248-270.

LEE, C. E. & B. W. FROST, 2002. Morphological stasis in the *Eurytemora affinis* species complex (Copepoda: Temoridae). Hydrobiologia, **480**: 111-128.

LEE, C. E., M. KIERGAARD, G. W. GELEMBIUK, B. D. EADS & M. POSAVI, 2011. Pumping ions: rapid parallel evolution of ionic regulation following habitat invasions. Evol., **65**: 2229-2244.

LEE, C. E., W. E. MOSS, N. OLSON, K. F. CHAU, Y. M. CHANG & K. E. JOHNSON, 2013. Feasting in freshwater: impacts of food concentration on freshwater tolerance and the evolution of food salinity response during the expansion from saline into fresh water habitats. Evol. App., **6**: 673-689.

LEE, C. E. & C. H. PETERSEN, 2003. Effects of developmental acclimation on adult salinity tolerance in the freshwater-invading copepod *Eurytemora affinis*. Physiol. Biochem. Zool., **76**: 296-301.

LEE, C. E., M. POSAVI & G. CHARMANTIER, 2012. Rapid evolution of body fluid regulation following independent invasions into freshwater habitats. J. Evol. Biol., **25**: 625-633.

LEE, C. E., J. L. REMFERT & Y. M. CHANG, 2007. Response to selection and evolvability of invasive populations. Genet., **129**: 179-192.

LEE, C. E., J. L. REMFERT & G. W. GELEMBIUK, 2003. Evolution of physiological tolerance and performance during freshwater invasions. Integr. Comp. Biol., **43**: 439-449.

LEPAGE, G. & A. ROY, 1984. Improved recovery of fatty acid through direct transesterification without prior extraction or purification. J. Lip. Res., **25**: 1391-1396.

MARTY, Y., F. DELAUNAY, J. MOAL & J. F. SAMAIN, 1992. Changes in the fatty acid composition of *Pecten maximus* during larval development. J. Exp. Mar. Biol. Ecol., **163**: 221-234.

MÜLLER-NAVARRA, D., 1995. Evidence that a highly unsaturated fatty acid limits *Daphnia* growth in nature. Archiv für Hydrobiol., **132**: 297-297.

PARRISH, C. C., 2009. Essential fatty acids in aquatic food webs. In: M. T. ARTS, M. T. BRETT & M. KAINZ (eds.), Lipids in aquatic ecosystems: 309-326. (Springer, New York).

PERNET, F., R. TREMBLAY, L. COMEAU & H. GUDERLEY, 2007. Temperature adaptation in two bivalve species from different thermal habitats: energetics and remodelling of membrane lipids. J. Exp. Biol., **210**: 2999-3014.

PIERSON, J. J., D. G. KIMMEL & M. R. ROMAN, 2016. Temperature impacts on *Eurytemora carolleeae* size and vital rates in the upper Chesapeake Bay in winter. Estuar. Coast., **39**: 1122-1132.

POLIS, G. A., C. A. MYERS & R. D. HOLT, 1989. The ecology and evolution of intraguild predation: potential competitors that eat each other. Ann. Rev. Ecol. Evol. Syst., **20**: 297-330.

RAHLFF, J., J. PETERS, M. MOYANO, O. PLESS, C. CLAUSSEN & M. A. PECK, 2017. Short-term molecular and physiological responses to heat stress in neritic copepods *Acartia tonsa* and *Eurytemora affinis*. Comp. Biochem Physiol., (A, Molecular & Integrative Physiology), **203**: 348-358.

SCHLECHTRIEM, C., M. T. ARTS & I. D. ZELLMER, 2006. Effect of temperature on the fatty acid composition and temporal trajectories of fatty acids in fasting *Daphnia pulex* (Crustacea, Cladocera). Lip., **41**: 397-400.

SIROIS, P. & J. J. DODSON, 2000. Influence of turbidity, food density and parasites on the ingestion and growth of larval rainbow smelt *Osmerus mordax* in an estuarine turbidity maximum. Mar. Ecol. Prog. Ser., **193**: 167-179.

SOUISSI, A., S. SOUISSI & B. W. HANSEN, 2016. Physiological improvement in the copepod *Eurytemora affinis* through thermal and multi-generational selection. Aquacult. Res., **47**: 2227-2242.

SOUISSI, A., S. SOUISSI & J.-S. HWANG, 2013. The effect of epibiont ciliates on the behavior and mating success of the copepod *Eurytemora affinis*. J. Exp. Mar. Biol. Ecol., **445**: 38-43.

ST-ONGE-DROUIN, S., G. WINKLER, J. F. DUMAIS & S. SENNEVILLE, 2014. Hydrodynamics and spatial separation between two clades of a copepod species complex. J. Mar. Syst., **129**: 334-342.

STILLWELL, W. & S. R. WASSALL, 2003. Docosahexaenoic acid: membrane properties of a unique fatty acid. Chem. Phys. Lip., **126**: 1-27.

SUKHIKH, N., A. SOUISSI, S. SOUISSI & V. R. ALEKSEEV, 2013. Invasion of *Eurytemora* sibling species (Copepoda: Temoridae) from North America into the Baltic Sea and European Atlantic coast estuaries. J. Nat. Hist., **47**: 753-767.

SUKHIKH, N., A. SOUISSI, S. SOUISSI, G. WINKLER, V. CASTRIC, A. C. HOLL & V. ALEKSEEV, 2016. Genetic and morphological heterogeneity among populations of *Eurytemora affinis* (Crustacea: Copepoda: Temoridae) in European waters. C. R. Biol., **339**: 197-206.

TOWLE, D. W. & D. WEIHRAUCH, 2001. Osmoregulation by gills of euryhaline crabs: molecular analysis of transporters. Am. Zool., **41**: 770-780.

TURNER, N., P. L. ELSE & A. J. HULBERT, 2003. Docosahexaenoic acid (DHA) content of membranes determines molecular activity of the sodium pump: implications for disease states and metabolism. Naturwissenschaften, **90**: 521-523.

WINKLER, G., J. CABROL & J.-B. FAVIER, 2016. Diversity, repartition and ecology of the *Eurytemora* cryptic complex species in the nursery area of the St. Lawrence Upper Estuary (trad). Nat. Can., **140**: 7-18.

WINKLER, G., J. J. DODSON, N. BERTRAND, D. THIVIERGE & W. F. VINCENT, 2003. Trophic coupling across the St. Lawrence River estuarine transition zone. Mar. Ecol. Prog. Ser., **251**: 59-73.

WINKLER, G., J. J. DODSON & C. E. LEE, 2008. Heterogeneity within the native range: population genetic analyses of sympatric invasive and non-invasive clades of the freshwater invading *Eurytemora affinis*. Mol. Ecol., **17**: 415-430.

First received 27 August 2019.
Final version accepted 28 October 2019.

[When citing this volume, please refer to *Crustaceana* 93 (2020), 241-547]

*Proceedings of the* Eurytemora *conference, St. Petersburg, 2019*

# THE ECOLOGICAL PREFERENCES OF *EURYTEMORA AFFINIS* (POPPE, 1880) IN THE WATER BODIES OF KAZAKHSTAN (CENTRAL ASIA) AND SOME NOTES ABOUT *EURYTEMORA CASPICA* SUKHIKH & ALEKSEEV, 2013

BY

ELENA G. KRUPA[1,2,3])

[1]) Kazakh Agency of Applied Ecology, Amangeldy 70a, Almaty 050012, Kazakhstan
[2]) Institute of Zoology, Ministry of Education and Science, al Faraby 93a, Almaty 050060, Kazakhstan

## ABSTRACT

The purpose of this work is to analyse the influence of external factors on quantitative variables of *Eurytemora affinis* in water bodies of Kazakhstan. In 1997-2018, more than 150 water bodies located in various regions of the country were surveyed. The results obtained and analysis of the published literature showed that *E. affinis* primarily inhabits small shallow water bodies of Central and Eastern Kazakhstan. Populations from the North-Eastern Caspian Sea, originally identified as "*Eurytemora affinis*", could be attributed to the recently described species *Eurytemora caspica*. In the water bodies examined, *E. affinis* was found at water temperatures from 12 to 25°C and at salinities from 0.2 to 7.1 g/dm$^3$. The highest abundance of *Eurytemora* was recorded at 21-22°C and a salinity of 1.0-3.0 g/dm$^3$. The optimum temperature range in the water bodies examined in an extreme continental climate is higher than the values given for *E. affinis* in water bodies in a moderately temperate climate. The optimal TDS [Total Dissolved Solids] value (1.0-3.0 g/dm$^3$) has shifted towards more truly fresh waters than was stated for *E. affinis* collected from the sea (6.2-8.2‰), and this was confirmed in experiments (from 3.0-10.0‰ to 10.0-15.0‰). In the majority of the examined water bodies of Kazakhstan, males dominated the *Eurytemora* populations. Male dominance in *E. affinis* populations might be due to the adverse effect of elevated temperature on the lifespan of females, and it was also connected with the level of toxic pollution of the water bodies. It seems that *E. affinis* endures a relatively high content of heavy metals, but it does not inhabit water bodies with very high toxic pollution. The revealed features of *E. affinis'* biology in Kazakhstan water bodies might be associated with both the heterogeneity of its populations from different parts of the range, and the existence of a complex of closely related species that at present is collectively named "*Eurytemora affinis*".

Key words. — Extreme continental climate, optimum salinity range, optimum temperature range, sex structure

[3]) e-mail: elena_krupa@mail.ru

 | DOI:10.1163/9789004465565_012

## RÉSUMÉ

Le but de ce travail est d'analyser l'influence des facteurs externes sur les variables quantitatives de *Eurytemora affinis* dans les collections d'eau du Kazakhstan. De 1997 à 2018, plus de 150 collections d'eau situées dans diverses régions du pays ont été étudiées. Les résultats obtenus et l'analyse de la littérature ont montré que *E. affinis* vit principalement dans les petites collections d'eau peu profondes du Kazakhstan central et oriental. Les populations de la mer Caspienne nord-orientale, identifiées à l'origine comme "*Eurytemora affinis*", pourraient être attribuées à l'espèce récemment décrite *Eurytemora caspica*. Dans les collections d'eau examinées, *E. affinis* a été trouvée à des températures de l'eau de 12 à 25°C et à des salinités de 0,2 à 7,1 g/dm$^3$. La plus forte abondance d'*Eurytemora* a été trouvée à 21–22°C et à une salinité de 1,0–3,0g/dm$^3$. L'intervalle optimal de température dans les collections d'eau examinées dans un climat continental extrême est supérieur aux valeurs données pour *E. affinis* dans les collections d'eau de climat modérément tempéré. La valeur optimale de TDS [Total Dissolved Solids] (1,0–3,0 g/dm$^3$) correspond à des eaux vraiment plus douces par rapport à ce qui avait été établi pour *E. affinis* collectée en mer (6,2–8,2‰), et ceci a été confirmé au cours d'expérimentations (de 3,0–10,0‰ à 10,0–15,0‰). Dans la majorité des collections d'eau du Kazakhstan étudiées, les mâles étaient dominants dans les populations de *Eurytemora*. La dominance des mâles dans les populations de *E. affinis* pourrait être due à l'effet néfaste de la température élevée sur la durée de vie des femelles, et elle est aussi reliée au niveau de pollution toxique des collections d'eau. Il semble que *E. affinis* supporte une concentration relativement élevée de métaux lourds, mais ne peut pas vivre dans les collections d'eau présentant une pollution toxique élevée. Les caractéristiques de la biologie de *E. affinis* dans les collections d'eau du Kazakhstan pourraient être associées à la fois à l'hétérogénéité de ses populations dans différentes régions de sa répartition et à l'existence d'un complexe d'espèces proches actuellement nommé collectivement "*Eurytemora affinis*."


Mots clés. — Climate continental extrême, optimum salinity range, optimum temperature range, sex structure

## INTRODUCTION

Information about *Eurytemora* from the water bodies of Kazakhstan is exceptionally scarce. According to literature data (Krupa et al., 2016), five species of this genus were known from this country: *Eurytemora affinis* (Poppe, 1880), *E. velox* (Lilljeborg, 1853), *E. lacustris* (Poppe, 1887), *E. grimmi* (Sars, 1897), and *E. minor* Behning, 1938. *Eurytemora caspica* Sukhikh & Alekseev, 2013 was recently described from the Russian part of the Northern Caspian Sea and the delta of the River Volga (Sukhikh & Alekseev, 2013), and can potentially be found in the Kazakhstan sector of the Caspian Sea.

Until the early 1990s, *E. grimmi* and *E. minor* were a constant component of the zooplankton in the Middle and Southern Caspian (Hossieni et al., 1996). The occurrence of the ctenophore *Mnemiopsis leidyi* A. Agassiz, 1865 in the Caspian Sea led to a reduction in the abundance of *Eurytemora* (Roohi et al., 2008). In recent decades, there is no reliable information about the findings of these species in the Caspian Sea. Two other species, *E. velox* and *E. lacustris*, are also rarely found in shallow brackish water bodies of West Kazakhstan.

Compared with other *Eurytemora* species, *E. affinis* is a most common component of the zooplankton in Kazakhstan (Krupa et al., 2016) as well as in large parts of the Northern Hemisphere (Borutzky et al., 1991). Like other species of the genus, *E. affinis* mostly inhabits estuaries (Heinle & Flemer, 1975) and more or less semi-enclosed coastal areas of the sea, like bays or gulfs (Semenova et al., 2010). The species also lives in rivers (Croskery, 1978) and lakes (Torke, 2001), and less often in temporary water bodies (Evdokimov & Ermokhin, 2009). *E. affinis* occurs mostly at low temperatures in winter and early spring (Oghenekaro et al., 2018), but can quickly adapt to temperature stress (Bradley et al., 1988). *E. affinis* further is a widely euryhaline species (Viitasalo et al., 1994; Paturej & Gutkowska, 2015; Karlsson et al., 2018). Only the most extreme salinities (i.e., 0 and 35‰) detrimentally affect naupliar survival (Devreker et al., 2007). In marine regions with oceanic salinity (30.0-35.0‰), *E. affinis* prefers the desalinated coastal areas (Borutzky et al., 1991). It consumes bacterial plankton (Boak & Goulder, 1983), phytoplankton (Lehtinen et al., 2010), and detritus (Heinle & Flemer, 1975; Heinle et al., 1977). The presence of food facilitates the adaptation of this copepod to salinity (Lee et al., 2014; Hammock et al., 2015), and thus food apparently is a more important factor than temperature (Ban, 1994).

The combination of the critical factors mentioned (Diekmann et al., 2012) varies significantly, depending on the regional-climatic conditions, which determine the peculiarities of the biology of *Eurytemora affinis* in different parts of its vast range. The above mentioned information on the biology of *E. affinis* refers to populations in water bodies in a moderate climate (Ban, 1994; Viitasalo et al., 1994; Irigoien et al., 1996; Ishikawa et al., 1999; Devreker et al., 2007; Paturej & Gutkowska, 2015; Souissi et al., 2016; Karlsson et al., 2018). Kazakhstan is mainly located in an extreme continental climate (Uteshev, 1959). High summer temperatures of air and water, the predominance of water with a high content of dissolved solids (Zhevago, 1989), and relatively low contents of nutrients (Nikolayenko, 1989; Frumin & Krashanovskaya, 2014) determine the distinctness of water bodies in this region.

In the published literature, there is no information about the environmental preferences of *E. affinis* in the extreme continental climate: as specifically in Kazakhstan (Krupa et al., 2016). This work fills this gap to some extent. Its goal is to analyse the influence of critical factors (temperature, salinity, nutrients and heavy metal contents of the water) on the distribution and the quantitative variables of *E. affinis* populations in the water bodies of Kazakhstan.

## STUDY SITE

There are four natural zones on the territory of Kazakhstan: forest-steppe, steppe, semi-desert, and desert. Altitudinal zonality is observed in the mountains along the southern and eastern borders of the republic. Water resources are unevenly distributed. Except for the far west, peripheral areas have the highest density of the hydrographic network. The largest water bodies are the Caspian Sea, the Balkhash Lake, and the Small Aral Sea. All major rivers (Ural, Syr-Darya, Ili, Irtysh) have their origins on the territory of neighbouring states.

The climate of most of Kazakhstan is "extreme continental", with large amplitudes of temperatures, from −45°C in winter to +45°C in summer (Uteshev, 1959). The amount of precipitation varies from 100 mm in desert areas to 1600-3000 mm in the north of the country and the mountainous regions. In the central areas, the amount of precipitation does not exceed 180-250 mm. Maximum water temperatures vary considerably. In Southern and South-Western Kazakhstan, water bodies warm up in the spring to 20-26°C, in summer to 28-36°C. A relatively high temperature of water lasts until the end of September and mid-October in these regions. In the north of the country, the maximum water temperatures do not exceed 17-20°C, in the central and eastern parts 23-25°C, in mountainous areas 8-12°C. Except for the far south, water bodies freeze in winter, especially the shallow ones.

Most water bodies of Kazakhstan are characterized by a high content of total dissolved solids (TDS). In plain lotic water bodies, the TDS value varies from 0.3 to 5.0 g/dm$^3$ (Krupa, 2012a). Highly mineralized water bodies are widespread, with a total dissolved solids content from 10-80 to 40-50 g/dm$^3$, in some cases up to 500 g/dm$^3$ (Zhevago, 1989). Ultrafresh water bodies, with TDS less than 0.1-0.2 g/dm$^3$, are located mainly in the mountainous and foothill areas. Generally, the content of nutrients is low (Nikolayenko, 1989; Frumin & Krashanovskaya, 2014). This leads to lower productivity of Kazakhstan's water bodies, compared to similar water bodies in humid territories (Krupa, 2012a).

## MATERIAL AND METHODS

### Field work

In the spring, summer, and autumn of 1997-2018, more than 150 water bodies in Kazakhstan were surveyed (fig. 1A). Zooplankton samples were collected using a Juday net with an inlet diameter of 12 cm, by pulling it through from bottom to surface. These samples were fixed with 4% formaldehyde. A total of 3528 zooplankton samples were selected and processed. The physicochemical properties of the water at the sampling stations were measured simultaneously with zooplankton sampling. The measurements of the temperature and pH values of the surface water layers were taken during field trips by using Hanna HI 98129 instruments. The transparency of the water was determined with a Secchi disk.

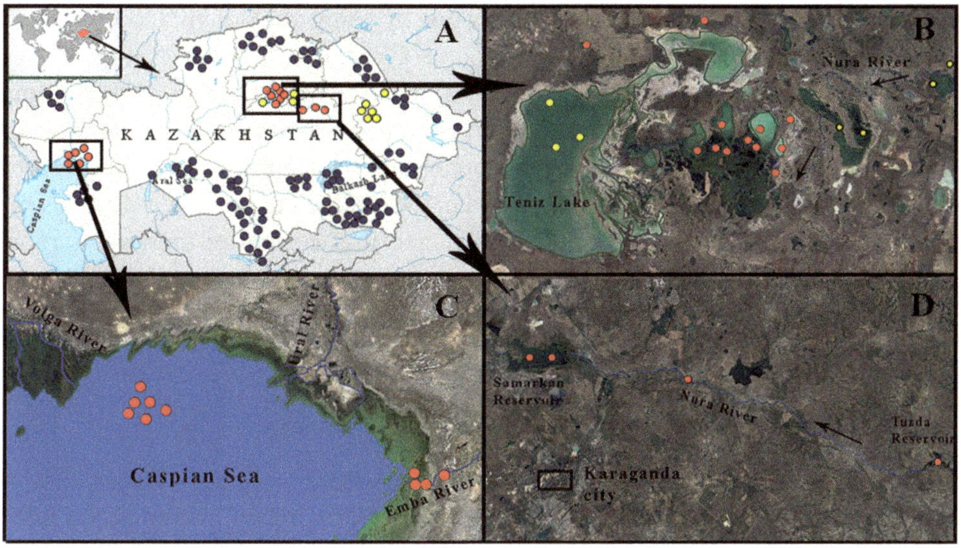


Fig. 1. A, Schematic map of the water bodies surveyed in Kazakhstan; filled, dark blue circles, surveyed water bodies; B-C, red (author's data) and yellow filled circles (data from Stuge, 1995, 2000, 2007, 2008) represent the sites of finds of: B, D, *Eurytemora affinis* (Poppe, 1880); and, C, (potentially) *E. caspica* Sukhikh & Alekseev, 2013.

We selected water samples to determine the Total Dissolved Solids (TDS), heavy metals, and nutrient concentrations from each water body. In small water bodies (less than 1 km$^2$ area), three water samples were taken, which were then mixed. From this mixed sample, one integrated subsample was selected for subsequent analysis. In large water bodies, water samples were taken from a grid of stations evenly spread throughout the water area. A total more than 1000 samples were taken to determine the values of TDS; 589 water samples were taken to determine the content of nutrients and the same number of samples to determine the contents of Cd, Cu, Zn and Pb. The samples for heavy metals were fixed on site by adding nitric acid; samples for nutrients were fixed with chloroform. All samples collected were transported to the laboratory in an icebox.

## Methods of chemical analysis

For the chemical analysis of the water (TDS, nutrients) conventional methods were used (Semenova, 1977; Fomin, 1995). Water samples were analysed in tri- or quadruplicate. The estimated error for the contents of major ions in the water was 0.5-5.0%, depending on the type of analysis.

The analysis of water samples for heavy metals was carried out in the analytical laboratory "Kazekoanaliz" (accreditation certificate No. KZ.I.02.1017) according to the Interstate Standard (2013). Heavy metal measuring was performed by mass spectrometry with inductively coupled plasma by using an Agilent 7500 A device, manufactured by Agilent Technologies (Santa Clara, CA, U.S.A.) (National Standard RK ISO 17294-2-2006). The device allows for the detection of the various chemical elements in complex matrices, including those in the sea and grey water, and the biological objects in micro-trace quantities. Concentrated nitric acid (based on 1 $cm^3$ of nitric acid per 200 $cm^3$ of water) was added to the water samples before the analysis. The water sample was heated in a current of inert gas (argon) according to a programme including drying, ashing, atomization, and annealing of the furnace. Abundance Sensitivity of the Agilent 7500 A: Low Mass $< 5 \times 10^{-7}$, High Mass $< 1 \times 10^{-7}$.

## Methods of analysis of zooplankton

Zooplankton samples were processed by standard methods (Balushkina & Winberg, 1979; Winberg & Lavrenteva, 1984). The numbers of females, males, copepodites and naupliar stages were separately calculated for each sample to characterize the structure of the *Eurytemora* population in that sample. Then, the average values of *Eurytemora* abundance were calculated for each water body. Also, in each sample, the ratio of males and females was found to characterize the sex structure of the population.

## Statistical data analysis

To determine the influence of environmental factors on the quantitative variables of *Eurytemora affinis* populations, Spearman's rank correlation coefficients ($R$, at $p<0.05$) were calculated in the Statistica 10 program. Based on the analysis of the matrix of pair correlation coefficients, insignificant factor variables were excluded from the further statistical analysis. A forward multiple step-by-step regression analysis (Glantz, 1998) was done in the Statistica 10 program to identify the main factors affecting the quantitative variables of *Eurytemora* in the water bodies of Kazakhstan. A forward multiple step-by-step regression analysis starts with no variables in the model, testing the addition of each variable using a chosen model fit criterion, adding the variable whose inclusion gives the most statistically significant improvement of the fit, and repeating this process until none further improves the model to a statistically significant extent. The number of steps depends on the number of predictors. The last step is the final assessment of the significance of predictors. The Fisher test ($F$-test) was used to assess the significance of the regression model. To analyse the distribution of population abundance within a gradient of external factors, Scatterplot Diagrams were built using the program Excel. In that same program, the descriptive statistics were calculated for all biological parameters.

TABLE I
Physico-chemical characteristics of the surveyed water bodies in Kazakhstan

| Water body | Water area ($km^2$) | Maximum depth (m) | Water temperature (°C) | TDS ($g/dm^3$) |
|---|---|---|---|---|
| Caspian Sea* | 371 000 | 138.0 | 23.1-24.1 | 0.2-1.4 |
| Lakes | | | | |
| Bozaral | 12.0 | 1.3 | 20.0-20.2 | 3.6-3.7 |
| Esey | 42.0 | 2.5 | 18.0-18.5 | 5.4-7.1 |
| Kokay | 24.3 | 3.5 | 21.0-21.3 | 1.9-2.8 |
| Koktal | 6.9 | 3.2 | 21.0-21.3 | 0.9-1.0 |
| Sultankeldy | 38.0 | 3.5 | 19.5-21.0 | 1.5-2.8 |
| Taban | 5.0 | 2.5 | 21.6-22.0 | 2.9-4.9 |
| Tabankazy | 3.6 | 2.5 | 17.5-23.5 | 1.5-1.6 |
| Sholak | 10.0 | 4.0 | 20.0-20.2 | 0.9-1.0 |
| Reservoirs | | | | |
| Kunovsky | 0.2 | 2.5 | 18.9-19.4 | 1.1-1.2 |
| Sandybay | 0.005 | 1.5 | 18.9-20.3 | 2.2-2.3 |
| Samarkan | 51.5 | 9.0 | 22.6-23.7 | 0.8-2.1 |
| Tuzda | 8.0 | 5.0 | 23.5-24.2 | 0.5-0.6 |
| Rivers and canals | | | | |
| Nura River | – | 3.0 | 13.0-23.0 | 1.1-1.8 |
| Canal Karaozek* | – | 1.5 | 20.8-21.6 | – |
| Canal Sultankeldy | – | 1.7 | 18.0-18.6 | 1.8-1.9 |
| Canal Tabankazy | – | 4.0 | 21.8-22.0 | 1.0-1.1 |

*Western Kazakhstan; all other water bodies are located in Central Kazakhstan.

## RESULTS

### Distribution of *Eurytemora affinis* in Kazakhstan

*Eurytemora affinis* was found in 15 water bodies of Kazakhstan, including lakes, reservoirs, rivers, and canals (table I). Most of the water bodies belong to the Nura River basin (Central Kazakhstan) (fig. 1A, B, D). Delta lakes of the lower reaches of the river (Bozaral, Kokai, Koktal, Sultankeldy, Taban, Tabankazy, Sholak) are located on the territory of the Korgalzhyn Nature Reserve. The shallow ponds are located north of Teniz Lake, the largest lake in Central Kazakhstan. The Tuzda and Samarkan reservoirs are formed in the upper and middle reaches of the Nura River, respectively. In addition to Central Kazakhstan, individuals of *Eurytemora* originally identified as "*Eurytemora affinis*" were found in the Emba River and in the Northern Caspian Sea (Western Kazakhstan) (fig. 1C).

## Physical and chemical characteristics of *Eurytemora affinis* habitats

*Eurytemora* was found mainly in small, plain, shallow water bodies (table I). In the Caspian Sea, individuals of *Eurytemora* previously identified as "*Eurytemora affinis*" were found at depths of 1.2-4.3 m in the zone of influence of the Volga River flow (fig. 1C). All water bodies are warm, with maximum summer temperatures from 23.0 to 28.0°C. During the study period, the water temperature was below the maximum and varied within relatively small limits (table I). All water bodies have alkaline water, with pH values from 8.2 to 9.2.

The salinity of the water bodies examined ranged from 0.2 to 7.1 g/dm$^3$ (table I). In some lakes, the average value of this parameter varied in different years of research. The content of nutrients was at a relatively low level, although it varied greatly depending on the year and water body (table II). The maximum concentrations of $NH_4^+$ were found in Tabankazy Lake in 2013. The content of $NO_3^-$ varied from 0.001 to 3.42 mg/dm$^3$, with maximum value in Sultankeldy Lake in 2014. In most of the water bodies examined, the concentrations of $NO_2^-$ were below 0.03 mg/dm$^3$. A relatively high content of $NO_2^-$ (0.100-0.400 mg/dm$^3$) was recorded in the lakes of the Korgalzhyn Nature Reserve in 2013 and 2014. The phosphate content was low. Relatively high concentrations of phosphates were established in 2013 in the lakes of the Korgalzhyn Nature Reserve and in the Samarkan reservoir in 2006.

The Cd and Zn content in all examined water bodies was very low and did not exceed 0.002-0.004 m/dm$^3$ (table II). The amount of Cu in the water reached 0.003-0.150 mg/dm$^3$, with the maximum in the lakes of the Korgalzhyn Nature Reserve (Taban, Esey, Sultankeldy) and the Nura River in 2014. In 2012, the Korgalzhyn lakes were characterized by an increased Pb content (0.014-0.035 mg/dm$^3$).

## Quantitative variables of *Eurytemora affinis*

The highest abundance of the crustacean was recorded in small water bodies: Sandybai pond and Tabankazy Lake (table III). In these water bodies, *E. affinis* formed 71.8% and 32.6% of the zooplankton abundance, respectively. Approximately in one-third of the surveyed water bodies, *Eurytemora* accounted for 10 to 20% of the total abundance of zooplankton. In other water bodies, *E. affinis* did not have any significant impact on zooplankton abundance. The number of mature individuals of *Eurytemora* was mostly low, and did not exceed 0.02-16.0% of the total abundance of its populations.

In the North Caspian Sea (May 2015), the abundance of *Eurytemora* previously identified as "*Eurytemora affinis*" was 9900 $\pm$ 400 ind./m$^3$, female abundance: 190 $\pm$ 110, males: 90 $\pm$ 50 ind./m$^3$.

TABLE II

The content of nutrients and heavy metals in the water of the surveyed water bodies of Kazakhstan

| Water body | Year | Concentration ($mg/dm^3$) | | | | | | | |
|---|---|---|---|---|---|---|---|---|---|
| | | $NH_4^+$ | $NO_3^-$ | $NO_2^-$ | $PO_4$ | Cd | Cu | Zn | Pb |
| Lakes | | | | | | | | | |
| Bozaral | 2012 | 0.630 | 0.000 | 0.019 | 0.000 | 0.002 | 0.010 | 0.002 | 0.035 |
| Esey | 2013 | 0.100 | 0.500 | 0.400 | 0.100 | 0.000 | 0.047 | 0.000 | 0.003 |
| | 2014 | 0.200 | 0.010 | 0.100 | 0.020 | 0.000 | 0.146 | 0.000 | 0.000 |
| Kokay | 2012 | 0.320 | 0.000 | 0.022 | 0.003 | 0.002 | 0.005 | 0.004 | 0.025 |
| | 2013 | 0.040 | 0.180 | 0.009 | 0.080 | 0.000 | 0.019 | 0.000 | 0.003 |
| | 2014 | – | – | – | – | 0.000 | 0.048 | 0.000 | 0.000 |
| Koktal | 2013 | 0.040 | 1.300 | 0.009 | 0.060 | 0.000 | 0.009 | 0.000 | 0.003 |
| Sultankeldy | 2012 | 0.320 | 0.000 | 0.026 | 0.000 | 0.001 | 0.004 | 0.003 | 0.014 |
| | 2013 | 0.040 | 0.200 | 0.100 | 0.080 | 0.000 | 0.012 | 0.000 | 0.003 |
| | 2014 | 0.070 | 3.420 | 0.100 | 0.016 | 0.000 | 0.083 | 0.000 | 0.000 |
| Taban | 2012 | 0.500 | 0.000 | 0.022 | 0.000 | 0.001 | 0.003 | 0.004 | 0.022 |
| | 2013 | 0.040 | 0.200 | 0.009 | 0.080 | 0.000 | 0.023 | 0.000 | 0.003 |
| | 2014 | 0.040 | 0.070 | 0.100 | 0.016 | 0.000 | 0.150 | 0.000 | 0.000 |
| Tabankazy | 2012 | 0.32 | 0.00 | 0.02 | 0.00 | 0.001 | 0.007 | 0.002 | 0.016 |
| | 2013 | 9.10 | 1.60 | 0.01 | 0.09 | 0.000 | 0.009 | 0.000 | 0.003 |
| Sholak | 2014 | 0.03 | 0.01 | 0.01 | 0.02 | 0.000 | 0.024 | 0.000 | 0.000 |
| Reservoirs | | | | | | | | | |
| Samarkan | 2006 | 0.26 | 0.01 | 0.01 | 0.07 | 0.001 | – | – | 0.004 |
| | 2007 | 0.60 | 0.02 | 0.00 | 0.01 | 0.002 | – | – | 0.016 |
| | 2008 | 0.78 | 0.12 | 0.01 | 0.01 | – | – | – | – |
| Tuzda | 2001 | – | – | – | – | 0.001 | 0.006 | 0.001 | 0.004 |
| Rivers and canals | | | | | | | | | |
| Nura River | 2014 | 0.04 | 0.01 | 0.22 | 0.03 | 0.000 | 0.142 | 0.000 | 0.000 |
| | 2006 | 0.20 | 0.11 | 0.00 | 0.07 | – | – | – | – |
| Canal Sultankeldy | 2013 | 0.10 | 0.50 | 0.01 | 0.07 | 0.000 | 0.021 | 0.000 | 0.003 |
| Canal Tabankazy | 2014 | 0.03 | 0.01 | 0.10 | 0.02 | – | – | – | – |

–, no data.

## Scatterplot diagrams

Analysis of scatterplot diagrams showed that the abundance of *E. affinis* populations was higher in water bodies with a high abundance of zooplankton (fig. 2A). Males were present in 57.8% of the processed samples and were more numerous than females (fig. 2B). There were fewer females, i.e., in 45.0% of samples. The ratio of males and females in the populations varied from 0.4 : 1 to 7.1 : 1. Males thus dominated the majority of the *E. affinis* populations.

Analysis of the scatterplot diagrams showed that the highest abundance of *E. affinis* was recorded at a water temperature of 21.0-22.0°C (fig. 3A) at depths of 1.0-3.0 m (fig. 3B). In the same narrow temperature range, the number of mature

TABLE III

Abundance of total zooplankton and *Eurytemora affinis* (Poppe, 1880) populations in the water bodies of Kazakhstan surveyed

| Water body | Date | Abundance (×1000 ind./$m^3$) | | | | |
|---|---|---|---|---|---|---|
| | | Total zooplankton | *E. affinis* | | | |
| | | | Females | Egg bearing females | Males | Total |
| Lakes | | | | | | |
| Bozaral | July 2012 | 141.8 ± 1.3 | 0.0 | 0.32 ± 0.32 | 0.07 ± 0.04 | 20.8 ± 6.1 |
| Esey | July 2013 | 92.2 ± 33.9 | 0.008 ± 0.008 | 0.0 | 0.0 | 3.1 ± 1.8 |
| | July 2014 | 284.4 ± 34.6 | 0.0 | 0.0 | 0.0 | 19.4 ± 8.3 |
| Kokay | July 2012 | 311.5 ± 106.6 | 0.07 ± 0.05 | 0.11 ± 0.05 | 0.30 ± 0.08 | 19.0 ± 1.4 |
| | July 2013 | 104.7 ± 11.1 | 0.006 ± 0.006 | 0.0 | 0.02 ± 0.01 | 8.9 ± 1.2 |
| | July 2014 | 345.4 ± 72.6 | 0.06 ± 0.03 | 0.05 ± 0.03 | 0.33 ± 0.15 | 16.2 ± 7.2 |
| Koktal | July 2013 | 157.1 ± 65.5 | 0.026 ± 0.016 | 0.0 | 0.06 ± 0.03 | 34.4 ± 16.6 |
| Sultankeldy | July 2012 | 464.6 ± 302.6 | 0.51 ± 0.48 | 0.22 ± 0.11 | 0.87 ± 053 | 60.0 ± 25.6 |
| | July 2013 | 156.5 ± 14.5 | 0.0 | 0.0 | 0.0 | 11.0 ± 3.7 |
| | July 2014 | 339.0 ± 176.9 | 0.03 ± 0.03 | 0.44 ± 0.35 | 1.07 ± 0.36 | 23.7 ± 7.3 |
| Taban | July 2012 | 269.6 ± 13.8 | 0.02 ± 0.02 | 0.07 ± 0.04 | 0.20 ± 0.12 | 38.0 ± 27.8 |
| | July 2013 | 396.8 ± 158.6 | 0.03 ± 0.02 | 0.04 ± 0.03 | 0.19 ± 0.09 | 65.9 ± 26.0 |
| | July 2014 | 134.5 ± 26.2 | 0.01 ± 0.01 | 0.0 | 0.03 ± 0.01 | 19.5 ± 11.3 |
| Tabankazy | July 2012 | 342.3 ± 100.9 | 0.0 | 0.0 | 0.03 ± 0.02 | 111.6 ± 54.6 |
| | July 2013 | 142.9 ± 50.9 | 0.0 | 0.0 | 0.0 | 34.2 ± 11.9 |
| Sholak | July 2014 | 1614.0 ± 36.9 | 0.007 ± 0.007 | 0.015 ± 0.007 | 0.62 ± 0.43 | 20.2 ± 4.0 |

TABLE III
(Continued)

| Water body | Date | Abundance (×1000 ind./m$^3$) | | | | |
|---|---|---|---|---|---|---|
| | | Total zooplankton | *E. affinis* | | | |
| | | | Females | Egg bearing females | Males | Total |
| Ponds and reservoirs | | | | | | |
| Kunovsky | July 2014 | 448.3 ± 300.0 | 0.0 | 0.0 | 0.0 | 66.4 ± 33.0 |
| Sandybay | July 2014 | 337.3 ± 250.0 | 0.0 | 0.0 | 0.0 | 242.3 ± 25.3 |
| Samarkan | June 2006 | 324.5 ± 135.9 | 0.01 ± 0.007 | 0.007 ± 0.007 | 0.02 ± 0.01 | 12.1 ± 5.9 |
| | July 2007 | 156.1 ± 44.4 | 0.007 ± 0.004 | 0.010 ± 0.006 | 0.02 ± 0.007 | 2.4 ± 2.2 |
| | July 2008 | 153.5 ± 14.3 | 0.001 ± 0.001 | 0.003 ± 0.002 | 0.02 ± 0.007 | 19.9 ± 2.5 |
| Tuzda | August 2001 | 70.5 ± 37.3 | 0.0 | 0.01 ± 0.01 | 0.0 | 1.3 ± 1.3 |
| Rivers and canals | | | | | | |
| Nura River | June 2006 | 28.1 ± 7.2 | 0.09 ± 0.04 | 0.004 ± 0.004 | 0.04 ± 0.02 | 1.6 ± 0.6 |
| | September 2007 | 24.7 ± 20.7 | 0.33 ± 0.33 | 0.0 | 0.35 ± 0.32 | 4.3 ± 3.7 |
| | July 2014 | 1189.6 ± 526.3 | 0.04 ± 0.04 | 0.0 | 0.24 ± 0.24 | 4.0 ± 3.0 |
| Canal Karaozek | June 2011 | 165.4 ± 16.1 | 1.8 ± 1.3 | 1.9 ± 0.4 | 4.82 ± 1.10 | 40.4 ± 13.5 |
| Canal Sultankeldy | July 2013 | 70.5 ± 34.6 | 0.0 | 0.0 | 0.10 ± 0.10 | 10.1 ± 0.4 |
| Canal Tabankazy | July 2012 | 264.9 ± 128.6 | 0.0 | 0.0 | 0.0 | 107.3 ± 80.5 |
| | July 2014 | 7.6 ± 3.5 | 0.0 | 0.0 | 0.0 | 2.0 ± 2.0 |

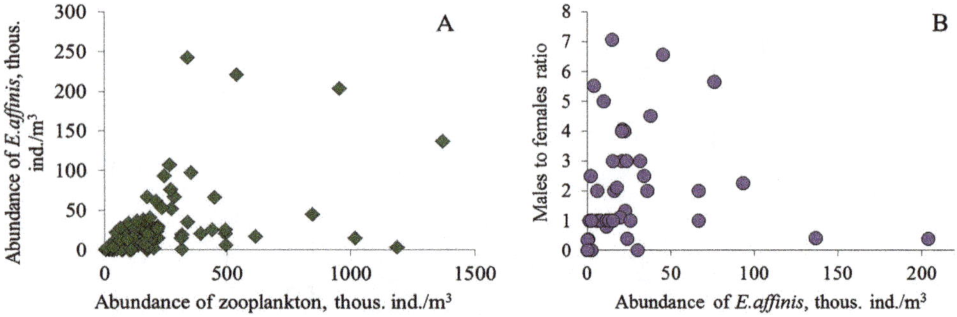


Fig. 2. A, Changes in the populations of *Eurytemora affinis* (Poppe, 1880), depending on the abundance of zooplankton; B, changes in the sex ratio (male/female) when the population abundance of *E. affinis* is changing.

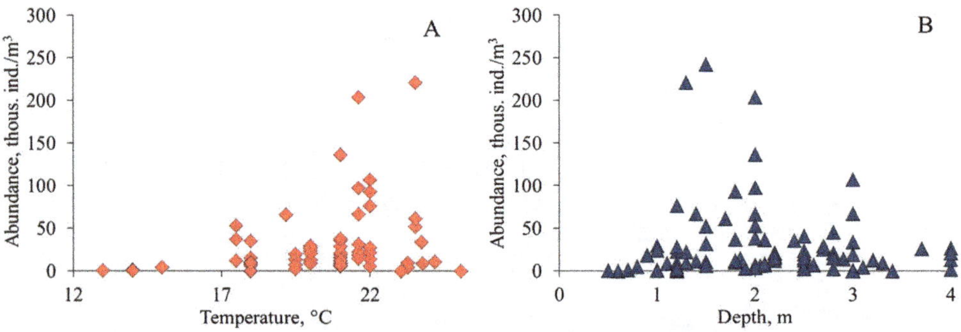


Fig. 3. Distribution of the abundance of *Eurytemora affinis* (Poppe, 1880) populations, as observed in gradients of A, temperature; and, B, depth.

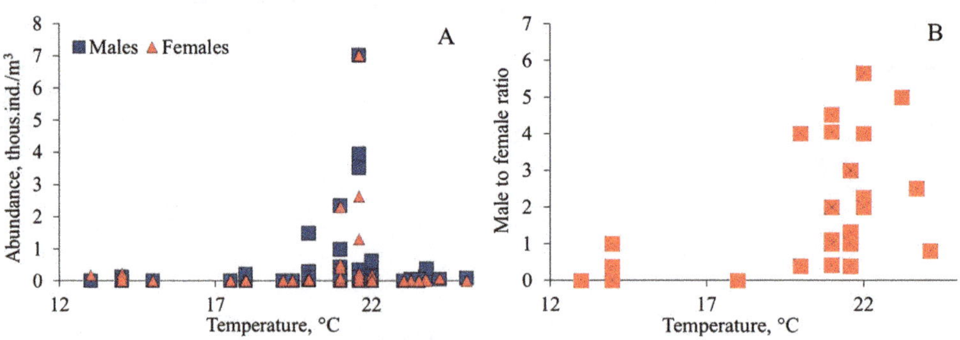


Fig. 4. The distribution in populations of *Eurytemora affinis* (Poppe, 1880) within a water temperature gradient of: A, the abundance of mature individuals; and, B, the ratio of males and females.

individuals of *Eurytemora* was maximal (fig. 4A). There was a slight tendency of male domination to grow with the increase in the water temperature (fig. 4B).

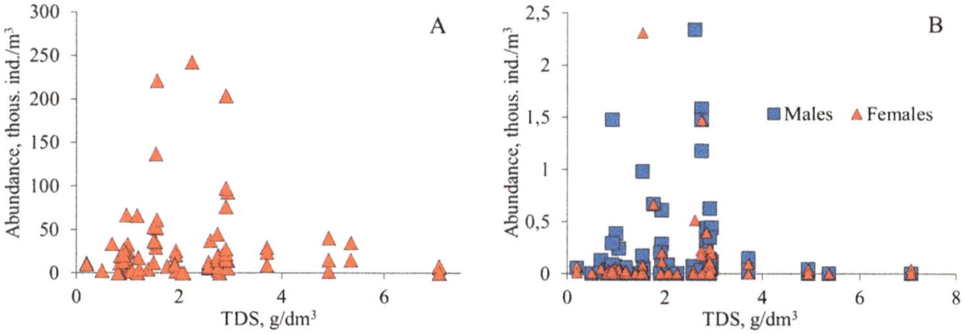


Fig. 5. The distribution in populations of *Eurytemora affinis* (Poppe, 1880) within a TDS gradient of: A, the total abundance of the species; and, B, the abundance of mature individuals.

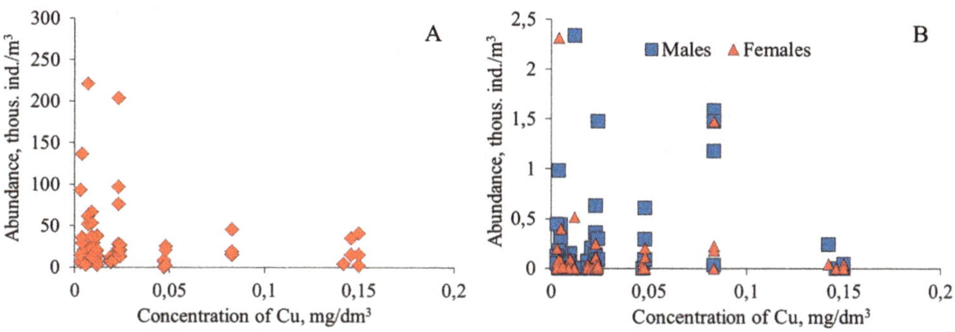


Fig. 6. The distribution in populations of *Eurytemora affinis* (Poppe, 1880), depending on the Cu content of the water, of: A, total abundance of the species; and, B, the abundance of mature individuals; in the examined water bodies of Kazakhstan.

*E. affinis* was found at salinities from 0.2 to 7.1 g/dm$^3$. The highest abundance of *Eurytemora* (fig. 5A), including mature individuals (fig. 5B), was recorded at salinities from 1.0 to 3.0 g/dm$^3$.

The heavy metals Cd, Pb and Zn were present in the water of the examined water bodies in low concentrations and did not have any significant effect on the abundance of *Eurytemora* populations. It was observed that both the total abundance of populations (fig. 6A) and the number of mature individuals (fig. 6B) tended to decrease when the Cu content in the water rose above 0.02 mg/dm$^3$.

## Correlation analysis

According to the results of correlation analysis, the relationship between the abundance of *E. affinis* and environmental factors (temperature, depth, TDS, the content of nitrates, nitrites, and ammonium) was either statistically insignificant or very weak ($R$ = 0.210-0.270, with $p$<0.05). Statistically significant negative relationships were established between the concentration of phosphates and the number of adult *E. affinis* ($R = -0.477$, $p<0.05$), including males ($R = -0.452$, $p<0.05$) and females with egg sacs ($R = -0.542$, $p<0.05$). The number of females with egg sacs increased with increasing concentrations of Zn in the water ($R = 0.537$, $p<0.05$). Statistically significant positive relationships were established between the number of males and the water temperature ($R = 0.454$, with $p<0.05$).

## Multiple step-by-step regression analysis

According to the results of multiple step-by-step regression analysis, the influence of external factors on the abundance of females and the total abundance of *Eurytemora affinis* was statistically insignificant (table IV). The abundance of males and the total abundance of mature individuals of *E. affinis* increased under the influence of nitrates and decreased under the influence of phosphates. The proportion of males in *Eurytemora* populations decreased with an increase in TDS but enhanced with an increase in Cu concentrations in water. The total abundance of zooplankton was under the negative influence of phosphates and lead.

# DISCUSSION

In addition to the water bodies listed in table I, *Eurytemora affinis* is found in 14 other localities in Kazakhstan (Stuge, 1995, 2000, 2007, 2008a, b). Taking into account all the available data, it is possible to outline the distribution area of *E. affinis* in this country. The species occupies predominantly the central and partly north-eastern parts of the territory (fig. 1). The *Eurytemora* population of the Caspian Sea can preliminarily be assigned to the recently described species *E. caspica*, although further studies are required for a definitive conclusion. In the Kazakhstan sector of the Caspian Sea, *E. caspica* individuals were found in the zone of influence of the Volga River, i.e., almost in the same locality from where the new species was described (Sukhikh & Alekseev, 2013). An indirect confirmation of this assumption is also the fragmentation of the range of *Eurytemora affinis* within Kazakhstan (fig. 1A). It should be noted that during many years of seasonal research in the Caspian Sea (spring, summer, autumn of 1997-2018), *E. caspica* was found only at the end of spring, and at a water temperature not higher than 24.0°C. In summer, with temperatures of the surface water of 28-30°C (Skolskaya & Uvarov, 2018), this species was never found in the zooplankton. Perhaps, like other species of this genus (Souissi et al., 2016), *E. caspica* prefers moderate and low temperatures. With a salinity of the open water area of the Caspian Sea of 9.0-13.0‰, *E. caspica* occurred locally in the desalinated part of the area, near the mouth of the Volga River (fig. 1C). The

TABLE IV

Assessment of the influence of external factors on the *Eurytemora affinis* (Poppe, 1880) populations from water bodies of Kazakhstan, based on the results of multivariate step-by-step regression analysis

| Variable | Step 1 | Step 2 | Step 3 | Final solution |
|---|---|---|---|---|
| Total zooplankton abundance | $PO_4$ b = −0.28 | $PO_4$ b = −0.57*<br>Pb b = −0.52* | $PO_4$ b = −0.57*<br>Pb b = −0.61*<br>$NO_3^-$ b = −0.28 | $PO_4$ b = −0.81*<br>Pb b = −0.95*<br>$NO_3^-$ b = −0.33<br>Cu b = −0.39 |
| Total *Eurytemora* abundance | $NO_2^-$ b = −0.41 | $NO_2^-$ b = −0.32 | $NO_2^-$ b = −0.34 | $NO_2^-$ b = −0.34 |
| Adult *Eurytemora* abundance | – | – | $PO_4$ b = −0.48* | $PO_4$ b = −0.57*<br>$NO_3^-$ b = 0.49* |
| Female *Eurytemora* abundance | Zn b = 0.44 | Zn b = 0.948*<br>Pb b = −0.63 | Zn b = 0.192<br>Pb b = −1.4 | Zn b = 0.192<br>Pb b = −1.4<br>* |
| Male *Eurytemora* abundance | $NO_3^-$ b = 0.442 | $NO_3^-$ b = 0.537*<br>$PO_4$ b = −0.530* | $NO_3^-$ b = 0.591*<br>$PO_4$ b = −0.470* | $NO_3^-$ b = 0.591*<br>$PO_4$ b = −0.470*<br>$NH_4^+$ b = −0.210 |
| Sex ratio of *Eurytemora* | – | TDS b = −0.78* | TDS b = −0.61* | TDS b = −0.61*<br>Cu b = 0.97* |

*Significant b.

occurrence of *E. caspica* at low salinity (0.2-1.4‰) might be associated not only with the peculiar chemical composition of the Caspian Sea water (Alekin & Lyakhin, 1984) but also with favourable feeding conditions in the zone of influence of the river flow. According to experimental data (Hammock et al., 2015), the amount of food made a difference to this copepod's adaptation to salinity, but not the temperature.

Concerning *E. affinis*, one of the reasons for its rare occurrence in Kazakhstan is the elevated water temperatures in summer. In a temperate continental climate, *Eurytemora* inhabits water bodies at a temperature of 5-20°C (Evdokimov & Ermokhin, 2009), and the optimum temperature range is 12-14°C (Diekmann et al., 2012). A summary of the available data showed that in Kazakhstan *E. affinis* occurred at temperatures from 12.0°C (Stuge, 2000, 2008a) to 25.0°C (present author's data). According to the results of the correlation and regression analysis (table IV) in the above range of values, the temperature did not significantly affect the quantitative variables of the *E. affinis* population. Fig. 4 demonstrates that the maximum abundance of this crustacean was recorded at a temperature of 21.0-22.0°C. This is higher than the optimal values of temperature given for *E. affinis* populations from other regions (Diekmann et al., 2012). Over the entire study period (1997-2018), *E. affinis* was never discovered in the southern areas of Kazakhstan, where the water temperature reaches 28.0-32.0°C in summer. Evidently, a temperature of 25°C is the upper thermal limit for this species in Kazakhstan. The results agree with the experimental data, according to which, after a temperature increase to 24°C, the population remained viable, but copepod size was significantly decreased, just as well as female fecundity (Souissi et al., 2016). Freezing of shallow water bodies in winter at air temperatures down to −45°C (Uteshev, 1959) could be one of the reasons for the absence of this species in the northern and eastern regions of Kazakhstan.

Along with temperature, hydrochemical conditions have a decisive influence on the composition of the fauna of any water body (Khlebovich, 2013). In the Baltic Sea, *E. affinis* was constantly encountered at salinities of 1.8-9.3‰ (Paturej & Gutkowska, 2015). According to (Karlsson et al., 2018), *E. affinis* can tolerate conditions close to freshwater at high temperatures, but with a significant reduction in fitness. Under experimental conditions, *E. affinis* endured a salinity ranging from 2.5 to 25.0-30.0‰ (Devreker et al., 2007), but the hatching of eggs occurred at a salinity not higher than 20.0‰ (Ishikawa et al., 1999). The optimal salinity limits for this species established in experiments are 5.0-15.0‰ (Ishikawa et al., 1999; Cailleaud et al., 2007), or 3.0-10.0‰ according to Roddie et al. (1984). In the Baltic Sea, the maximum abundance of populations of this crustacean was observed at a salinity of 6.2-8.2‰ (Diekmann et al., 2012). According to Viitasalo et al. (1994), the upper limit of *E. affinis*' distribution in the Baltic Sea was at a salinity of 6.5‰.

As a part of this study, *E. affinis* was registered in the gradient of total dissolved solids (TDS) from 0.2 to 7.1 g/dm$^3$. According to the results of correlation and regression analysis, in the gradient mentioned above, the effect of salinity on the quantitative variables of *E. affinis* was statistically insignificant. The maximum number of this species was recorded with TDS at 1.0-3.0 g/dm$^3$ (fig. 5). The results indicate a shift of the optimal salinity range in the surveyed continental waters towards more fresh waters than reported for the seas (Cailleaud et al., 2007) and established in experiments (Ishikawa et al., 1999; Devreker et al., 2007) for this species. In Eastern Kazakhstan, *Eurytemora* was found in both fresh waters (with a salinity of 0.27 g/dm$^3$) and in hypersaline water bodies (40.0-343.0 g/dm$^3$) (Stuge, 2000). At extreme TDS values (less than 0.3 and more than 40.0 g/dm$^3$), only immature *E. affinis* were represented in the zooplankton, and their abundance did not exceed 0.1-0.5 thous. ind./m$^3$ (Stuge, 2000, 2007, 2008). Taking into account the obtained results and the analysis of the literature data, the occurrences of *E. affinis* in ultra-fresh and hypersaline water bodies are most likely due to random drift.

The content of nutrients is an indirect indicator of trophic conditions in a water body (Ansari et al., 2015). According to the results of the correlation and regression analysis, the number of sexually mature individuals of *E. affinis* increased under the influence of nitrates and decreased under the influence of phosphates (table IV). The relationship between the total abundance of *Eurytemora* and the nutrient content was statistically insignificant, possibly due to their low absolute concentrations in the water bodies examined. Nevertheless, indirect evidence indicates that, in Kazakhstan, *E. affinis* preferred water bodies with good trophic conditions. Total abundance of zooplankton can serve as an indicator of the number of feed resources. As can be seen from fig. 3A, the number of *E. affinis* was higher in the water bodies of Kazakhstan with a high total abundance of planktonic invertebrates. Other researchers have noted similar preferences of *E. affinis*. In the bays of the Baltic Sea, *Eurytemora* dominated in spring during the period of high total abundance of zooplankton and maximum phytoplankton biomass (Semenova et al., 2010). In the Rhine delta, the decline in the abundance of *Eurytemora* from spring to summer was associated not only with temperature increase but also with a decreased amount of chlorophyll (Gulati & Doornekamp, 1991). Obviously, this is due to the fact that *E. affinis* has wide feeding spectra (Heinle & Flemer, 1975; Heinle et al., 1977; Boak & Goulder, 1983; Lehtinen & Sopanen, 2010), and it also needs more food for growth than for some other species of Copepoda (Escaravage & Soetaert, 1995).

Information about the sex structure of *E. affinis* populations is scarce. Along with other Calanoida species (Meshkova, 1975), females dominate the *Eurytemora* populations. Short-term dominance of males can be observed at the beginning of the breeding season, as males mature a few days earlier (Bazarkina, 2008). Predators selectively eat females of copepods (Castonguay & FitzGerald, 1990), which have larger body sizes than males; therefore, predators can influence the sex ratio in populations of *E. affinis* (cf. Heuschele et al., 2014). When the temperature rises, the lifespan of *E. affinis* females decreases (Katona, 1970), which may also be the reason for male dominance. According to Heinle (1970), males dominated *E. affinis* populations at a temperature of 15°C and 20°C in the laboratory. In populations of *E. affinis* from the Gulf of Finland (Baltic Sea), males dominated in 2008 and 2010 when summer temperatures were very high; females dominated at unusually low temperatures in 2015 (Sukhikh et al., 2019). Given the above, one of the reasons for the almost constant dominance of males in *E. affinis* populations (fig. 2B) might be the higher temperatures of Kazakhstan's water bodies, compared to water bodies of a temperate climate. Statistically significant positive relationships, although weak in strength, were established between the number of males of *Eurytemora* and the water temperature ($R = 0.454$, at $p<0.05$).

Another reason for the dominance of males in the *Eurytemora* populations might be the toxic pollution of the examined water bodies in the Nura River basin. In some periods, elevated concentrations of heavy metals were recorded both in the Samarkan reservoir (middle reaches) (Krupa et al., 2009; Slivinsky et al., 2009) and in lakes of the lower reaches of the Nura River (Slivinsky& Krupa, 2013). In the surveyed water bodies, inhabited by *E. affinis*, the content of copper and partly lead was elevated in some cases (table II). According to the results of multiple step-by-step regression analysis, copper caused an increase in the proportion of males in *Eurytemora* populations. The effect of toxic pollution on the sex structure of copepods in the direction of male dominance has been shown previously (Krupa, 2005, 2012b).

Information on the relationship of *E. affinis* to toxic pollution is not available in the literature. Findings of this species in the Samarkan reservoir and the Nura River indicate that *Eurytemora* can tolerate a relatively high level of toxic pollution. Industrial and municipal wastewater is discharged into the Samarkan reservoir and the Nura River for more than 75 years (Malinovskaya & Ten, 1983; Ilyushchenko et al., 2000; Matmuratov et al., 2005). Chronic pollution of the reservoir causes the accumulation of heavy metals in soils (Matmuratov et al., 2005), pronounced interannual changes in the composition and quantitative indicators of zooplankton communities (Stuge & Bartosh, 1998; Stuge et al., 2002; Matmuratov et al., 2005; Krupa et al., 2009; Slivinsky et al., 2009), as well as the constant presence of malformed specimens in Cyclopoida populations (Krupa, 2005, 2012a, b). The variable composition of wastewater, along with the climatic and hydrological conditions of the year, cause significant changes in the concentrations of heavy metals (mainly Cd, Zn, Cu) in the water of the reservoir and the river at different periods of research. An analysis of the published literature (Stuge & Bartosh, 1998; Stuge et al., 2002; Matmuratov et al., 2005; Krupa et al., 2009; Slivinsky et al., 2009) showed that *Eurytemora* was absent in the zooplankton of the Samarkan reservoir when the concentration of Cd in the water was more than 0.01 mg/dm$^3$, Zn more than 0.02 mg/dm$^3$ and Cu more than 0.005 mg/dm$^3$.

As can be seen from the above, the environmental preferences of *E. affinis* in Kazakhstan's water bodies under conditions of the extreme continental climate differ from those described in the literature sources. This might be due both to the heterogeneity of *Eurytemora* populations from different regions and to the existence of a complex of related species called *"Eurytemora affinis"* (Lee, 2000; Lee & Frost, 2002; Alekseev & Soussi, 2011).

## CONCLUSION

Taking into account both my own data and data from the literature, *Eurytemora affinis* is currently known from 29 water bodies of Central and Eastern Kazakhstan. Most water bodies refer to the Nura River basin. Populations from the North-Eastern Caspian Sea originally identified as "*Eurytemora affinis*" can be attributed to the recently described species *Eurytemora caspica*. In Kazakhstan, *Eurytemora affinis* inhabits predominantly small, shallow water bodies. It prefers water bodies with temperatures from 12 to 25°C and a salinity from 0.2 to 7.1 g/dm$^3$. Findings of *Eurytemora* at salinities less than 0.3 and more than 40.0 g/dm$^3$ can be considered accidental, since under such regimes the abundance of its populations was very low, and no mature individuals were present in those conditions. The optimum temperature range is 21-22°C, which is higher than the values given for *E. affinis* populations in water bodies of the temperate climate. The maximum abundance of *Eurytemora* populations was registered at a salinity of 1.0-3.0 g/dm$^3$. The results obtained indicated a shift of the optimal salinity range in continental water bodies towards more freshwater conditions than was reported for coastal sea areas (6.2-8.2‰), and as established in experiments (from 3.0-10.0 to 10.0-15.0‰). In Kazakhstan, *E. affinis* preferred water bodies with good trophic conditions. It seems that *E. affinis* endures relatively high contents of heavy metals, but it does not inhabit water bodies with very high toxic pollution. Male dominance in *E. affinis* populations might due both to the adverse effect of elevated temperature on the lifespan of females, and was also estsablished to be connected with toxic pollution of water bodies.

The described features of the biology of *E. affinis* biology in conditions of an extreme continental climate can be associated both with the heterogeneity of *Eurytemora* populations from different regions, and the existence of a complex of related species, still provisionally called "*Eurytemora affinis*" [but then sensu lato]. Detailed future studies should be undertaken to elucidate the actual situation.

## REFERENCES

ALEKIN, O. A. & Y. I. LYAKHIN, 1984. Chemistry of the ocean: 1-344. (Gidrometeoizdat, Leningrad). [In Russian.]

ALEKSEEV, V. R. & A. SOUISSI, 2011. A new species within the *Eurytemora affinis* complex (Copepoda: Calanoida) from the Atlantic coast of USA, with observations on eight morphologically different European populations. Zootaxa, **2767**: 41-56.

ANSARI, A. A., S. A. GHANIM, S. TRIVEDI, H. REHMAN, Z. K. ABBAS & S. SAGGU, 2015. Seasonal dynamics in the trophic status of water, floral and faunal density along some selected coastal areas of the Red Sea, Tabuk, Saudi Arabia. Int. Aquat. Res., **7**: 337-348. DOI:10.1007/s40071-015-0118-6.

BALUSHKINA, E. V. & G. G. WINBERG, 1979. The relationship between the length and weight of planktonic crustaceans. In: G. G. WINBERG (ed.), Experiments and field studies of the biological base of lake productivity: 58-79. (Academy of Sciences of the USSR, Leningrad [now St. Petersburg]). [In Russian.]

BAN, S., 1994. Effect of temperature and food concentration on post-embryonic development, egg production and adult body size of calanoid copepod *Eurytemora affinis*. Journal of Plankton Research, **16**(6): 721-735. DOI:10.1093/plankt/16.6.721.

BAZARKINA, L. A., 2008. Trophic relationships of aquatic organisms in the pelagial of Azabachie Lake in 2006. Studies of Aquatic Biological Resources of Kamchatka and the Northwestern Part of the Pacific Ocean, **11**: 32-41. [In Russian.]

BOAK, A. C. & R. GOULDER, 1983. Bacterioplankton in the diet of the calanoid copepod *Eurytemora* sp. in the Humber estuary. Mar. Biol., **73**(2): 139-149. DOI:10.1007/BF00406881.

BORUTZKY, E. V., L. A. STEPANOVA & M. S. KOS, 1991. A handbook of Calanoida from the fresh waters of the USSR: 1-503. (Zoological Institute, Nauka, Leningrad [now St. Petersburg]). [In Russian.]

BRADLEY, B. P., R. HAKIMZADEH & J. S. VINCENT, 1988. Rapid responses to stress in *Eurytemora affinis*. Hydrobiologia, **167**: 197. DOI:10.1007/BF00026305.

CAILLEAUD, K., G. MAILLET, H. BUDZINSKI, S. SOUISSI & J. FORGET-LERAY, 2007. Effects of salinity and temperature on the expression of enzymatic biomarkers in *Eurytemora affinis* (Calanoida, Copepoda). Comp. Biochem. Physiol., **147**: 841-849.

CASTONGUAY, M. & G. J. FITZGERALD, 1990. The ecology of the calanoid copepod *Eurytemora affinis* in salt marsh tide pools. Hydrobiologia, **202**(3): 125-133. DOI:10.1007/BF00006839.

CROSKERY, P., 1978. The freshwater co-occurrence of *Eurytemora affinis* (Copepoda: Calanoida) and *Manayunkia speciosa* (Annelida: Polychaeta): possible relicts of a marine incursion. Hydrobiologia, **59**(3): 237-241. DOI:10.1007/BF00036503.

DEVREKER, D., S. SOUISSI, J. FORGET-LERAY & F. LEBOULENGER, 2007. Effects of salinity and temperature on the post-embryonic development of *Eurytemora affinis* (Copepoda; Calanoida) from the Seine estuary: a laboratory study. J. Plankton. Res., **29**: 117-133.

DIEKMANN, A. B. S., C. CLEMMESEN, M. A. ST. JOHN, M. PAULSEN & M. A. PECK, 2012. Environmental cues and constraints affecting the seasonality of dominant calanoid copepods in brackish, coastal waters: a case study of *Acartia*, *Temora* and *Eurytemora* species in the south-west Baltic. Mar. Biol., **159**(11): 2399-2414. DOI:10.1007/s00227-012-1955-0.

ESCARAVAGE, V. & K. SOETAERT, 1995. Secondary production of the brackish copepod communities and their contribution to the carbon fluxes in the Westerschelde estuary (The Netherlands). Hydrobiologia, **311**(1-3): 103-114. DOI:10.1007/BF00008574.

EVDOKIMOV, N. A. & M. V. ERMOKHIN, 2009. Crustaceans of zooplankton in temporary waterbodies of Saratov district on territory of different natural zones. Biology of Inland Waters, **1**: 62-69. [In Russian.]

FOMIN, G. S. (ed.), 1995. Water. Control of chemical, bacterial, and radiation safety according to international standards: 1-618. (NGO "Alternative", Moscow). [In Russian.]

FRUMIN, G. T. & Y. V. KRASHANOVSKAYA, 2014. Assessment of the trophic status of the lakes of Kazakhstan. Ecological Chemistry, **23**(1): 8-12. [In Russian.]

GLANTZ, S. A., 1998. Primer of BIOSTATISTICS: 1-459. (Practice, Moscow). [In Russian.]

GULATI, R. D. & A. DOORNEKAMP, 1991. The spring-time abundance and feeding of *Eurytemora affinis* (Poppe) in Volkerak-Zoommeer, a newly-created freshwater lake system in the Rhine delta (The Netherlands). Hydrobiological Bulletin, **25**(1): 51-60. DOI:10.1007/BF02259589.

HAMMOCK, B. G., S. LESMEISTER, I. FLORES, G. S. BRADBURD, F. H. HAMMOCK & S. J. THE, 2015. Low food availability narrows the tolerance of the copepod *Eurytemora affinis* to salinity, but not to temperature. Estuaries and Coasts, **39**(1): 189-200. DOI:10.1007/s12237-015-9988-5.

HEINLE, D. R., 1970. Population dynamics of exploited cultures of calanoid copepods. Helgoländer Wissenschaftliche Meeresuntersuchungen, **20**: 360-372.

HEINLE, D. R. & D. A. FLEMER, 1975. Carbon requirements of a population of the estuarine copepod *Eurytemora affinis*. Mar. Biol., **31**(3): 235-247. DOI:10.1007/BF00387152.

HEINLE, D. R., R. P. HARRIS, J. F. USTACH & D. A. FLEMER, 1977. Detritus as food for estuarine copepods. Mar. Biol., **40**(4): 341-353. DOI:10.1007/BF00395727.

HEUSCHELE, J., S. CEBALLOS, C. M. ANDERSEN BORG, O. BJÆRKE, S. ISARI, R. LASLEY-RASHER, E. LINDEHOFF, A. SOUISSI, S. SOUISSI & J. TITELMAN, 2014. Non-consumptive effects of predator presence on copepod reproduction: insights from a mesocosm experiment. Mar. Biol., **161**(7): 1653-1666. DOI:10.1007/s00227-014-2449-z.

HOSSIENI, A., A. ROOHI, K. A. GANJIAN, M. ROSHANTABARI, A. HASHEMIAN, A. SOLIMANROUDI & H. S. NASROLLAZADEH, 1996. Hydrology and hydrobiology of the southern Caspian Sea: 1-510. (Iranian Fisheries Research Organization, Tehran [= Teheran]).

ILYUSHCHENKO, M. A., L. V. YAKOVLEVA, S. HAVEN & E. LAPSHIN, 2000. Mercury pollution of the Nura River. The Industry of Kazakhstan, **3**(6): 56-59. [In Russian.]

Interstate Standard 31 870-2012, 2013. Drinking water. Determination of elements content by atomic spectrometry methods. (Standardinform, Moscow). [In Russian.]

IRIGOIEN, X., J. CASTEL & S. GASPARINI, 1996. Gut clearance rate as predictor of food limitation situations. Application to two estuarine copepods: *Acartia bifilosa* and *Eurytemora affinis*. Mar. Ecol. Prog. Ser., **131**: 159-163.

ISHIKAWA, A., S. BAN & N. SHIGA, 1999. Effects of salinity on survival, and embryonic and postembryonic development of *Eurytemora affinis* from a freshwater lake. Plankton Biol. Ecol., **46**(2): 113-119.

KARLSSON, K., S. PUIAC & M. WINDER, 2018. Life-history responses to changing temperature and salinity of the Baltic Sea copepod *Eurytemora affinis*. Mar. Biol., **165**: art. 30. DOI:10.1007/s00227-017-3279-6.

KATONA, S. K., 1970. Growth characteristics of the copepods *Eurytemora affinis* and *E. herdmani* in laboratory cultures. Helgoländer Wissenschaftliche Meeresuntersuchungen, **20**: 373-384. DOI:10.1007/BF01609914.

KHLEBOVICH, V. V., 2013. Critical salinity – homeostasis – sustainable development. Proceedings of the Zoological Institute RAS, **3**: 3-6. [In Russian.]

KRUPA, E. G., 2005. Population densities, sex ratios of adults, and occurrence of malformations in three species of cyclopoid copepods in waterbodies with different degrees of eutrophy and toxic pollution. Journal of Marine Science and Technology, **13**(3): 226-237.

KRUPA, E. G., 2012a. Zooplankton of limnic and lotic ecosystems of Kazakhstan. Structure, patterns of formation: 1-346. (Palmarium Academic Publishing, Saarbrücken). [In Russian.]

KRUPA, E. G., 2012b. Sex structure of the copepod populations (Copepoda: Cyclopoida, Calanoida) in water bodies of Kazakhstan. In: A. M. MELDEBEKOV (ed.), Fauna of Kazakhstan and adjacent territories. Materials of the international scientific conference: 127-128. (Institute of Zoology, Almaty). [In Russian.]

KRUPA, E. G., O. V. DOBROKHOTOVA & T. S. STUGE, 2016. Calanoida fauna (Crustacea, Copepoda) of Kazakhstan and adjacent territories: 1-208. (EtalonPrint, Almaty). [In Russian.]

KRUPA, E. G., T. S. STUGE & G. G. SLIVINSKY, 2009. Zooplankton of the Samarkan reservoir under conditions of complex pollution by sewage. Bulletin of Pavlodar State University Named after CM. Toraigyrova, **4**: 114-125. [In Russian.]

LEE, C. E., 2000. Global phylogeography of a cryptic copepod species complex and reproductive isolation between genetically proximate «populations». Evolution, **54**(6): 2014-2027.

LEE, C. E. & B. W. FROST, 2002. Morphological stasis in the *Eurytemora affinis* species complex (Copepoda: Temoridae). Hydrobiologia, **480**: 111-128.

LEE, C. E., K. E. JOHNSON, L. PERREAU, G. CHARMANTIER & M. CHARMANTIER-DAURES, 2014. Without gills: rapid evolution of osmoregulatory function in the copepod *Eurytemora affinis* during habitat invasions. Copepoda and the global environment. In: 12th international conference on Copepoda in Korea: 47. (Hanyang University, Seoul).

LEHTINEN, S., S. SOPANEN & T. TAMMINEN, 2010. Community-level and species-specific responses of coastal phytoplankton to the presence of two mesozooplanktonic grazers. Mar. Biol., **157**(7): 1555-1565. DOI:10.1007/s00227-010-1429-1.

MALINOVSKAYA, A. S. & V. A. TEN, 1983. Hydrofauna of reservoirs of Kazakhstan: 1-208. (Science of the Kazakh SSR, Alma-Ata). [In Russian.]

MATMURATOV, S. A., T. S. STUGE, T. T. TROSHINA, E. G. KRUPA, G. ZH. AKBERDINA & N. S. AINABAEVA, 2005. Zooplankton ponds of the river basin Nura in conditions of chronic pollution. Proceedings of the Institute of Zoology, **49**: 130-145.

MESHKOVA, T. M., 1975. Patterns of development of zooplankton Sevan Lake: 1-237. (Academy of Sciences of the Armenian SSR, Yerevan). [In Russian.]

NIKOLAYENKO, V. A., 1989. The hydrochemical regime and water quality of reservoirs in Central Asia. Hydrobiological Journal, **25**(1): 78-83. [In Russian.]

OGHENEKARO, E. U., P. CHIGBU, O. F. OSEJI & K. W. TANG, 2018. Seasonal factors influencing copepod abundance in the Maryland coastal bays. Estuaries and Coasts, **41**(2): 495-506. DOI:10.1007/s12237-017-0285-3.

PATUREJ, E. & A. GUTKOWSKA, 2015. The effect of salinity levels on the structure of zooplankton communities. Arch. Biol. Sci., **67**(2): 483-492.

RODDIE, B. D., R. J. G. LEAKEY & A. J. BERRY, 1984. Salinity-temperature tolerance and osmoregulation in *Eurytemora affinis* (Poppe) (Copepoda: Calanoida) in relation to its distribution in the zooplankton of the upper reaches of the Forth Estuary. Journal of Experimental Marine Biology and Ecology, **79**(2): 191-211. DOI:10.1016/0022-0981(84)90219-3.

ROOHI, A., Z. YASIN, A. E. KIDEYS, A. T. HWAI, A. G. KHANARI & E. EKER-DEVELI, 2008. Impact of a new invasive ctenophore (*Mnemiopsis leidyi*) on the zooplankton community of the Southern Caspian Sea. Marine Ecology, **29**: 421-434.

SEMENOVA, A. D. (ed.), 1977. Guideline for chemical analysis of surface water: 1-541. (Gidrometeoizdat, Leningrad [now St. Petersburg]). [In Russian.]

SEMENOVA, V. S., A. S. SEMENOVA & O. A. DMITRIEVA, 2010. Seasonal dynamics and trophic relationships of phyto- and zooplankton in the Vistula Lagoon of the Baltic Sea. Ecology of aquatic invertebrates. In: Collection of materials of the international conference dedicated to the 100th anniversary of the birth of F. D. Mordukhay-Boltovsky: 289-291. (IBVV RAS, Borok).

SKOLSKAYA, E. & V. UVAROV, 2018. Natural conditions of the North-East Caspian. In: V. A. SKOLSKY (ed.), Environmental monitoring studies of the environment of the Northeast Caspian Sea during the development of oil fields by NCOC N.V. in the period from 2006 to 2016: 33-57. (NCOC N.V., KAPE, Almaty). [In Russian.]

SLIVINSKY, G. G. & E. G. KRUPA, 2013. The current state of Teniz-Korgalzhyn lakes according to hydrochemical and toxicological variables. Bulletin of the Kazakh National University, (Environmental Series) **1**(37): 74-81. [In Russian.]

SLIVINSKY, G. G., E. G. KRUPA & G. ZH. AKBERDINA, 2009. Characterization of the Nura river basin in the zone of influence of the Temirtau-Karaganda industrial complex according to hydrochemical and toxicological variables. Bulletin of the Kazakh National University, (Environmental Series) **3**(26): 82-90. [In Russian.]

SOUISSI, A., S. SOUISSI & J.-S. HWANG, 2016. Evaluation of the copepod *Eurytemora affinis* life history response to temperature and salinity increases. Zoological Studies, **55**: art. 4, 1-9. DOI:10.6620/ZS.2016.55-04.

STUGE, T. S., 1995. About zooplankton of salt lakes of Kazakhstan. Reservoirs of Pavlodar and Kokshetau regions. Kazakhstan Zoological Journal “Selevinia”, **1**: 35-41. [In Russian.]

STUGE, T. S., 2000. On the autumn zooplankton zone of influence of the Semipalatinsk test site. Bulletin of NNC Rk. Radioecology, (Environmental Protection) **3**: 73-77. [In Russian.]

STUGE, T. S., 2007. The species composition and quantitative characteristics of the zooplankton of Lake Teniz. In: Proceedings of the international scientific conference “Biological diversity of the Asian steppes”, Kostanay: 291-295. [In Russian.]

STUGE, T. S., 2008a. Studies of the zooplankton of the Uyaly-Shalkar and Shalkar-Birtaban lakes of the r. Nury. In: Proceedings of the Korgalzhyn State Nature Reserve. Korgalzhyn: 44-64. (Korgalzhyn State Nature Reserve, Korgalzhyn). [In Russian.]

STUGE, T. S., 2008b. Summer zooplankton of water bodies of the Semipalatinsk testing range based on the materials of 2001. Tethys Aqua Zoological Research, **4**: 93-104. [In Russian.]

STUGE, T. S. & E. S. BARTOSH, 2003. Species diversity of planktonic crustaceans in ponds of the river basin Nura and assessment of the state of the community and water quality based on 1998 materials. Tethys Aqua Zoological Research, **2**: 73-91. [In Russian.]

STUGE, T. S., S. A. MATMURATOV & E. G. KRUPA, 2002. Mesozooplankton of the Nura River in the zone of industrial pollution (summer 2000). In: A. M. MELDEBEKOV (ed.), Zoological research in Kazakhstan: current status and prospects. Materials of the international scientific conference: 186-188. (Institute of Zoology, Almaty). [In Russian.]

SUKHIKH, N. M. & V. R. ALEKSEEV, 2013. *Eurytemora caspica* sp. nov. from the Caspian Sea – one more new species within the *E. affinis* complex (Copepoda: Calanoida: Temoridae). Proceedings of the Zoological Institute RAS, **317**(1): 85-100.

SUKHIKH, N., A. SOUISSI, S. SOUISSI, A.-C. HOLL, N. V. SCHIZAS & V. ALEKSEEV, 2019. Life in sympatry: coexistence of native *Eurytemora affinis* and invasive *Eurytemora carolleeae* in the Gulf of Finland (Baltic Sea). Oceanologia, **61**: 227-238. DOI:10.1016/j.oceano.2018.11.002.

TORKE, B., 2001. The distribution of calanoid copepods in the plankton of Wisconsin lakes. Hydrobiologia, **453**: 351-365. DOI:10.1023/A:1013185916287.

UTESHEV, A. S. (ed.), 1959. The climate of Kazakhstan: 1-370. (Hydrometeorological Publishing House, Leningrad [now St. Petersburg]). [In Russian.]

VIITASALO, M., T. KATAJISTO & I. VUORINEN, 1994. Seasonal dynamics of *Acartia bifilosa* and *Eurytemora affinis* (Copepoda: Calanoida) in relation to abiotic factors in the northern Baltic Sea. Hydrobiologia, **292**(293): 415-422.

WINBERG, G. G. & G. M. LAVRENTEVA (eds.), 1984. Zooplankton and its products. Guidelines for the collection and processing of materials in hydrobiological research in freshwater water bodies: 1-34. (GosNIORH, Leningrad [now St. Petersburg]). [In Russian.]

ZHEVAGO, V. S. (ed.), 1989. Hydrogeochemistry of Kazakhstan: 1-190. (Science of the Kazakh SSR, Almaty). [In Russian.]

First received 3 May 2019.
Final version accepted 20 September 2019.

[When citing this volume, please refer to *Crustaceana* 93 (2020), 241-547]

*Proceedings of the* Eurytemora *conference, St. Petersburg, 2019*

# SALINITY EFFECTS ON EGG PRODUCTION, HATCHING, AND SURVIVAL OF *EURYTEMORA AFFINIS* (COPEPODA, CALANOIDA)

BY

LAURI KUISMANEN[1]), LOUISE FORSBLOM[1]), JONNA ENGSTRÖM-ÖST[2,4]), ULF BÅMSTEDT[3]) and OLIVIER GLIPPA[2])

[1]) Åbo Akademi University, Environmental and Marine Biology, Faculty of Science and Engineering, Åbo, Finland

[2]) Novia University of Applied Sciences, Ekenäs, Finland

[3]) Umeå Marine Sciences Centre, University of Umeå, Norrbyn, Hörnefors, Sweden

## ABSTRACT

Salinity is an important biodiversity regulating factor in the Baltic Sea, forming a physiological dispersal barrier for species. The salinity in the Baltic Sea has been predicted to decline due to increased precipitation and fewer saline water inflows from the ocean. This causes stress to species already living on the edge of their tolerances and can alter species compositions and interactions in ecosystems. Calanoid copepod resting eggs, originating from a known egg bank on the seabed in the western Gulf of Finland, were incubated in the laboratory. We monitored the hatching of the calanoid copepods *Acartia* sp. and *Eurytemora affinis*, as well as the survival to maturity of hatched *Eurytemora affinis* nauplii in salinities ranging from 0 to 25. Further, we also investigated salinity-related effects on body size and egg production. Based on the results of our generalized linear mixed model, peak hatching occurred within the salinity range 5-20 (6.3 at the study site). Body size was not affected by salinity and most eggs were produced in salinities of 5, 7.5 and 15. The results suggest that *E. affinis* lives on the edge of an optimal salinity and that a decline of salinity could affect the fitness of the local populations of the species.

Key words. — Egg production, hatching success, survival to adulthood, resting eggs

## RÉSUMÉ

La salinité est un important facteur de régulation de la biodiversité dans la mer Baltique, formant une barrière de dispersion physiologique pour les espèces. Il a été prévu une diminution de la salinité de la mer Baltique en raison de l'augmentation des précipitations et de la diminution des apports d'eau salée provenant de l'océan. Cela peut provoquer un stress pour les espèces qui vivent déjà à la limite de leur tolérance et peut modifier la composition des espèces et les interactions dans les écosystèmes. Des œufs de résistance de copépodes calanoïdes, provenant d'une banque d'œufs connu sur le fond marin dans l'ouest du golfe de Finlande, ont été incubés en laboratoire. Nous avons

[4]) Corresponding author; e-mail: jonna.engstrom-ost@novia.fi

 | DOI:10.1163/9789004465565_013

surveillé l'éclosion des copépodes calanoïdes *Acartia* sp. et *Eurytemora affinis*, ainsi que la survie des nauplii *E. affinis* jusqu'à l'âge adulte dans des salinités allant de 0 à 25. En outre, nous avons également étudié les effets de la salinité sur le taille du corps et la production d'œufs. D'après les résultats de notre modèle linéaire mixte généralisé, le pic d'éclosion s'est produit dans la plage de salinité de 5 à 20 (6,3 sur le site de l'étude). La taille du corps n'a pas été affectée par la salinité et la plupart des œufs ont été produits à des salinités de 5, 7,5 et 15. Les résultats suggèrent qu'*E. affinis* vit à la limite d'une salinité optimale et qu'une baisse de la salinité pourrait affecter la valeur adaptative des populations locales de l'espèce.

## INTRODUCTION

Climate change is regionally leading to increased precipitation, which, via higher river runoff, affects the estuarine salinity level in many coastal zones (Graham, 2004; Andersson et al., 2015; Thompson et al., 2015). Increased precipitation-caused river runoff affects the inter-annual variability in surface salinity along coastlines globally, e.g., in the Arabian Sea, the Arctic Ocean, the Gulf of Mexico, the North Sea and the Norwegian Sea (Blindheim et al., 2000; Kim et al., 2014; Núñez-Riboni & Akimova, 2017; Khare, 2018; Tang et al., 2018). Enclosed brackish-water basins, such as the Baltic Sea, are also dependent on saline water inflows from the ocean, as salinity is an important regulating factor for the ecosystem (Leppäkoski et al., 2002). Major marine inflows have become more unpredictable and less frequent, down to circa one inflow per decade (Mohrholz et al., 2015), due to the anomalous barotropic pressure gradient that hampers saltwater inflows (Schrum, 2001).

The Baltic Sea will likely be subject to profound changes in hydrography in the future, with important consequences for both taxonomic and functional biodiversity to be expected (Ojaveer et al., 2010; Mäkinen et al., 2017). Especially organisms living in brackish conditions on the edge of their salinity tolerance, are sensitive to freshening, as salinity can affect reproductive output, oxygen consumption, feeding, growth and survival (Lee & Petersen, 2002; Beyrend-Dur et al., 2009; Devreker et al., 2009; Lee, 2016). Copepod populations have regionally declined in the Baltic Sea, and this phenomenon has been suggested to be due to eutrophication, warming or overfishing, leading to an increased abundance of planktivores (Suikkanen et al., 2013; Kuosa et al., 2017).

Among the copepod populations, *Eurytemora affinis* (Poppe, 1880) is a common euryhaline grazer in many estuaries and other brackish watersheds throughout the Northern Hemisphere, withstanding a large range of salinities (Lee, 1999; Devreker et al., 2009). The species has a central role in estuarine zooplankton communities globally (Gasparini et al., 1999; Winkler et al., 2005), and its tolerance to different salinities as well as its ability to invade freshwater habitats are widely studied (reviewed by Lee, 2016). While the species has been shown to persist in many communities, Karlsson et al. (2018) found little variation between genotypes in terms of response to salinity, implying that adaptation can be difficult, and a salinity decrease could thus reduce fitness. The species produces resting eggs that are deposited in the sediment and the hatching of the eggs is at least partially influenced by temperature (Katajisto et al., 1998). As resting eggs can survive multiple years in the sediment, this creates a reservoir of individuals adapted to fluctuating environmental conditions (Viitasalo & Katajisto, 1994). These resting stages can be important for the population, as overlapping generation times can provide migrants from the past, and thus influence adaptation (Yamamichi et al., 2019). As salinity, apart from temperature, is one of the most important factors regulating the zooplankton diversity in the Baltic Sea (Ojaveer et al., 2010), salinity can be expected to influence the species. Therefore, the current work aimed to study the effect of a salinity gradient from 0 to 25, using calanoid copepod resting eggs from the sediment. After collection, these eggs (and the individuals they produced) were tested on hatching success, survival to maturity, body size, and egg production. We expected *Eurytemora* to be best adapted to intermediate salinities (Beyrend-Dur et al., 2009; Devreker et al., 2009), and that its fitness will decrease above and below the optimum.

## MATERIAL AND METHODS

### Sampling

Sampling took place in a pelagic area close to Tvärminne Zoological Station (TZS) in the western Gulf of Finland (59°51′21″N 23°15′56″E) in March 2017. Resting eggs of *Eurytemora affinis* and other calanoid copepods have earlier been studied at this site (Katajisto et al., 1998; Katajisto, 2006). The depth is 33 m, and most of the water column consists of surface water of the open Baltic proper (Katajisto et al., 1998). Surface salinity has varied between 5.8 and 7.3 since 1927 (Rousi et al., 2013). The water temperature at the bottom can be below 1°C during winter (CTD measurement; this study). Salinity and temperature profiles were measured using a CTD probe (Citadel CTD-NV, Teledyne RD Instruments, Poway, CA, U.S.A.). Two sediment samples were collected using an Ekman grab, pooled, and stored at 4°C.

### Extraction of resting eggs

The sediment was thoroughly mixed before extracting 5 ml of sediment for each replicate and the replicates were subsequently weighed (Sartorius CP32025 scale). Resting eggs were extracted as described in Glippa et al. (2011) according to the method of Onbé (1978). During the egg extraction process, the temperature was kept at 4°C to avoid hatching. The sediment was sieved through 250 $\mu$m and 45 $\mu$m mesh sieves, using water of the treatment-specific salinity, before washing remaining particles into centrifugation tubes, using a sugar solution (1.43 : 1) and centrifuged for 5 min at 3000 rpm and 4°C. The supernatant containing the eggs was rinsed through a 45 $\mu$m sieve using water of treatment-specific salinity to remove the sugar. The eggs were counted, and the type determined from the remaining material, using a counting chamber (Bogorov design tray) with filtered seawater.

## Preparation of salinity treatments

Water was produced for the eight different salinities between 0 and 25. Fresh water was collected from Gennarbyviken Bay, an embayment that previously was connected to the sea (Lignell, 2015). The fresh water was filtered through a 0.45 $\mu$m filter (Sartobran 300 sterile capsule, Sartorius Stedim Biotech, Göttingen, Germany), using a peristaltic pump (Heidolph Pumpdrive 5006). Sterile filtered seawater of salinity 5 from Tvärminne Zoological Station was used to produce salinity 2.5-25 water. The salt was a commercial sea salt mix (Instant Ocean® synthetic sea salt, Aquarium Systems, Loessnitz, Germany). The desired salinity of the water was obtained either by diluting with fresh water or by adding salt to TZS water (salinity 5) and monitored by using a salinometer (WTW inoLab Cond 720). Salinity in the washing bottles was measured before use and adjusted when needed. Water was kept cold and stored in the dark at approx. 10°C. Some of the water was stored at the same temperature as in the climate chamber where the zooplankton was held, to avoid temperature stress when adding or changing the water in the zooplankton containers.

## Experimental set-up

Copepods were fed with a mixture of high-quality algae (Engström-Öst et al., 2017) *Rhinomonas nottbecki* Majaneva, 2014 (07B6) and *Brachiomonas submarina* Bohlin, 1897 (TV15), grown in a temperature-controlled climate chamber in 2-litre Erlenmeyer glass flasks, using an f/2 medium with silicate omitted, according to Guillard (1975). Both cultures were non-axenic batch monocultures and kept at 17°C under a 16 : 8 h light : dark cycle and 14 $\mu$mol photons/m$^2$ per s in seawater of salinity 6. Cell concentrations in the cultures were determined twice a week by microscopy (Leica DMIL, using a 10 $\times$ 12.5 magnification) with a Bürker chamber. Before nauplii reached the copepodite stage, 200 $\mu$l of algae was added to the copepod containers. As the copepods developed, the dose was increased to 400 $\mu$l. The mean algae concentration was 12 200 $\pm$ 2271 cells/$\mu$l.

## Hatching, survival and reproduction

Eggs (numbers ranging between 21-59) of each replicate were incubated in 20 ml crystallizing dishes creating 5 replicates for each salinity treatment (table I). The dishes were placed on a tray, according to treatment, and covered with plastic film (Parafilm® M, Bemis, Oshkosh, WI, U.S.A.), to minimize evaporation. The dishes were then stored in a climate chamber at 18.6 ± 0.3°C, in a 16 : 8 h light : dark cycle, at a light intensity of 24.14 ± 4.43 $\mu$mol/m$^2$ per s. The copepod

TABLE I

Hatching of nauplii from *Eurytemora affinis* (Poppe, 1880) and *Acartia* sp. Eggs, extracted from the sediment (eggs not distinguished between the two species), and number of adult survivors of *E. affinis* from hatched eggs

| Salinity (Mean ± SD for replicates 1-3 and 4-5) | Eggs (Mean ± SD) | Hatched (Mean ± SD) | Adult *E. affinis* of replicates | | | | |
|---|---|---|---|---|---|---|---|
| | | | 1 | 2 | 3 | 4 | 5 |
| 0<br>1-3: 0 ± 0<br>4-5: 0 ± 0 | 28.4 ± 3.6 | 11.6 ± 2.6 | | | 2♀2♂ | | |
| 2.5<br>1-3: 2.6 ± 0.1<br>4-5: 2.6 ± 0.04 | 35.8 ± 4.4 | 24.2 ± 1.6 | | 5♂ | 4♀2♂ | 2♀1♂ | 4♀1♂ |
| 5<br>1-3: 5.1 ± 0.1<br>4-5: 5.1 ± 0.1 | 36.4 ± 5.3 | 33.4 ± 4.6 | 2♀3♂ | 3♀2♂ | 4♀5♂ | | 3♀4♂ |
| 7.5<br>1-3: 7.7 ± 0.2<br>4-5: 7.8 ± 0.2 | 36.2 ± 13.8 | 29.6 ± 10.6 | 2♀4♂ | 3♀2♂ | 3♀6♂ | 3♂ | |
| 10<br>1-3: 10.1 ± 0.1<br>4-5: 10.1 ± 0.1 | 44.2 ± 8.2 | 37 ± 5.7 | 6♀3♂ | 2♀2♂ | 3♀4♂ | 1♀5♂ | 5♀2♂ |
| 15<br>1-3: 15.2 ± 0.2<br>4-5: 15.2 ± 0.2 | 33 ± 6.2 | 29 ± 6.0 | 2♀2♂ | 1♂ | 5♀5♂ | | 3♀4♂ |
| 20<br>1-3: 20.2 ± 0.2<br>4-5: 20.2 ± 0.2 | 24.2 ± 4.4 | 20.4 ± 6.8 | 1♂ | | | | |
| 25<br>1-3: 25.2 ± 0.2<br>4-5: 25.2 ± 0.3 | 27.6 ± 2.4 | 17.8 ± 5.3 | | | | 1♂ | |

Eggs, eggs extracted from the sediment; Hatched, hatched individuals from eggs (hatching success of *E. affinis* and *Acartia* sp. eggs); Adults, *E. affinis* adults, survivors to adulthood from Hatched of replicates 1-5.

eggs were repositioned on the shelf once per day to ensure similar conditions. The zooplankton eggs were monitored daily using stereo-microscopes (Leica MZ12 and Leica M125) for the first ten days, and subsequently every two days, until the experiment was terminated after 22 $\pm$ 4 days. During the monitoring, the water in the dishes was changed and the salinity was measured by pooling water from replicates 1-3 and 4-5 (due to small water amount, table I). Then we calculated a weighted mean salinity for each treatment that was used in the subsequent analysis. All hatched nauplii of each replicate were transferred to 700-ml glass jars, pre-filled with 100 ml water of the corresponding treatment. As both the animal number and their size increased, the amount of water was increased close to full volume. As the eggs originated from the sediment in the natural environment, the total numbers of calanoid eggs (*E. affinis* and *Acartia* sp.) were counted per sample.

*E. affinis* survival was monitored approximately once a week, to observe the number of survivors. Salinity was measured from each jar, and all, or part of the water, was changed during the check. An average salinity value was calculated for each replicate during the survival experiment. The sex of adult *E. affinis* was also determined. Algae were added to the jars daily (as described previously). The jars were repositioned on the shelf daily and covered with plastic film (Parafilm® M). The prosome length was measured from the anterior tip of the cephalosome to the posterior end of the metasome, i.e., up to the joint with the first abdominal somite, using the software ImageJ (Version 1.51j8; Abràmoff et al., 2004).

Individual *E. affinis* females and males from the same replicate were paired and transferred into false-bottom chambers (120 $\mu$m mesh size) immersed in separate beakers and filled with water of the corresponding salinity for the egg production (EP) experiment. This was done for each female in every replicate where at least one female and male had survived (table I). The beakers were covered with Parafilm to avoid evaporation and food was supplied daily. The females were preserved in a 4% formalin solution after the termination of the experiment and subsequently measured as previously described. The EP was monitored once or twice daily and during these checks, all or part of the water was changed to keep the salinity constant. As some treatments only had few survivors and no males, the EP was only investigated for salinities 0-15. As the resulting number of females was low, and as the females sometimes originated from the same replicate (according to the original experimental set-up with 5 replicates), the results from the EP experiment are only presented visually (precluded from formal statistical analysis). The EP was standardized to the size of the female by dividing the EP with the prosome length of the female. We also investigated the relationship between EP and the prosome length of the female.

## Data analyses

We studied both the hatching success and survival from nauplii to sexual maturity along a salinity gradient using separate, generalized linear mixed models. The response variables in the two analyses were hatching success (hatched, not hatched) and survival (alive, dead) respectively, using a logit link function. Both models were fitted using maximum likelihood estimation, using the 'glmer' function from the 'lme4' package (Bates et al., 2015). The continuous salinity variables were standardized to zero by demeaning in both analyses. To account for the expected quadratic relationship in the salinity response observed in a pilot study (fig. 1; and see below), both salinity and squared salinity were used as predictors. To account for overdispersion, the unique ID was added as an observation level random effect in both analyses, and the hatching analysis additionally had a random effect of treatment level included. Initially, the random effect of treatment was included in both analyses, but it led to a singular fit in the survival analysis and was dropped. We used a likelihood ratio test for both models, comparing them to null models without salinity terms, to investigate the significance of the salinity

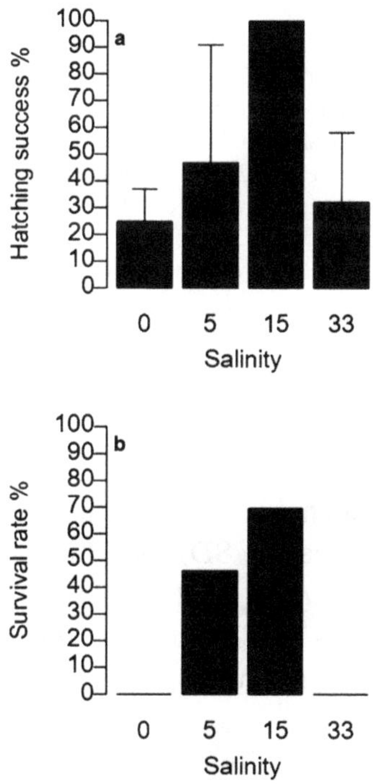


Fig. 1. a, Hatching success (%) of resting calanoid copepod eggs (*Eurytemora affinis* (Poppe, 1880) and *Acartia* sp.); and, b, survival to adulthood (%) of *E. affinis* after 20 days; in the pilot study, in the northern Bothnian Sea. Error bars are standard deviations within each treatment. (See "Supplementary information" for details on methodology.).

terms. Statistical significance was defined at $p<0.05$ in all analyses and the 95% confidence intervals for the model predictions were calculated using a bootstrap procedure with 10 000 resampling events. Model fit was investigated visually by looking at quantile-quantile graphs of Dunn-Smythe residuals.

We also investigated the salinity effect on the prosome length of adult individuals using a linear model. We used the deviance from mean prosome length within sexes as the response variable. The predictor variables were salinity, squared salinity, and the sex. All analyses were conducted using R version 3.4.2 (R Core Team, 2017).

### Supplementary information

Sediment samples were taken for the pilot study on 11 May 2012 in the Öre River (northern Bothnian Sea, depth 18 m) using a modified Kajak gravity corer. The upper oxygenated sediment was removed and kept at 5°C in darkness. One month later, 12 subsamples were taken (1.43 $cm^3$/subsample) and the extraction procedure was performed in triplicate for treatments of salinity 0, 5, 15 and 33. Filtered UV-treated seawater (1 $\mu$m) and autoclaved fresh water were mixed to produce waters of different salinities for extraction and incubation of resting eggs. Subsamples of sediment were washed through a 33 $\mu$m mesh using waters of the mentioned salinities. The pH was adjusted to 7.0 in all units. The residue remaining on the sieve was transferred into centrifuge tubes with saturated solution (1 : 1) of sucrose and water (salinity 0, 5, 15 or 33). Centrifugation was done twice at 5000 rpm for 6 min. The supernatant was rinsed on a 33 $\mu$m sieve to remove the sugar. Calanoid copepod eggs were quantified under a stereoscopic microscope and transferred into a 5-ml crystallizing dish. Eggs were incubated under optimal conditions (16°C, 16 : 8 h light : dark) to determine hatching and next the species. Hatching was monitored every 2 days; nauplii were identified and transferred into one 2-litre beaker per treatment. The copepods were fed at a concentration of approx. 20 000 cells/ml using *Rhodomonas minuta* Skuja, 1948 and *R. marina* (P. A. Dangeard) Lemmermann, 1899. The experiment was terminated 18 days after the extraction, and all copepods were checked (fig. 1).

## RESULTS

### Hatching success

The average number of calanoid eggs (*Acartia* sp. and *Eurytemora affinis*) per sample was 33.23 ± 9.23 (mean ± SD), of which 25.35 ± 10.03 hatched, and 3.20 ± 3.25 developed into *E. affinis* adults (table I). The mean temperature during the hatching experiment was ~18°C. In total, 129 *E. affinis* individuals survived: 59 females and 70 males. The mean temperature in the survival experiment was 18.6 ± 0.14 °C.

There was a clear effect of salinity on hatching success (table II) as indicated by the likelihood ratio test (LRT = 10.525, df = 1, 3, $p$ = 0.005). Both salinity terms in the model were significant, thus indicating a quadratic effect of salinity as expected. The hatching success was relatively high, between 70 and 90%, over a broad range of salinities from 5 to 20, below and above which it decreases (fig. 2a).

TABLE II

Parameter estimates, standard errors (SE), $z$-test statistics and statistical significance ($p$) for the fixed effects of the generalized linear mixed models of the hatching success and survivorship used in the present study on hatching success and subsequent survival of *Eurytemora affinis* (Poppe, 1880) eggs and hatchlings

| Fixed effects | Estimate | SE | $z$-value | $p$ |
|---|---|---|---|---|
| Hatching success | | | | |
| (Intercept) | 2.133 | 0.236 | 9.022 | <0.001*** |
| Salinity | 0.069 | 0.020 | 3.376 | <0.001*** |
| Salinity$^2$ | −0.012 | 0.003 | −4.665 | <0.001*** |
| Survival to adulthood | | | | |
| (Intercept) | −1.681 | 0.231 | −7.283 | <0.001*** |
| Salinity | −0.070 | 0.030 | −2.336 | 0.01948* |
| Salinity$^2$ | −0.014 | 0.004 | −3.279 | 0.00104** |

*** $p<0.001$, ** $p<0.01$, * $p<0.05$.

## Survival to adulthood

The salinity also influenced the survival to adulthood (table II) as indicated by the likelihood ratio test (LRT = 21.113, df = 1, 3, $p<0.001$); however, the overall survival remained low, generally well below 20% (fig. 2b). Compared to the hatching analysis, the survival success levelled off already above salinities of 10 and subsequently remained low. Based on the model prediction, the highest survival from nauplii to adult occurred at salinities between 5 and 10 (fig. 2b). The observed survival displayed relatively large variation, ranging from approximately 0 to 30% in salinities 2.5-7.5 and 15.

## Prosome length and clutch size

There was no clear effect of salinity on the prosome length of adult *E. affinis* after controlling for the sex, $F_{(3, 82)} = 0.4841$, $p = 0.6942$. Females were larger than males within the same treatments. The mean prosome length (as defined above) of female and male *E. affinis* was 743.8 ± 53.9 $\mu$m ($n$ = 39), and 669.8 ± 42.9 $\mu$m ($n$ = 47), respectively. As we detected a positive tendency ($p = 0.06$) between the female prosome length and the number of eggs produced, we standardized the EP to the length of the female (fig. 3). Comparatively high numbers of eggs were produced in salinities of 5, 7.5 and 15. Since we had only one female in the 7.5 salinity treatment, the EP obtained (0.027, EP/prosome length) was lower than the highest observed EP within the 15 salinity treatments (0.032, EP/prosome length).

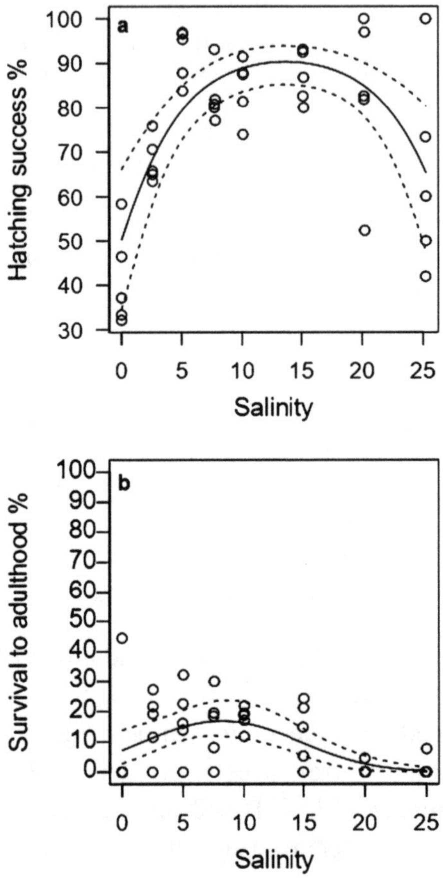


Fig. 2. The expected relationships, a, between salinity and hatching success (%) of calanoid resting eggs (*Eurytemora affinis* (Poppe, 1880) and *Acartia* sp.); and, b, between salinity and survival to adulthood (%) of *E. affinis*, as predicted using the respective models (solid lines). Presented with 95% confidence intervals (dashed lines) and the raw data on hatching and survival percentages (open circles).

## DISCUSSION

This study, focusing on *Eurytemora affinis*, demonstrates how hatching success and survival to adulthood of calanoid copepods are affected by different salinities for individuals originating from the sediment. The salinity during sampling was 6.3 and has been measured since 1927 in the Tvärminne area, varying between 5.8 and 7.3 (Rousi et al., 2013). Nevertheless, salinity is predicted to decrease in the study area before the year 2100 (Meier et al., 2012). Based on the model predictions, our results indicate that the optimal salinity conditions are 5-20 for hatching (*E. affinis*, *Acartia* sp.), and 5-10 for survival (*E. affinis*; fig. 2, table II), including the salinity level at the sampling site. We did not find an effect of salinity on the prosome

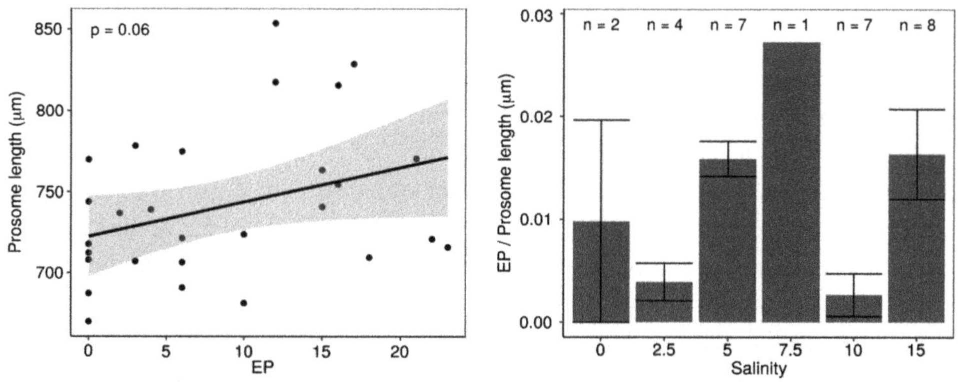


Fig. 3. The relationship between egg production and female prosome length in *Eurytemora affinis* (Poppe, 1880): Left panel, the relationship between egg production (EP) and female prosome length in $\mu$m; right panel, the average number of eggs produced in the first clutch (EP), divided by the prosome length in $\mu$m, in relation to salinity. The error bars show standard errors, and n indicates the number of individuals.

length of *E. affinis*. The largest clutch sizes (*E. affinis*) were also produced within a wide salinity range, i.e., in the 5, 7.5 and 15 salinity treatments, and there was a positive correlation between egg production and female size (fig. 3). Clutch size dropped at salinity 10. This is difficult to explain apart from stochastic processes, as experimental conditions were similar in all treatments.

## Hatching success

Our results suggest that salinity 5 and higher, up to salinity 20, are beneficial for the hatching success of *E. affinis* and *Acartia* sp. (fig. 2a). The hatching peak salinities include salinity levels in the Kattegat, Belt Sea, and Arkona Sea in the south-western Baltic Sea, comprising the transition zone between the North Sea and the Baltic Sea, after which the salinity gradient decreases further into the Baltic Sea. The hatching success, however, seems quite robust in the different salinity levels, compared to the survival to adulthood (fig. 2b). The hatching success varied between 32-100% and was higher than 80% in salinities between 5 and 20 in the current work. It is comparable to observations made in the Seine estuary, where all reproductive variables were negatively affected by salinity 25 (Devreker et al., 2009). Our results are supported by pilot data of hatching success (*E. affinis* and *Acartia* sp.) that peaked at salinity 15, even though the salinity in the region was 3 (off Umeå in the Bothnian Bay). Karlsson et al. (2018) found that salinities higher than the ambient in the Gulf of Riga (salinity 5), and the Bothnian Bay (salinity 3), were in general beneficial for the hatching success and survival of *E. affinis*. Hatching success in the current study started to decrease in the lowest salinity treatments (2.5 and 0) and suggests that *E. affinis* fitness could decline in a freshening Baltic Sea (cf. Karlsson et al., 2018).

## Survival

Survival to adulthood can be regarded as one of the most central components of an organism's fitness (Posavi et al., 2014), as it usually means that the organism is ready for reproduction. The survival to adulthood of *E. affinis* at different salinities is an indicator of its tolerance to the different salinities. The survival to adulthood was highest between salinities 5 and 15. The ambient salinity of 6.3 at the sampling site is thus at the lower end of the salinity range. Survival was highest in the salinity 15 treatment in a pilot study conducted in the Bothnian Bay, approximately 20% higher than in the salinity 5 treatment (fig. 1). Diekmann et al. (2012) reported that *E. affinis* is usually abundant in the Kiel Canal with salinity ranging between 4.9 and 12.4, but not in the Kiel Bight with salinities ranging from 11.9 to 21.4. Even though salinity is considered as the most important biodiversity-shaping factor for brackish animals (Ojaveer et al., 2010), many other factors, such as competition, food quantity and quality, predation pressure, and fecundity have a major effect on *E. affinis* survival to adulthood (Lee et al., 2013). Copepods tolerate environmental stress better when food is abundant compared to starving copepods. Lee et al. (2013) found that low salinity tolerance was significantly increased by high food concentrations. The slightly fluctuating quality of the *Rhinomonas nottbecki* culture during the experiment might have affected the survival of *E. affinis*, but as we fed the copepods with a mixture of two different algae, we assume they obtained enough nutrition at all times.

Lee et al. (2012) have shown that *E. affinis* regulates the osmolality of the haemolymph. In decreasing salinity, organisms must maintain essential ions against steep concentration gradients, and this can be energetically stressful and costly (Morris, 2001). During non-optimal salinity conditions, copepods experience osmotic stress, which affects their metabolism, respiration, moulting, and osmoregulation (Roddie et al., 1984; Devreker et al., 2004; Beyrend-Dur et al., 2009), which may manifest as reduced egg production (Mauchline, 1998; Diekmann et al., 2012). However, as *E. affinis* is originally an estuarine and salt marsh species that successfully has invaded fresh water repeatedly (Lee et al., 2016), it is expected to be well adapted in higher salinities. The developmental stages could also have different salinity tolerances (Devreker et al., 2004). Karlsson et al. (2018) found that the nauplii of *E. affinis* were more sensitive to lower salinities than the copepodites, suggesting that survival to adulthood is mainly determined by nauplii survival. Devreker et al. (2004) observed that only the most extreme salinities in their study, 0 and 35, had negative effects on nauplii survival of *E. affinis*, and that the optimal salinity was 15 at 15°C, whereas a lower temperature, 10°C, increased the mortality rates. Nauplii survived for longer, but development terminated at the second nauplius stage at 10°C (Devreker et al., 2004).

Our results show that *E. affinis* seems to perform better at salinities higher than ambient, but its fitness would probably decrease in decreasing salinities. This is also supported by the result of the pilot study conducted in the northernmost Baltic Sea (fig.1), where the species seems to perform better at salinities higher than ambient.

There was no effect of salinity on prosome length in *E. affinis*, which was supported by other studies and could be due to the euryhaline nature of this species (Katona, 1970; Souissi et al., 2016).

Salinity changes in the Baltic Sea are predicted to be the largest in the Danish straits region, and get gradually smaller further into the Baltic Sea, toward the Bothnian Bay and Gulf of Finland (Meier, 2015). Marine species are likely to suffer from the lower salinities, driving their distributions toward higher salinities in the southern parts of the Baltic Sea, while more freshwater-tolerant species are becoming more dominant, leading to changes in the food webs. Our results suggest that *E. affinis* benefits from ambient or higher salinity but could experience energetically costly osmotic stress in lower salinity, potentially leading to increased fitness costs, such as lower survival, as well as lower hatching and offspring numbers (cf. Karlsson et al., 2018).

Individuals inhabiting estuaries and coastal zones may experience rapid short-term changes in abiotic factors, such as salinity, pH, or temperature and sometimes the changes exceeding the normal ranges as predicted to occur due to climate change (Diekmann et al., 2012; Meier et al., 2012; Almén et al., 2014). This is considered to be one reason why copepods are fairly robust to environmental changes (Fabry et al., 2012; Lewis et al., 2013). Coastal species, such as *E. affinis*, are exposed to relatively wide environmental conditions, that potentially could increase the selective advantage of adaptation to future salinity decline (cf. Mäkinen et al., 2017). While this study explored responses from hatching to egg production by one generation under different salinities, future studies should focus on salinities close to current conditions but integrating a multigenerational approach.

## ACKNOWLEDGEMENTS

The pilot study is a contribution to the Danish National Strategic Research Council project IMPAQ grant no. 10-093522 and to the French Seine-Aval program. It was co-financed by the Copepod Selection WP of IMPAQ and ZOOSEINE and ZOOGLOBAL projects of GIP Seine-Aval, France. There are neither conflicts of interest, nor human or animal rights to declare. Sampling took place within the protected area of Tvärminne Zoological Station (University of Helsinki, Finland) with permission from the personnel. Two reviewers are thanked for valuable comments, and A. Lindén and S. Souissi for discussions. The data collection was supported by the Academy of Finland (project nr. 276947), Onni Talaan säätiö, Oskar Öflunds stiftelse, Svenska Kulturfonden, Societas pro Fauna et Flora Fennica, Walter och Andrée de Nottbecks stiftelse, Waldemar von Frenckells stiftelse, Mary och Ragnar Numelins stipendiefond, and Bergsrådinnan Sophie von Julins stiftelse.

## REFERENCES

ABRÀMOFF, M. D., P. J. MAGALHÃES & S. J. RAM, 2004. Image processing with ImageJ. Biophotonics Int., **11**: 36-42. https://doi.org/10.1017/CBO9781107415324.004.

ALMÉN, A.-K., A. VEHMAA, A. BRUTEMARK & J. ENGSTRÖM-ÖST, 2014. Coping with climate change? Copepods experience drastic variations in their physicochemical environment on a diurnal basis. J. Exp. Mar. Biol. Ecol., **460**: 120-128. https://doi.org/10.1016/j.jembe.2014.07.001.

ANDERSSON, A., H. E. M. MEIER, M. RIPSZAM, O. ROWE, J. WIKNER, P. HAGLUND, K. EILOLA, C. LEGRAND, D. FIGUEROA, J. PACZKOWSKA, E. LINDEHOFF, M. TYSKLIND & R. ELMGREN, 2015. Projected future climate change and Baltic Sea ecosystem management. Ambio, **44**: 345-356. https://doi.org/10.1007/s13280-015-0654-8.

BATES, D., M. MÄCHLER, B. BOLKER & S. WALKER, 2015. Fitting linear mixed-effects models using lme4. J. Stat. Softw., **67**: 1-48. https://doi.org/10.18637/jss.v067.i01.

BEYREND-DUR, D., S. SOUISSI, D. DEVREKER, G. WINKLER & J.-S. HWANG, 2009. Life cycle traits of two transatlantic populations of *Eurytemora affinis* (Copepoda: Calanoida): salinity effects. J. Plankton Res., **31**: 713-728. https://doi.org/10.1093/plankt/fbp020.

BLINDHEIM, J., V. BOROVKOV, B. HANSEN, S. A. MALMBERG, W. R. TURRELL & S. ØSTERHUS, 2000. Upper layer cooling and freshening in the Norwegian Sea in relation to atmospheric forcing. Deep-Sea Res., (Part I, Oceanogr. Res. Pap.) **47**: 655-680. https://doi.org/10.1016/S0967-0637(99)00070-9.

DEVREKER, D., S. SOUISSI & L. SEURONT, 2004. Development and mortality of the first naupliar stages of *Eurytemora affinis* (Copepoda, Calanoida) under different conditions of salinity and temperature. J. Exp. Mar. Biol. Ecol., **303**: 31-46. https://doi.org/10.1016/j.jembe.2003.11.002.

DEVREKER, D., S. SOUISSI, G. WINKLER, J. FORGET-LERAY & F. LEBOULENGER, 2009. Effects of salinity, temperature and individual variability on the reproduction of *Eurytemora affinis* (Copepoda; Calanoida) from the Seine estuary: a laboratory study. J. Exp. Mar. Biol. Ecol., **368**: 113-123. https://doi.org/10.1016/j.jembe.2008.10.015.

DIEKMANN, A. B. S., C. CLEMMESEN, M. A. ST. JOHN, M. PAULSEN & M. A. PECK, 2012. Environmental cues and constraints affecting the seasonality of dominant calanoid copepods in brackish, coastal waters: a case study of *Acartia*, *Temora* and *Eurytemora* species in the south-west Baltic. Mar. Biol., **159**: 2399-2414. https://doi.org/10.1007/s00227-012-1955-0.

ENGSTRÖM-ÖST, J., N. BARRETT, A. BRUTEMARK, A. VEHMAA, A. DWYER, A.-K. ALMÉN & B. T. DE STASIO, 2017. Feeding, survival, and reproduction of two populations of *Eurytemora* (Copepoda) exposed to local toxic cyanobacteria. J. Great Lakes Res., **43**: 1091-1100.

GASPARINI, S., J. CASTEL & X. IRIGOIEN, 1999. Impact of suspended particulate matter on egg production of the estuarine copepod, *Eurytemora affinis*. J. Mar. Syst., **22**: 195-205.

GLIPPA, O., S. SOUISSI, L. DENIS & S. LESOURD, 2011. Calanoid copepod resting egg abundance and hatching success in the sediment of the Seine estuary (France). Estuar. Coast. Shelf Sci., **92**: 255-262. https://doi.org/10.1016/j.ecss.2010.12.032.

GRAHAM, L. P., 2004. Climate change effects on river flow to the Baltic Sea. Ambio, **33**: 235-241. https://doi.org/10.1579/0044-7447-33.4.235.

GUILLARD, R. L. L., 1975. Culture of phytoplankton for feeding marine invertebrates. In: W. L. SMITH & M. H. CHANLEY (eds.), Culture of marine invertebrate animals (1st ed.): 29-60. (Springer Verlag, New York, NY).

KARLSSON, K., S. PUIAC & M. WINDER, 2018. Life-history responses to changing temperature and salinity of the Baltic Sea copepod *Eurytemora affinis*. Mar. Biol., **165**: 30. https://doi.org/10.1007/s00227-017-3279-6.

KATAJISTO, T., 2006. Benthic resting eggs in the life cycles of calanoid copepods in the northern Baltic Sea: 1-46. (Ph.D. Thesis, Walter Andrée Nottbeck Found. Sci. Rep., University of Helsinki, Finland).

KATAJISTO, T., M. VIITASALO & M. KOSKI, 1998. Seasonal occurrence and hatching of calanoid eggs in sediments of the northern Baltic Sea. Mar. Ecol. Prog. Ser., **163**: 133-143. https://doi.org/10.3354/meps163133.

KATONA, S. K., 1970. Growth characteristics of the copepods *Eurytemora affinis* and *E. herdmani* in laboratory cultures. Helgoländ. Wiss. Meeresunters., **20**: 373-384. https://doi.org/10.1007/BF01609914.

KHARE, N., 2018. Evidence of increased rainfall prior to 3500 years BP as revealed by river borne terrigenous flux: a study from west coast of India. Quat. Int., **479**: 100-105. https://doi.org/10.1016/j.quaint.2017.05.055.

KIM, H., S. SON, P. MONTAGNA, B. SPIERING & J. NAM, 2014. Linkage between freshwater inflow and primary productivity in Texas estuaries: downscaling effects of climate variability. J. Coast. Res., **68**: 65-73.

KUOSA, H., V. FLEMING-LEHTINEN, S. LEHTINEN, M. LEHTINIEMI, H. NYGÅRD, M. RAATEOJA, J. RAITANIEMI, J. TUIMALA, L. UUSITALO & S. SUIKKANEN, 2017. A retrospective view of the development of the Gulf of Bothnia ecosystem. J. Mar. Syst., **167**: 78-92. https://doi.org/10.1016/j.jmarsys.2016.11.020.

LEE, C. E., 1999. Rapid and repeated invasions of fresh water by the copepod *Eurytemora affinis*. Evolution, **53**: 1423-1434. https://doi.org/10.2307/2640889.

LEE, C. E., 2016. Evolutionary mechanisms of habitat invasions, using the copepod *Eurytemora affinis* as a model system. Evol. Appl., **9**: 248-270. https://doi.org/10.1111/eva.12334.

LEE, C. E., W. E. MOSS, N. OLSON, K. F. CHAU, Y.-M. CHANG & K. E. JOHNSON, 2013. Feasting in fresh water: impacts of food concentration on freshwater tolerance and the evolution of food × salinity response during the expansion from saline into freshwater habitats. Evol. Appl., **6**: 673-689. https://doi.org/10.1111/eva.12054.

LEE, C. E. & C. H. PETERSEN, 2002. Genotype-by-environment interaction for salinity tolerance in the freshwater-invading copepod *Eurytemora affinis*. Physiol. Biochem. Zool., **75**: 335-344. https://doi.org/10.1086/343138.

LEE, C. E., M. POSAVI & G. CHARMANTIER, 2012. Rapid evolution of body fluid regulation following independent invasions into freshwater habitats. J. Evol. Biol., **25**: 625-633.

LEPPÄKOSKI, E., S. GOLLASCH, P. GRUSZKA, H. OJAVEER, S. OLENIN & V. PANOV, 2002. The Baltic a sea of invaders. Can. J. Fish. Aquat. Sci., **59**: 1175-1188. https://doi.org/10.1139/f02-089.

LIGNELL, A., 2015. Gennarbyviken — En undersökning av vattenkvalité och sedimenten 2014, samt en jämförelse med vattenkvalitén 1950-1970: 1-79. (B.Sc. Thesis, Novia University of Applied Sciences, Finland). [In Swedish.]

MÄKINEN, K., I. VUORINEN & J. HÄNNINEN, 2017. Climate-induced hydrography change favours small-bodied zooplankton in a coastal ecosystem. Hydrobiologia, **792**: 83-96. https://doi.org/10.1007/s10750-016-3046-6.

MAUCHLINE, J., 1998. The biology of calanoid copepods. Adv. Mar. Biol., **33**: i-x, 1-710. (Academic Press, San Diego, CA). https://doi.org/10.1086/393440.

MEIER, H. E. M., 2015. Projected change — marine physics. In: THE BACC II AUTHOR TEAM (eds.), Second assessment of climate change for the Baltic Sea Basin. Regional Climate Studies: 243-252. (Springer, Cham). https://doi.org/10.1007/978-3-319-16006-1_13.

MEIER, H. E. M., H. C. ANDERSSON, B. ARHEIMER, T. BLENCKNER, B. CHUBARENKO, C. DONNELLY, K. EILOLA, B. G. GUSTAFSSON, A. HANSSON, J. HAVENHAND, A. HÖGLUND, I. KUZNETSOV, B. R. MACKENZIE, B. MÜLLER-KARULIS, T. NEUMANN, S. NIIRANEN, J. PIWOWARCZYK, U. RAUDSEPP, M. RECKERMANN, T. RUOHO-AIROLA, O. P. SAVCHUK, F. SCHENK, S. SCHIMANKE, G. VÄLI, J. M. WESLAWSKI & E. ZORITA, 2012. Comparing reconstructed past variations and future projections of the Baltic Sea ecosystem — first results from multi-model ensemble simulations. Environ. Res. Lett., **7**: 034005. https://doi.org/10.1088/1748-9326/7/3/034005.

MOHRHOLZ, V., M. NAUMANN, G. NAUSCH, S. KRÜGER & U. GRÄWE, 2015. Fresh oxygen for the Baltic Sea — an exceptional saline inflow after a decade of stagnation. J. Mar. Syst., **148**: 152-166. https://doi.org/10.1016/j.jmarsys.2015.03.005.

MORRIS, S., 2001. Neuroendocrine regulation of osmoregulation and the evolution of air-breathing in decapod crustaceans. J. Exp. Biol., **204**: 979-989.

NÚÑEZ-RIBONI, I. & A. AKIMOVA, 2017. Quantifying the impact of the major driving mechanisms of inter-annual variability of salinity in the North Sea. Prog. Oceanogr., **154**: 25-37. https://doi.org/10.1016/j.pocean.2017.04.004.

OJAVEER, H., A. JAANUS, B. R. MACKENZIE, G. MARTIN, S. OLENIN, T. RADZIEJEWSKA, I. TELESH, M. L. ZETTLER & A. ZAIKO, 2010. Status of biodiversity in the Baltic Sea. PLoS One, **5**: e12467. https://doi.org/10.1371/journal.pone.0012467.

ONBÉ, T., 1978. Sugar flotation method for sorting the resting eggs of marine cladocerans and copepods from sea-bottom sediment. Bull. Jap. Soc. Scient. Fish., **44**: 1411.

POSAVI, M., G. W. GELEMBIUK, B. LARGET & C. E. LEE, 2014. Testing for beneficial reversal of dominance during salinity shifts in the invasive copepod *Eurytemora affinis*, and implications for the maintenance of genetic variation. Evolution, **68**: 3166-3183. https://doi.org/10.1111/evo.12502.

R CORE TEAM, 2017. R: a language and environment for statistical computing. (R Foundation for Statistical Computing, Vienna, Austria). https://www.R-project.org/.

RODDIE, B. D., R. J. G. LEAKEY & A. J. BERRY, 1984. Salinity-temperature tolerance and osmoregulation in *Eurytemora affinis* (Poppe) (Copepoda: Calanoida) in relation to its distribution in the zooplankton of the upper reaches of the Forth Estuary. J. Exp. Mar. Biol. Ecol., **79**: 191-211.

ROUSI, H., A. O. LAINE, H. PELTONEN, P. KANGAS, A.-B. ANDERSIN, J. RISSANEN, E. SANDBERG-KILPI & E. BONSDORFF, 2013. Long-term changes in coastal zoobenthos in the northern Baltic Sea: the role of abiotic environmental factors. ICES J. Mar. Sci., **70**: 440-451. https://doi.org/10.1093/icesjms/fss197.

SCHRUM, C., 2001. Regionalization of climate change for the North Sea and Baltic Sea. Clim. Res., **18**: 31-37. https://doi.org/10.3354/cr018031.

SOUISSI, A., S. SOUISSI & J.-S. HWANG, 2016. Evaluation of the copepod *Eurytemora affinis* life history response to temperature and salinity increases. Zool. Stud., **4**. https://doi.org/10.6620/ZS.2016.55-04.

SUIKKANEN, S., S. PULINA, J. ENGSTRÖM-ÖST, M. LEHTINIEMI, S. LEHTINEN & A. BRUTEMARK, 2013. Climate change and eutrophication induced shifts in northern summer plankton communities. PLoS One, **8**: e66475. https://doi.org/10.1371/journal.pone.0066475.

TANG, W., S. YUEH, D. YANG, A. FORE, A. HAYASHI, T. LEE, S. FOURNIER & B. HOLT, 2018. The potential and challenges of using Soil Moisture Active Passive (SMAP) sea surface salinity to monitor Arctic Ocean freshwater changes. Remote Sens., **10**: 869. https://doi.org/10.3390/rs10060869.

THOMPSON, P. A., T. D. O'BRIEN, H. W. PAERL, B. L. PEIERLS, P. J. HARRISON & M. ROBB, 2015. Precipitation as a driver of phytoplankton ecology in coastal waters: a climatic perspective. Estuar. Coast. Shelf Sci., **162**: 119-129. https://doi.org/10.1016/j.ecss.2015.04.004.

WINKLER, G., P. SIROIS, L. E. JOHNSON & J. J. DODSON, 2005. Invasion of an estuarine transition zone by *Dreissena polymorpha* veligers had no detectable effect on zooplankton community structure. Can. J. Fish. Aquat. Sci., **62**: 578-592.

YAMAMICHI, M., N. G. HAIRSTON, JR., M. REES & S. P. ELLNER, 2019. Rapid evolution with generation overlap: the double-edged effect of dormancy. Theor. Ecol., **12**: 179-195.

First received 27 August 2019.
Final version accepted 13 December 2019.

[When citing this volume, please refer to *Crustaceana* 93 (2020), 241-547]

*Proceedings of the* Eurytemora *conference, St. Petersburg, 2019*

# *EURYTEMORA AFFINIS* (POPPE, 1880) (COPEPODA, CALANOIDA) IN THE GULF OF RIGA, BALTIC SEA — ELEMENTAL COMPOSITION AND DIURNAL VERTICAL MIGRATION

BY

ASTRA LABUCE[1,3]), SOLVITA STRAKE[1]) and INTA DIMANTE-DEIMANTOVICA[1,2])

[1]) Latvian Institute of Aquatic Ecology, Agency of Daugavpils University, 4 Voleru str., 1007, Riga, Latvia

[2]) Norwegian Institute for Nature Research, Gaustadalléen 21, 0349, Oslo, Norway

## ABSTRACT

The present paper reports the results of two surveys conducted in the Gulf of Riga (Baltic Sea) during 1997. Elemental composition (carbon and nitrogen) and diurnal vertical migration (DVM) throughout development of the calanoid copepod *Eurytemora affinis* (Poppe, 1880) were studied. This is a species with a high genetic and morphological heterogeneity, inhabiting a seasonally stratified, herring (*Clupea harengus* Linnaeus, 1758) rich ecosystem. The DVM of copepods is a widespread advantageous behaviour, although with several trade-offs. At the expense of time spent in the food-richer surface layer, copepods avoid visual predators (planktivorous fish) by descending to greater depths, thereby decreasing their mortality. Being a part of the food web, copepods themselves serve as a source of nutrients with specific elemental contents. In the present study, we observed differences both in DVM and elemental content between the developmental stages of the studied species. The majority (57-89%) of nauplii and 1st to 3rd copepodids were found near the thermocline during daytime, whereas at night they were detected above the thermocline. The 5th stage copepodids and adults displayed evident nocturnal ascent, when most (>90%) individuals moved above 15 m depth. During daytime, they tended to disperse throughout the water column. Carbon (C%) and nitrogen (N%) in dry weight of *E. affinis* increased with ontogenetic development, reaching the highest values in 4th to 5th copepodids and adults. Adult males showed higher values of C% and N%, as well as an increased C : N molar ratio, compared to adult non-ovigerous females. The amount of time spent at the food-richer surface layer appeared to be related to N% in *E. affinis*.

Key words. — Diel vertical migration, DVM, carbon, nitrogen, C : N ratio, ontogeny, copepod

## RÉSUMÉ

L'article présent donne les résultats de deux suivis menés dans le Golfe de Riga (mer Baltique) en 1997. La composition élémentaire (carbone et azote) et la migration verticale diurne (DVM) au cours du développement du copépode calanoïde *Eurytemora affinis* (Poppe, 1880) ont été étudiées. C'est

[3]) Corresponding author; e-mail: astra.labuce@lhei.lv

 | DOI:10.1163/9789004465565_014

une espèce à hétérogénéité génétique et morphologique élevée, vivant dans un écosystème riche à hareng (*Clupea harengus* Linnaeus, 1758), saisonnièrement stratifié. La DVM des copépodes est un comportement avantageux largement répandu, bien qu'avec plusieurs contreparties. Au détriment du temps passé dans la couche de surface riche en nourriture, les copépodes évitent les prédateurs visuels (poissons planctivores) en allant vers les grandes profondeurs, diminuant ainsi leur mortalité. Comme élément de la chaîne alimentaire, les copépodes eux-mêmes servent de source de nutriments avec un contenu élémentaire spécifique. Dans la présente étude, nous avons observé des différences à la fois dans la DVM et dans le contenu élémentaire entre les stades de développement des espèces étudiées. La majorité (57-89%) des nauplies et des $1^{ers}$ aux $3^{èmes}$ copépodites ont été trouvés près de la thermocline au cours de la journée alors que pendant la nuit, ils ont été détectés au-dessus de la thermocline. Les copépodites de stade 5 et les adultes ont présenté une montée nocturne évidente, lorsque la plupart (>90%) des individus se déplacent au-dessus de 15 m de profondeur. Au cours de la journée, ils tendent à se disperser partout dans la colonne d'eau. Le carbone (C%) et l'azote (N%) dans le poids sec de *E. affinis* augmente avec le développement ontogénique, atteignant les valeurs les plus élevées chez les copépodites 4 à 5 et les adultes. Les adultes mâles présentaient des valeurs plus élevées de C% et N%, ainsi qu'un rapport molaire C : N supérieur, par rapport aux valeurs des femelles adultes non-ovigères. Le temps passé dans la couche de surface la plus riche en nourriture apparaît corrélée à la valeur de N% chez *E. affinis*.

Mots clés. — Migration verticale diurne, DVM, carbone, azote, rapport C : N, ontogénie, copépode

## INTRODUCTION

*Eurytemora affinis* (Poppe, 1880) is prevalent in estuaries and other brackish waterbodies in the Northern Hemisphere. It is an egg-carrying species and primarily a filter feeder. This paper has a particular focus on the European clade, Baltic lineage *E. affinis* (see Lee & Frost, 2002; Sukhikh et al., 2016). In the Gulf of Riga (Baltic Sea), *E. affinis* is the dominant calanoid copepod with a key-role in the local pelagic ecosystem during summer. It is a phytoplankton grazer (Jurgensone et al., 2011) and is selectively preyed upon by the main planktivorous fish, i.e., herring, *Clupea harengus* Linnaeus, 1758 (cf. Livdāne et al., 2016). Thus, *E. affinis* plays an important role in the carbon cycling of the regional food web.

*Eurytemora affinis* is capable of distinct diel vertical migrations (DVM) — a flexible and highly variable behaviour of habitat selection. It can be viewed as a practical application of the animal's morphological characteristics, physiological state and responses to environmental conditions, especially as a way to avoid visual predators, e.g., planktivorous fish (Holliland et al., 2012; Andersson et al., 2017). Copepods also play an important role in export stoichiometry during DVM. As inter-zonal migrants, they actively transport material (Hannides et al., 2009). While the elemental composition of zooplankton itself remains relatively stable (Walve & Larsson, 1999), it can still change throughout ontogeny (Vrede et al., 1999, 2004; Carillo et al., 2001) regardless of the food quality (Villar-Argaiz et al., 2002). However, species-specific ontogenetic variability is rarely considered in DVM or studies focusing on elemental composition. Most of the conducted research in the Baltic Sea focuses on adult specimens, particularly females (Vuorinen, 1987; Koski, 1999; Pertola et al., 2002) as they are the most susceptible to visual predators. Hence, our aim was to fill the existing gap in the knowledge for this particular region. Therefore, we summarized the results of two so far unpublished surveys conducted in the Gulf of Riga during 1997. Carbon and nitrogen contents as well as DVM in *E. affinis* at all developmental stages were studied. In addition, by coupling the results of those two independently, but simultaneously conducted surveys, we discuss potential driving forces of DVM from the ontogenetic perspective and the changes of elemental composition from the viewpoint of deliberate habitat selection.

## MATERIAL AND METHODS

### Study area

The Gulf of Riga is a semi-enclosed, shallow brackish waterbody, located in the eastern Baltic Sea. The ecosystem is characterized by anthropogenic pressures, pronounced salinity and climate gradients, as well as seasonal thermal stratification (Kotta et al., 2008; Skudra & Lips, 2017; Snoeijs-Leijonmalm & Andrén, 2017). It is one of the most eutrophic sub-regions of the Baltic Sea (Andersen et al., 2011; HELCOM, 2014).

The phytoplankton succession in the Gulf of Riga follows the classical scenario of boreal phytoplankton development (Yurkovskis et al., 1999; Jurgensone et al., 2011). A pronounced diatom-dinoflagellate bloom occurs during spring. Summer is dominated by cyanobacteria accompanied by chlorophytes and cryptophytes, which is followed by an increase of diatoms in autumn. The composition and biomass of zooplankton in the Gulf of Riga change seasonally, but the zooplankton as a whole is always dominated by brackish water species. Rotifers of the genus *Synchaeta* Ehrenberg, 1832 together with copepod species of *Acartia* Dana, 1846 dominate in spring and autumn, whereas the cladoceran *Bosmina* (*Eubosmina*) *coregoni* Baird, 1857 and the copepod *Eurytemora affinis* dominate in the summer (Kotta et al., 2008).

### Vertical population structure during a 24-h period

Zooplankton samples were collected using a Judy net (mesh size 100 $\mu$m) at Station A (fig. 1). Vertical discrete net hauls from 6 depth strata (0-5 m, 5-10 m, 10-15 m, 15-20 m, 20-25 m, 25-30 m) were made with 4 h intervals from 14:00 h on 20 July to 14:00 h on 21 July 1997, resulting in 7 samplings per 24-hours (see

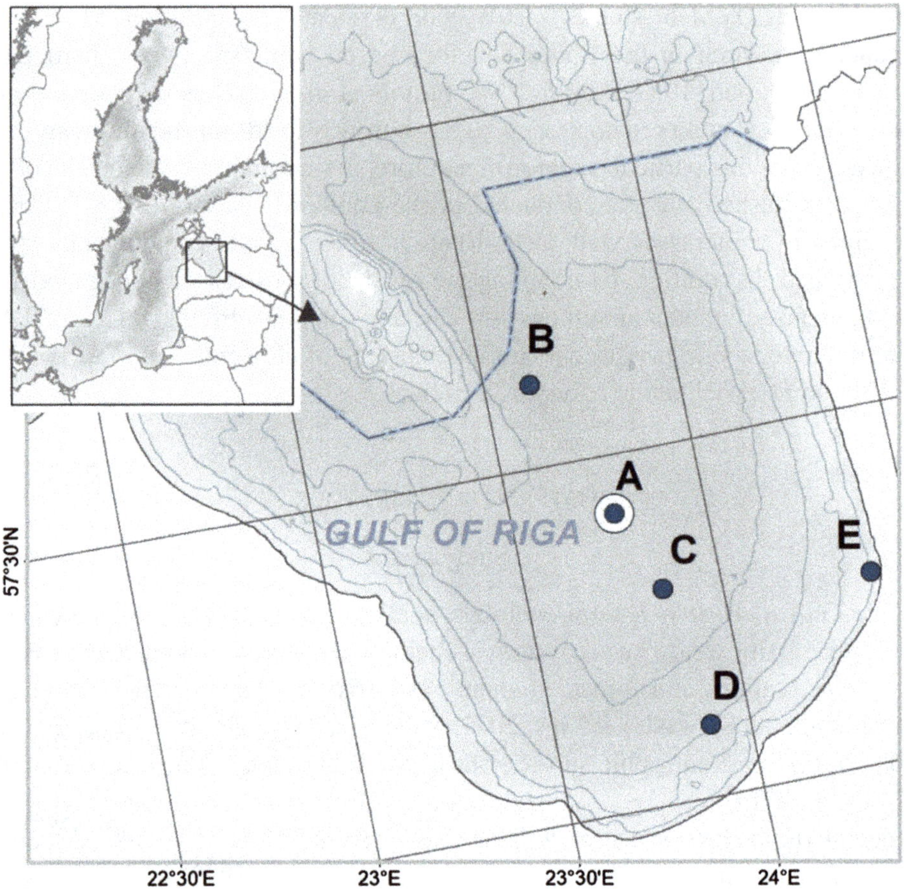


Fig. 1. Map of the study area — the Gulf of Riga, Baltic Sea. A, White circle represents location where zooplankton samples for analysing vertical population structure were obtained; A-E, dark dots represent locations where specimens for carbon and nitrogen content measurements were collected. Contour lines correspond to isobaths from 10 to 50 m depth (with an interval of 10 m).

Supplementary file: data table S1). The sunset on 20 July was at 22:00 h, followed by sunrise on 21 July at 5:01 h. Hence, samplings occurring at 14:00, 18:00, 6:00 and 10:00 h were conducted during daylight, while samplings at 22:00 and 2:00 h represent the night. Samples were preserved with a ~4% buffered formalin and analysed in the laboratory under a dissecting microscope. Each developmental stage of *E. affinis* was identified and counted.

Water samples for temperature, chlorophyll-a concentration, nutrients (DIN — nitrate, nitrite and ammonium; DIP — phosphates) and for an estimate of the phytoplankton community were collected with Nansen bottles from 0, 5, 10, 15, 20 and 40 m strata. The temperature was measured with a reversing thermometer attached to a Nansen bottle. Water temperature, DIN, DIP and phytoplankton

TABLE I
Amount of *Eurytemora affinis* (Poppe, 1880) individuals collected at each sampling site for carbon (C%) and nitrogen (N%) content measurements

| Station | Date | cop1 | cop2 | cop3 | cop4 | cop5 | n-ov F | M |
|---|---|---|---|---|---|---|---|---|
| A | 21 July 1997 | 0 | 0 | 50 | 50 | 50 | 0 | 50 |
| B | 25 July 1997 | 0 | 0 | 0 | 50 | 50 | 50 | 130 |
| C | 20 July 1997 | 85 | 72 | 80 | 0 | 0 | 60 | 0 |
| D | 19 July 1997 | 0 | 0 | 0 | 0 | 0 | 80 | 80 |
| E | 24 July 1997 | 50 | 144 | 60 | 0 | 0 | 0 | 0 |
| Total | | 134 | 216 | 189 | 99 | 99 | 189 | 260 |
| *n* | | 2 | 3 | 3 | 3 | 3 | 3 | 4 |
| ind per *n* | | 67 | 72 | 63 | 33 | 33 | 63 | 65 |

cop1-cop5, corresponding copepodid-stages; n-ovF, non-ovigerous females; M, males; *n*, number of replicates; ind per *n*, amount of individuals in the replicates.

community were assessed only at the starting point of zooplankton sampling at 14:00 on 20 July 1997. Secchi depth was also measured at the same time. Water samples for determination of chlorophyll-a concentration were collected at every zooplankton sampling occasion throughout a day-night cycle (7 times with 4 h in-between each sampling) to evaluate changes in the amount of available food sources. Preservation and processing of phytoplankton samples, and chlorophyll-a were done according to HELCOM recommendations (HELCOM, 1988), whereas measurements of DIN and DIP were conducted according to Grasshoff (1976).

To describe light conditions in the water column, Secchi depth ($d_S$) was converted to the attenuation coefficient ($k$) as

$$k = \frac{1.7}{d_S}$$

and the light irradiance at a specific depth ($I_d$) was calculated from

$$I_d = I_0 \times \mathrm{e}^{-kd}$$

where $I_0$ is light irradiance at the surface and $d$ is depth. As $I_0$ was not measured directly, it was assumed as 100% of the light available at the surface, consequently $I_d$ (expressed in percent) showed the attenuation of light respective to depth.

## Ontogenetic variability in carbon (C%) and nitrogen (N%) content

Animals for measuring C% and N% contents were collected from surface waters (0-10 m) with a vertical net haul (Judy net, mesh size 100 $\mu$m) during day-time. Sampling was conducted at stations A-E between 19 and 25 July 1997 (fig. 1; table I). Immediately after sample collection, individuals (copepodids, adult males and non-ovigerous females) of *E. affinis* were picked out under a stereomicroscope and temporally placed in 5 ml containers filled with filtered sea water. When a sufficient number of animals was obtained, they were collected on a fine plankton net and rinsed with deionised water, transferred to plastic vials and placed in a freezer (−18°C). In the laboratory, two to four replicates — each containing at least 33 animals — for each of the developmental stages were pooled together (table I). The individuals were transferred to delicate pre-weighed tin capsules for C% and N% analyses using a needle. Samples were dried at 60°C and weighed to obtain dry mass. The C% and N% content in dry mass was measured using a CHN analyser CHN-900 (Leco Corporation, St. Joseph, MI).

### Data analysis

Development-specific patterns of vertical population structure were analysed and compared using weighted time (WT). WT represents the part of a 24-h period spent at a specific depth ($d$), and it was calculated as follows. Copepod abundances at a certain depth ($ab_d$; ind/m$^3$) were expressed as the percentage ($ab_p$) of the summed abundances observed at every depth for every sampling time ($t$) and developmental stage ($s$):

$$ab_p(t, s, d) = \frac{ab_d}{\sum ab_d} \times 100\%. \tag{1}$$

Mean $ab_p$ ($\overline{ab_p}$; representing the mean abundance of a four-hours (4h) period) between two consecutive sampling times ($t + t1$) was calculated for every developmental stage at every depth:

$$\overline{ab_p}(t + t1, s, d) = \frac{ab_p(t) + ab_p(t + 1)}{2}. \tag{2}$$

WT spent at each depth for every *E. affinis* developmental stage was calculated:

$$\mathrm{WT}(s, d) = \frac{\sum(\overline{ab_p} \times 4\mathrm{h})}{24\mathrm{h}} \tag{3}$$

In order to evaluate the direct effects of habitat selection, the relationship between *E. affinis*' elemental composition (C% and N% contents, C : N molar ratio) and WT was analysed using linear regression analysis. Prior to the analysis, accordance of the data to the Gaussian distribution was checked. All data analyses were conducted using the software R 3.5.1 (R Core Team, 2018). The package "ggplot2" (Wickham, 2009) was used for data visualization.

## RESULTS

### Environmental characteristics

The vertical profile of the water column showed a typical mid-summer pattern on 20 July 1997 (fig. 2). The upper mixed layer reached down to 10 m and was

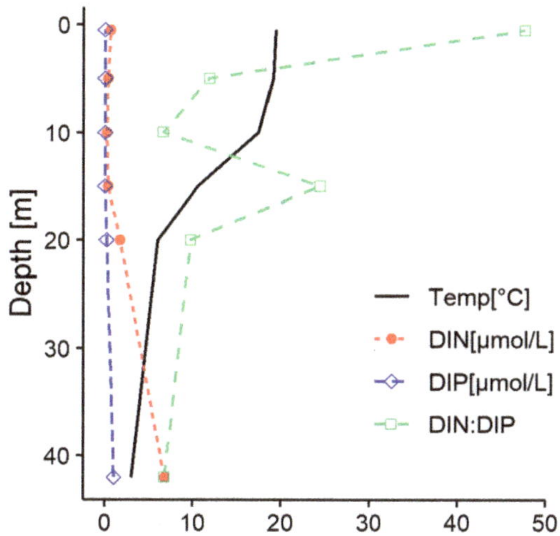


Fig. 2. Abiotic factors at 14:00 on 20 July 1997 in the Gulf of Riga. Temp, water temperature; DIN, dissolved inorganic nitrogen (nitrate, nitrite and ammonium); DIP, dissolved inorganic phosphorus (phosphates).

characterized by warm waters (mean 18.9°C). The DIN : DIP ratio was high at the surface (47.7), another peak appeared at 15 m depth (24.5), but overall the concentrations of DIN and DIP were low (close to 0 $\mu$mol/l). The thermocline was developed at the 10-20 m layer. Deeper waters (>20 m) had higher nutrient values (DIN exceeded 6.8 $\mu$mol/l, DIP reached 1.0 $\mu$mol/l) and low temperatures (3.0-6.0°C) (fig. 2). The Secchi depth was 3.2 m, indicating almost complete darkness below 10 m depth (fig. 3). The chlorophyll-a concentration followed the dynamics of the available light and was highest in the upper 10 m, but did not exceed 2 mg/m$^{-3}$ on average (fig. 3). The phytoplankton community was co-dominated by the cyanobacterium *Aphanizomenon flos-aquae* (Ralfs ex Bornet & Flahault, 1886), the chlorophytes *Oocystis* spp. Nägeli ex A. Braun and *Monoraphidium contortum* (Thuret) Komárková-Legnerová, and the dinoflagellate *Dinophysis acuminata* Claparède & Lachmann.

## Vertical population structure during a 24 h period

Ontogenetic differences of *Eurytemora affinis* were pronounced in the observed vertical population structure (fig. 4). The majority (57-89%) of nauplii and early stage copepodids (cop1-cop3) gathered at the thermocline (15-20 m) during the day, while during the night they ascended to the upper layers. Older copepodids (cop4-cop5) and adults did not cluster at a specific depth during the day — they tended to disperse throughout the water column. During the night, however, they displayed an evident nocturnal ascent, when the majority of individuals (>90%) moved above 15 m depth (fig. 4).

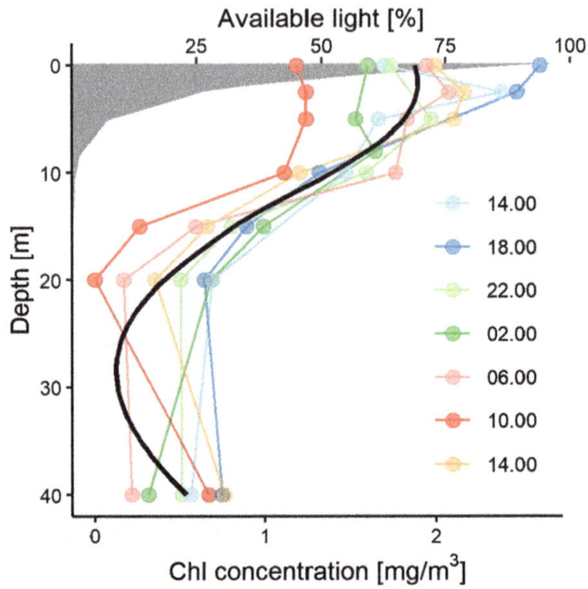


Fig. 3. Vertical profile of chlorophyll a concentration (Chl) throughout 24-hour period (starting at 14:00 on 20 July 1997). Line colours represent different timing. Black line portrays LOESS curve from all measurements. Grey area represents available light irradiance, calculated from Secchi depth at 14:00 on 20 July 1997 (as percentage of light available at the surface).

The proportion of time spent at each stratum (fig. 5) showed that adults tended to spend most time in the upper layers. Females dwelled in the 0-5 m layer for 0.16 parts of 24 h (≈4 h), while males spent 0.23 parts (≈5.5 h) in this layer. Simultaneously, cop1-cop3 lingered at a depth of 15-20 m almost half a day. Cop4 and cop5 specimens behaved similarly as adults, only with less time spent at the 0-5 m depth stratum (fig. 5).

## Ontogenetic variability in carbon (C%) and nitrogen (N%) content

The elemental composition showed ontogenetic variability. The C% and N% content in *E. affinis* increased gradually throughout development from cop1, reaching its maximum values in cop4 (table II). The exception is the N% content in males (10.73), which was higher than in cop4 (9.33). The C : N molar ratio showed a decreasing trend during ontogeny (cop1-cop5). The exception is the highest ratio observed in males (8.65), similar to that in cop1 (7.08).

The importance of *E. affinis* feeding in the euphotic layer (where the bulk of organic N and C production occurs) was evaluated by conducting linear regression analysis between elemental composition (table II) and the WT spent at particular depths: 0-5 m (fig. 6) and 0-10 m (not shown). The analysis pointed to a statistically significant ($R^2 = 0.63$, $p = 0.03$, $N = 7$) relationship between N% content in *E.*

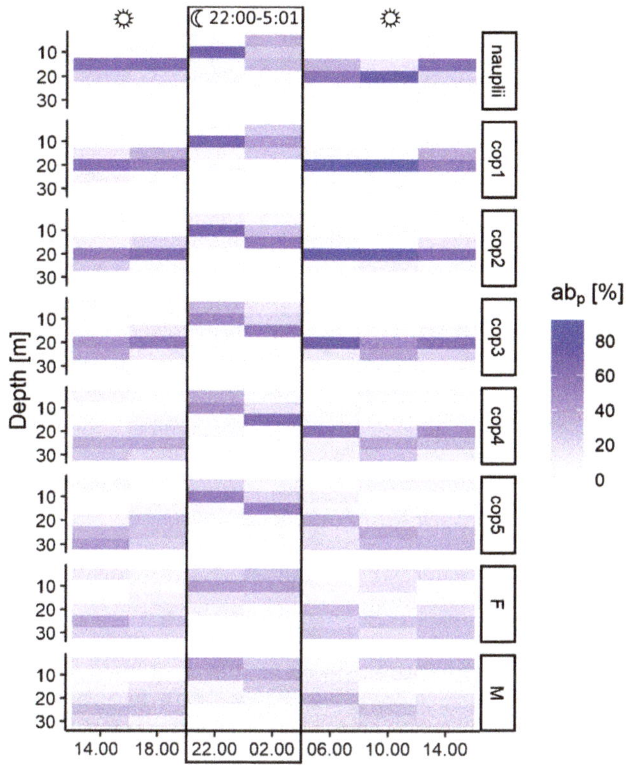


Fig. 4. Fraction (%) of observed abundance at study site A during 24-hour period (from 14:00 on 20 July 1997 to 14:00 on 21 July 1997) for every *Eurytemora affinis* (Poppe, 1880) developmental stage (cop1-cop5, corresponding copepodid-stages; F, females; M, males). Sunset (on 20 July) and sunrise (on 21 July) times are noted next to the 'moon' icon.

*affinis* and the time spent at 0-5 m depth (fig. 6). For 0-10 m depth, this relationship was not significant.

## DISCUSSION

The results of the present study on vertical distribution of *Eurytemora affinis* during a 24 h period coincided with the previously described DVM characteristics of Baltic *E. affinis* (Vuorinen, 1987; Holliland et al., 2012), as well as with other theoretical reviews (e.g., Cushing, 1951; Lampert, 1989; Mauchline, 1998). The *E. affinis* population in the Baltic Sea shows seasonal and ontogenetic differences, emphasizing the strategic importance of vertical migration for minimizing population loss (Vuorinen, 1987). The most pronounced DVM for the northern Baltic *E. affinis* population has been observed during July and August, and the reported dynamics clearly showed an increase in amplitude of the migrations with development and body growth (Holliland et al., 2012). The active migration usually is

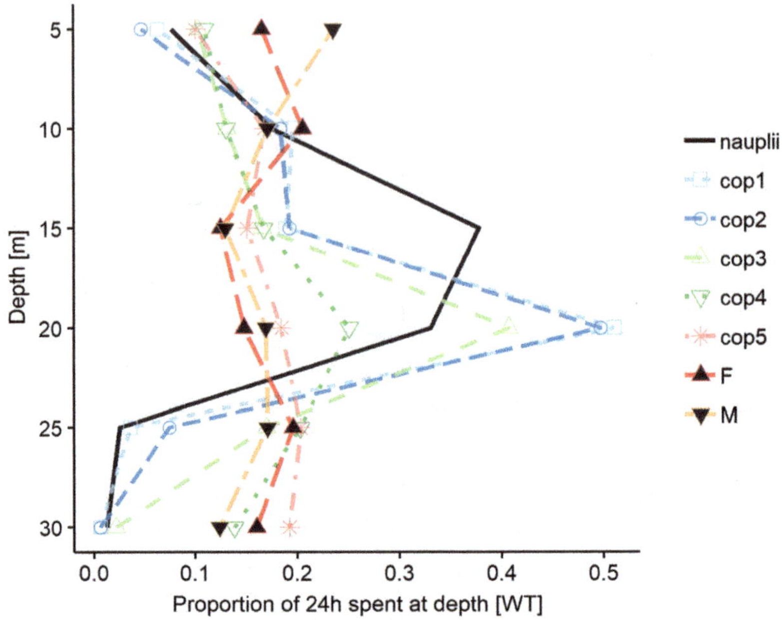


Fig. 5. Proportion of 24-hour period (from 14:00 on 20 July 1997 to 14:00 on 21 July 1997) spent (WT) at depth for every *Eurytemora affinis* (Poppe, 1880) developmental stage (cop1-cop5, corresponding copepodid-stages; F, females; M, males).

attributed to older copepodids and adults, whereas nauplii and early stage copepodids are considered as non- or weakly-migratory (Huntley & Brooks, 1982). In the present study a certain DVM pattern of nauplii and early stage copepodids was observed, however, most likely they were drifting passively, adjusting their vertical depth in the water column due to epilimnion turbulence, mixing patterns, and microstratification.

TABLE II

Carbon (C%) and nitrogen (N%) content and C : N molar ratio in dry weight of *Eurytemora affinis* (Poppe, 1880) collected in the Gulf of Riga during July 1997

| | C (%) | SE | N (%) | SE | C : N | SE |
|---|---|---|---|---|---|---|
| cop1 | 28.24 | ±2.02 | 2.50 | ±1.15 | 7.08 | ±0.65 |
| cop2 | 39.80 | ±10.94 | 4.51 | ±0.90 | 5.71 | ±1.39 |
| cop3 | 43.58 | ±1.04 | 6.09 | ±0.41 | 5.64 | ±1.11 |
| cop4 | 54.79 | ±8.36 | 9.33 | ±1.33 | 4.56 | ±0.23 |
| cop5 | 52.52 | ±1.67 | 8.98 | ±0.91 | 3.88 | ±0.09 |
| n-ov F | 43.36 | ±4.01 | 8.83 | ±0.93 | 4.67 | ±0.15 |
| M | 48.83 | ±4.16 | 10.73 | ±1.01 | 8.65 | ±2.99 |

SE, standard error; cop1-cop5, corresponding copepodid-stages; n-ovF, non-ovigerous females; M, males.

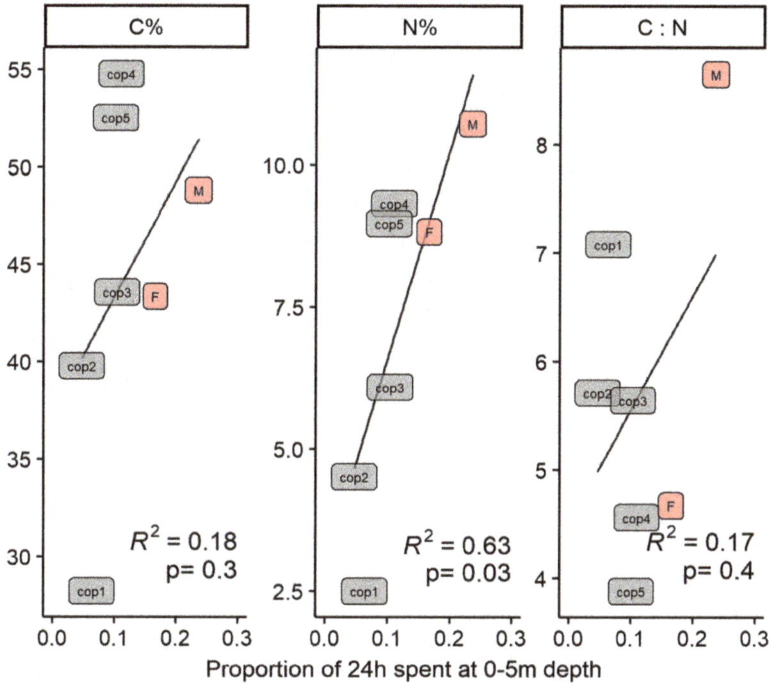


Fig. 6. Linear regression between proportion of time spent (WT) for every *Eurytemora affinis* (Poppe, 1880) developmental stage (cop1-cop5, corresponding copepodid-stages; F, females; M, males) at top surface layer (0-5 m) and their respective stoichiometric parameters (C%, carbon content (%); N%, nitrogen content (%)). $N = 7$ for all regression analysis.

During our study, most of the surface light was absorbed within the upper 5 m. Less than 10% penetrated deeper than 5 m, whereas less than 1% reached a depth below 10 m (fig. 3). Therefore, the bulk of the available, and relatively valuable (as an N source), prey for *E. affinis* was located in the top 5 m stratum, however, in variable amounts (fig. 3). The Gulf of Riga is deemed as N-limited during the summer period (Maestrini et al., 1999; Jurgensone et al., 2011). N-limitation favours development of diazotrophic cyanobacteria. They are considered as low in nutritional quality, especially low in lipids (Finkel et al., 2016; Jónasdóttir, 2019), but they have a high content of amino acids (Ahlgren et al., 1992), and a high N: P ratio (Klausmeier et al., 2004). Recent studies on Baltic copepods (*Acartia* spp. and *E. affinis*) show a positive impact of cyanobacterial consumption on egg reproduction and naupliar development (Hogfors et al., 2014; Motwani et al., 2017). While Koski et al. (1999) observed a decrease in reproduction and survival of the Baltic *E. affinis* after feeding on a toxic cyanobacterium strain (*Nodularia* sp. Mertens ex Bornet & Flahault, 1886, nom. cons.), when mixed with a chlorophyte (*Brachiomonas submarina* Bohlin) the negative effects faded (Koski et al., 1999). In the present study, a mixture of cyanobacteria, chlorophytes and dinoflagellates dominated the phytoplankton community. Hence, we concluded that the nutrients available in the water column provided sufficient nutritional diversity and quality (concentrated in the upper 5 m) for *E. affinis* in the Gulf of Riga.

The C : N molar ratio values reported in this study varied from 3.88 (cop5) to 8.65 (males). Other studies on copepods show a wide interval, ranging from 3 to 13, however, in the majority of species it varies between 3 and 5 (Mauchline, 1998). Copepods belong to N-demanding zooplankton (Andersen & Hessen, 1991; Sterner & Hessen, 1994; Gismervik, 1997), thus, their C : N ratio is predominantly low. The C% and N% content in the Baltic copepods, including *E. affinis*, have been analysed previously, but mainly in the northern Baltic or Baltic Proper (see data compilation from previous studies in table III), and so far not in the Gulf of Riga. Moreover, previous studies were mainly done on older copepodids or adults, and most frequently on females. The results of the present study fall into the range of the previous ones, e.g., the C : N ratio of *E. affinis* non-ovigerous females from the Gulf of Riga is similar to the summer values observed on the SW coast of Finland (Koski, 1999; table III), but slightly lower than in the Archipelago Sea and Baltic Proper (Pertola et al., 2002; table III). We observed differences in elemental composition throughout copepod ontogeny, and this is supported by already existing studies (e.g., Campbell et al., 2001; Methews et al., 2018). Meunier et al. (2016) revealed that nutritional demand and grazing selectivity differ among developmental stages in *Acartia tonsa* Dana, 1849. They observed that nauplii always selected P-rich algae, whereas copepodids favoured N-rich algae. This is also in accordance with the growth-rate hypothesis (GRH; Sterner & Elser, 2002) that relates P demand to the growth rate. Frost et al. (2005) suggested a modification to GRH by stating "that the N-intensive requirements of energy dominate at low absolute growth rates while the P-intensive requirements of protein synthesis machinery dominate at high growth rates". Indeed, nauplii grow at higher rates on average than copepodids (Kiørboe, 1997), confirming an increasing role of N during ontogeny. The pattern can be observed in the results of the present study as a decreasing C : N trend from cop1 to cop5, however, values observed for adults do not comply with the assumption (table II). We observed substantial stoichiometric differences between males and non-ovigerous females in *E. affinis* from the Gulf of Riga. To the best of our knowledge, there are no other supporting studies on sexual differences in *E. affinis* stoichiometric parameters for comparison. One of the reason for the sexual differences might be time spent at the 0-5 m layer. WT spent for females and males at the 0-5 m layer was ~4 h and ~5.5 h, respectively (fig. 6). Still, reproduction costs or/and post-reproduction costs cannot be neglected. Koski (1999) reported on the elemental composition of Baltic *E. affinis* females and their produced eggs. The C% and N% composition in eggs proportional to the composition found in

TABLE III

Stoichiometric studies done in situ on Baltic *Eurytemora affinis* (Poppe, 1880)

| Date | Location | Station | Layer | Stage | C (unit) | N (unit) | C : N | Reference |
|---|---|---|---|---|---|---|---|---|
| Oct. 1995 | BS, Himmerfjärden Bay | H4 | h.t. 5-10 m | 4-6 | 49.1 (%) | 12.0 (%) | 4.8 | Walve&Larsson (1999) |
| Sep. 1996 | BS, Landsort Depth | BY31 | 0-100 m | 6 | 48.5 (%) | 12.7 (%) | 4.5 | Walve&Larsson (1999) |
| Jul. 1997 | BS, Gulf of Riga | 120 | 0-10 m | 6* | 46.1 (%) | 9.78 (%) | 6.7 | Present study |
| Jul. 1997 | BS, Gulf of Riga | 120 | 0-10 m | n-ov F | 43.4 (%) | 8.83 (%) | 4.7 | Present study |
| May 1998 | BS, SW coast of Finland | | | F | 3.4 ($\mu$g/ind) | 0.8 ($\mu$g/ind) | 4.2 | Koski (1999) |
| Jun. 1998 | BS, SW coast of Finland | | | F | 1.8 ($\mu$g/ind) | 0.4 ($\mu$g/ind) | 4.1 | Koski (1999) |
| Jul. 1998 | BS, SW coast of Finland | | | F | 3.6 ($\mu$g/ind) | 0.8 ($\mu$g/ind) | 4.5 | Koski (1999) |
| Aug. 1998 | BS, SW coast of Finland | | | F | 3.0 ($\mu$g/ind) | 0.7 ($\mu$g/ind) | 4.5 | Koski (1999) |
| May 1998 | BS, SW coast of Finland | | | eggs | 0.02 ($\mu$g/ind) | 0.007 ($\mu$g/ind) | 3.1 | Koski (1999) |
| Jun. 1998 | BS, SW coast of Finland | | | eggs | 0.02 ($\mu$g/ind) | 0.008 ($\mu$g/ind) | 2.8 | Koski (1999) |
| Jul. 1998 | BS, SW coast of Finland | | | eggs | 0.03 ($\mu$g/ind) | 0.009 ($\mu$g/ind) | 3.3 | Koski (1999) |
| Aug. 1998 | BS, SW coast of Finland | | | eggs | 0.03 ($\mu$g/ind) | 0.007 ($\mu$g/ind) | 3.9 | Koski (1999) |
| Jun.-Jul. 1999 | BS, Bothnian Bay | F2 | 0-40 m | F | 2.4 ($\mu$g/ind) | 0.5 ($\mu$g/ind) | 5.8 | Pertola et al. (2002) |
| Jun.-Jul. 1999 | BS, Bothnian Sea | SR5 | 0-40 m | F | 1.6 ($\mu$g/ind) | 0.3 ($\mu$g/ind) | 5.8 | Pertola et al. (2002) |
| Jun.-Jul. 1999 | BS, Archipelago Sea | IU5 | 0-40 m | F | 1.7 ($\mu$g/ind) | 0.3 ($\mu$g/ind) | 5.8 | Pertola et al. (2002) |
| Jun.-Jul. 1999 | BS, Gulf of Finland | LL3A | 0-40 m | F | 2.3 ($\mu$g/ind) | 0.4 ($\mu$g/ind) | 6.2 | Pertola et al. (2002) |
| Jun.-Jul. 1999 | BS, northern Baltic Proper | LL12 | 0-40 m | F | 1.7 ($\mu$g/ind) | 0.3 ($\mu$g/ind) | 6.1 | Pertola et al. (2002) |

C, carbon; N, nitrogen; C : N, molar carbon: nitrogen ratio; BS, Baltic Sea; h.t., horizontal trawl; F, female; n-ov F, non-ovigerous female.

*, Mean from non-ovigerous females and males

females were approx. 0.6-1% and 0.9-2%, respectively, suggesting egg production as N-demanding (Koski, 1999). In the present study, non-ovigerous females were chosen for stoichiometric analysis. However, most likely, the majority of analysed females were in a latency period (LT; time between the hatching of nauplii and extrusion of a new clutch), as the time needed to produce the first clutch is on average shorter (0.5-1.5 days, depending on temperature) than the mean LT (Devreker et al., 2009). Kiørboe (1989) suggested N as a fecundity limiting constituent of the food for *A. tonsa*, whereas Checkley (1980) reported N-limited egg production for *Paracalanus parvus parvus* (Claus, 1863). Hence, we assume that females of *E. affinis* could be more N-demanding than males. Furthermore, egg-carrying females are known to spend more energy per se, as the brood adds to the body mass (Gilbert & Williamson, 1983). A clear conclusion on whether nutritional demand drives the behaviour of vertical migration or whether increase in N% is a result of predatory avoidance behaviour, cannot be drawn. Ontogenetic differences in metabolic activity and characteristics (Mauchline, 1998) might affect the ontogenetic variability in elemental composition of *E. affinis*.

## ACKNOWLEDGEMENTS

We thank the anonymous reviewers for careful reading of our manuscript and many reasoned comments and suggestions, and we are also grateful to Dr. Natalia Sukhikh for organizing the conference on *Eurytemora*. We would also like to express our gratitude to all those who took part in the cruise and field work back in 1997, and who commented on the manuscript during its development

## SUPPLEMENTARY FILE

Raw data on the vertical population structure of *Eurytemora affinis* (Poppe, 1880) during a 24-hour period are made accessible in table S1.

# Appendix

TABLE A1

Raw data on the vertical population structure of *Eurytemora affinis* (Poppe, 1880) during a 24-h period. Please note that the decimal values in columns 7-14 are in fact futile, as abundances of whole individuals are at issue

| Date | Station | Lat | Lon | Time | Depth | nauplii | cop1 | cop2 | cop3 | cop4 | cop5 | F | M | Parameter | Unit |
|---|---|---|---|---|---|---|---|---|---|---|---|---|---|---|---|
| 20-July-1997 | A | 57.4167 | 23.7667 | 14.00 | 5 | 512.8 | 352.6 | 1153.8 | 2500.0 | 1570.5 | 2435.9 | 3044.9 | 833.3 | *Eurytemora affinis* abundance | ind/m$^3$ |
| 20-July-1997 | A | 57.4167 | 23.7667 | 14.00 | 10 | 256.4 | 64.1 | 64.1 | 512.8 | 192.3 | 320.5 | 833.3 | 833.3 | *Eurytemora affinis* abundance | ind/m$^3$ |
| 20-July-1997 | A | 57.4167 | 23.7667 | 14.00 | 15 | 4423.1 | 3653.8 | 1153.8 | 576.9 | 384.6 | 576.9 | 1538.5 | 30 384.6 | *Eurytemora affinis* abundance | ind/m$^3$ |
| 20-July-1997 | A | 57.4167 | 23.7667 | 14.00 | 20 | 16 057.7 | 15 961.5 | 10 673.1 | 4903.8 | 2980.8 | 2403.8 | 2884.6 | 11 730.8 | *Eurytemora affinis* abundance | ind/m$^3$ |
| 20-July-1997 | A | 57.4167 | 23.7667 | 14.00 | 25 | 3141.0 | 7435.9 | 9102.6 | 8397.4 | 7756.4 | 5897.4 | 5256.4 | 1730.8 | *Eurytemora affinis* abundance | ind/m$^3$ |
| 20-July-1997 | A | 57.4167 | 23.7667 | 14.00 | 30 | 461.5 | 576.9 | 1230.8 | 6538.5 | 9903.8 | 3615.4 | 3615.4 | 615.4 | *Eurytemora affinis* abundance | ind/m$^3$ |
| 20-July-1997 | A | 57.4167 | 23.7667 | 18.00 | 5 | 549.5 | 576.9 | 1098.9 | 989.0 | 1016.5 | 1318.7 | 2554.9 | 1730.8 | *Eurytemora affinis* abundance | ind/m$^3$ |
| 20-July-1997 | A | 57.4167 | 23.7667 | 18.00 | 10 | 192.3077 | 576.9231 | 576.9231 | 769.2308 | 769.2308 | 1153.846 | 769.2308 | 1346.154 | *Eurytemora affinis* abundance | ind/m$^3$ |
| 20-July-1997 | A | 57.4167 | 23.7667 | 18.00 | 15 | 11 410.26 | 7820.513 | 3333.333 | 2051.282 | 2435.897 | 1538.462 | 2435.897 | 43 461.54 | *Eurytemora affinis* abundance | ind/m$^3$ |
| 20-July-1997 | A | 57.4167 | 23.7667 | 18.00 | 20 | 18 717.95 | 23 333.33 | 11 794.87 | 5000 | 7179.487 | 1923.077 | 3333.333 | 14 871.79 | *Eurytemora affinis* abundance | ind/m$^3$ |
| 20-July-1997 | A | 57.4167 | 23.7667 | 18.00 | 25 | 1076.923 | 2576.923 | 3153.846 | 7884.615 | 7051.282 | 3269.231 | 4038.462 | 1923.077 | *Eurytemora affinis* abundance | ind/m$^3$ |
| 20-July-1997 | A | 57.4167 | 23.7667 | 18.00 | 30 | 144.2308 | 120.1923 | 240.3846 | 4855.769 | 5961.538 | 2572.115 | 1682.692 | 673.0769 | *Eurytemora affinis* abundance | ind/m$^3$ |
| 20-July-1997 | A | 57.4167 | 23.7667 | 22.00 | 5 | 1846.154 | 4096.154 | 4730.769 | 4500 | 4442.308 | 1903.846 | 5942.308 | 4903.846 | *Eurytemora affinis* abundance | ind/m$^3$ |
| 20-July-1997 | A | 57.4167 | 23.7667 | 22.00 | 10 | 20 769.23 | 20 576.92 | 7307.692 | 5961.538 | 10 769.23 | 3269.231 | 5000 | 55 000 | *Eurytemora affinis* abundance | ind/m$^3$ |

TABLE A1
(Continued)

| Date | Station | Lat | Lon | Time | Depth | nauplii | cop1 | cop2 | cop3 | cop4 | cop5 | F | M | Parameter | Unit |
|---|---|---|---|---|---|---|---|---|---|---|---|---|---|---|---|
| 20-July-1997 | A | 57.4167 | 23.7667 | 22.00 | 15 | 3205.128 | 3012.821 | 1153.846 | 833.3333 | 2051.282 | 1474.359 | 1538.462 | 6602.564 | *Eurytemora affinis* abundance | ind/m$^3$ |
| 20-July-1997 | A | 57.4167 | 23.7667 | 22.00 | 20 | 2564.103 | 1089.744 | 833.3333 | 1025.641 | 1410.256 | 897.4359 | 1602.564 | 6153.846 | *Eurytemora affinis* abundance | ind/m$^3$ |
| 20-July-1997 | A | 57.4167 | 23.7667 | 22.00 | 25 | 158.371 | 101.81 | 56.56109 | 67.8733 | 169.6833 | 203.6199 | 169.6833 | 1085.973 | *Eurytemora affinis* abundance | ind/m$^3$ |
| 20-July-1997 | A | 57.4167 | 23.7667 | 22.00 | 30 | 96.15385 | 48.07692 | 19.23077 | 38.46154 | 52.88462 | 120.1923 | 134.6154 | 475.9615 | *Eurytemora affinis* abundance | ind/m$^3$ |
| 21-July-1997 | A | 57.4167 | 23.7667 | 02.00 | 5 | 9935.897 | 3205.128 | 3525.641 | 1282.051 | 3205.128 | 3525.641 | 3205.128 | 42 307.69 | *Eurytemora affinis* abundance | ind/m$^3$ |
| 21-July-1997 | A | 57.4167 | 23.7667 | 02.00 | 10 | 15 384.62 | 10 000 | 4615.385 | 3076.923 | 6730.769 | 4807.692 | 4423.077 | 31 730.77 | *Eurytemora affinis* abundance | ind/m$^3$ |
| 21-July-1997 | A | 57.4167 | 23.7667 | 02.00 | 15 | 9423.077 | 19 038.46 | 12 115.38 | 11 346.15 | 15 384.62 | 2500 | 3076.923 | 37 788.46 | *Eurytemora affinis* abundance | ind/m$^3$ |
| 21-July-1997 | A | 57.4167 | 23.7667 | 02.00 | 20 | 432.6923 | 625 | 432.6923 | 625 | 1923.077 | 576.9231 | 865.3846 | 4423.077 | *Eurytemora affinis* abundance | ind/m$^3$ |
| 21-July-1997 | A | 57.4167 | 23.7667 | 02.00 | 25 | 168.2692 | 120.1923 | 48.07692 | 288.4615 | 841.3462 | 240.3846 | 432.6923 | 2620.192 | *Eurytemora affinis* abundance | ind/m$^3$ |
| 21-July-1997 | A | 57.4167 | 23.7667 | 02.00 | 30 | 49.45055 | 98.9011 | 32.96703 | 24.72527 | 49.45055 | 74.17582 | 140.1099 | 1346.154 | *Eurytemora affinis* abundance | ind/m$^3$ |
| 21-July-1997 | A | 57.4167 | 23.7667 | 06.00 | 5 | 0 | 0 | 288.4615 | 288.4615 | 576.9231 | 0 | 576.9231 | 0 | *Eurytemora affinis* abundance | ind/m$^3$ |
| 21-July-1997 | A | 57.4167 | 23.7667 | 06.00 | 10 | 0 | 961.5385 | 320.5128 | 641.0256 | 2243.59 | 641.0256 | 1282.051 | 641.0256 | *Eurytemora affinis* abundance | ind/m$^3$ |
| 21-July-1997 | A | 57.4167 | 23.7667 | 06.00 | 15 | 2884.615 | 2500 | 769.2308 | 961.5385 | 1730.769 | 576.9231 | 1730.769 | 23 653.85 | *Eurytemora affinis* abundance | ind/m$^3$ |
| 21-July-1997 | A | 57.4167 | 23.7667 | 06.00 | 20 | 32 115.38 | 24 615.38 | 12 692.31 | 8653.846 | 6730.769 | 1923.077 | 3846.154 | 41 538.46 | *Eurytemora affinis* abundance | ind/m$^3$ |
| 21-July-1997 | A | 57.4167 | 23.7667 | 06.00 | 25 | 1510.989 | 1785.714 | 2609.89 | 1620.879 | 3269.231 | 1483.516 | 1950.549 | 2115.385 | *Eurytemora affinis* abundance | ind/m$^3$ |

TABLE A1
(Continued)

| Date | Station | Lat | Lon | Time | Depth | nauplii | cop1 | cop2 | cop3 | cop4 | cop5 | F | M | Parameter | Unit |
|---|---|---|---|---|---|---|---|---|---|---|---|---|---|---|---|
| 21-July-1997 | A | 57.4167 | 23.7667 | 06.00 | 30 | 230.7692 | 134.6154 | 173.0769 | 1711.538 | 3519.231 | 1711.538 | 2096.154 | 1038.462 | *Eurytemora affinis* abundance | ind/m$^3$ |
| 21-July-1997 | A | 57.4167 | 23.7667 | 10.00 | 5 | 0 | 0 | 288.4615 | 865.3846 | 1730.769 | 2019.231 | 3461.538 | 0 | *Eurytemora affinis* abundance | ind/m$^3$ |
| 21-July-1997 | A | 57.4167 | 23.7667 | 10.00 | 10 | 0 | 641.0256 | 0 | 320.5128 | 320.5128 | 2243.59 | 1282.051 | 0 | *Eurytemora affinis* abundance | ind/m$^3$ |
| 21-July-1997 | A | 57.4167 | 23.7667 | 10.00 | 15 | 1384.615 | 346.1538 | 807.6923 | 576.9231 | 576.9231 | 1153.846 | 461.5385 | 12 576.92 | *Eurytemora affinis* abundance | ind/m$^3$ |
| 21-July-1997 | A | 57.4167 | 23.7667 | 10.00 | 20 | 25 769.23 | 15 512.82 | 7179.487 | 2435.897 | 2435.897 | 1025.641 | 1410.256 | 53 846.15 | *Eurytemora affinis* abundance | ind/m$^3$ |
| 21-July-1997 | A | 57.4167 | 23.7667 | 10.00 | 25 | 1442.308 | 2500 | 5721.154 | 4759.615 | 6538.462 | 3365.385 | 3653.846 | 1442.308 | *Eurytemora affinis* abundance | ind/m$^3$ |
| 21-July-1997 | A | 57.4167 | 23.7667 | 10.00 | 30 | 256.4103 | 160.2564 | 705.1282 | 3429.487 | 5625 | 2307.692 | 2852.564 | 961.5385 | *Eurytemora affinis* abundance | ind/m$^3$ |
| 21-July-1997 | A | 57.4167 | 23.7667 | 14.00 | 5 | 480.7692 | 961.5385 | 0 | 961.5385 | 1442.308 | 1923.077 | 4807.692 | 961.5385 | *Eurytemora affinis* abundance | ind/m$^3$ |
| 21-July-1997 | A | 57.4167 | 23.7667 | 14.00 | 10 | 288.4615 | 288.4615 | 0 | 0 | 576.9231 | 288.4615 | 865.3846 | 1153.846 | *Eurytemora affinis* abundance | ind/m$^3$ |
| 21-July-1997 | A | 57.4167 | 23.7667 | 14.00 | 15 | 12 820.51 | 4487.179 | 1538.462 | 1025.641 | 384.6154 | 256.4103 | 641.0256 | 30 961.54 | *Eurytemora affinis* abundance | ind/m$^3$ |
| 21-July-1997 | A | 57.4167 | 23.7667 | 14.00 | 20 | 19 615.38 | 17 500 | 11 538.46 | 7500 | 2692.308 | 1923.077 | 2403.846 | 11 346.15 | *Eurytemora affinis* abundance | ind/m$^3$ |
| 21-July-1997 | A | 57.4167 | 23.7667 | 14.00 | 25 | 1057.692 | 2259.615 | 4471.154 | 4759.615 | 4711.538 | 3028.846 | 3125 | 1490.385 | *Eurytemora affinis* abundance | ind/m$^3$ |
| 21-July-1997 | A | 57.4167 | 23.7667 | 14.00 | 30 | 128.2051 | 224.359 | 1121.795 | 1955.128 | 4903.846 | 2916.667 | 3012.821 | 833.3333 | *Eurytemora affinis* abundance | ind/m$^3$ |

## REFERENCES


AHLGREN, G., I. B. GUSTAFSSON & M. BOBERG, 1992. Fatty acid content and chemical composition of freshwater microalgae. Journal of Phycology, **28**(1): 37-50. DOI:10.1111/j.0022-3646.1992.00037.x.

ANDERSEN, J. H., P. AXE, H. BACKER, J. CARSTENSEN, U. CLAUSSEN, V. FLEMING-LEHTINEN, M. JÄRVINEN, H. KAARTOKALLIO, S. KNUUTTILA, S. KORPINEN, A. KUBILIUTE, M. LAAMANEN, E. LYSIAK-PASTUSZAK, G. MARTIN, C. MURRAY, F. MØHLENBERG, G. NAUSCH, A. NORKKO & A. VILLNÄS, 2011. Getting the measure of eutrophication in the Baltic Sea: towards improved assessment principles and methods. Biogeochemistry, **106**: 137-156. DOI:10.1007/s10533-010-9508-4.

ANDERSEN, T. & D. O. HESSEN, 1991. Carbon, nitrogen and phosphorus content of freshwater zooplankton. Limnology and Oceanography, **36**: 807-814. DOI:10.4319/lo.1991.36.4.0807.

ANDERSSON, A., T. TAMMINEN, S. LEHTINEN, K. JURGENS, M. LABRENZ & M. VIITASALO, 2017. The pelagic food web. In: P. SNOEIJS-LEIJONMALM, H. SCHUBERT & T. RADZIEJEWSKA (eds.), Biological oceanography of the Baltic Sea: 281-333. (Springer, Dordrecht, Netherlands).

CAMPBELL, R. G., M. W. WAGNER, G. J. TEEGARDEN, C. A. BOUDREAU & E. G. DURBIN, 2001. Growth and development rates of the copepod *Calanus finmarchicus* reared in the laboratory. Marine Ecology Progress Series, **221**: 161-183. DOI:10.3354/meps221161.

CARILLO, P., M. VILLAR-ARGAIZ & J. M. MEDINA-SÀNCHEZ, 2001. Relationship between N: P ratio and growth rate during the life cycle of calanoid copepods: an in situ measurement. Journal of Plankton Research, **23**: 537-547. DOI:10.1093/plankt/23.5.537.

CHECKLEY, D. M., 1980. The egg production of a marine planktonic copepod in relation to its food supply: laboratory studies. Limnology and Oceanography, **25**: 430-446. DOI:10.4319/lo.1980.25.3.0430.

CUSHING, D. H., 1951. The vertical migration of planktonic Crustacea. Biological Reviews, **26**: 158-192. DOI:10.1111/j.1469-185X.1951.tb00645.x.

DEVREKER, D., S. SOUISSI, G. WINKLER, J. FORGET-LERAY & F. LEBOULENGER, 2009. Effects of salinity, temperature and individual variability on the reproduction of *Eurytemora affinis* (Copepoda; Calanoida) from the Seine estuary: a laboratory study. Journal of Experimental Marine Biology and Ecology, **368**: 113-123. DOI:10.1016/j.jembe.2008.10.015.

FINKEL, Z. V., M. J. FOLLOWS, J. D. LIEFER, C. M. BROWN, I. BENNER & A. J. IRWIN, 2016. Phylogenetic diversity in the macromolecular composition of microalgae. PLoS ONE, **11**(5): e0155977. DOI:10.1371/journal.pone.0155977.

FROST, P. C., M. A. EVANS-WHITE, Z. V. FINKEL, T. C. JENSEN & V. MATZEK, 2005. Are you what you eat? Physiological constraints on organismal stoichiometry in an elementally imbalanced world. Oikos, **109**: 18-28. DOI:10.1111/j.0030-1299.2005.14049.x.

GILBERT, J. J. & C. E. WILLIAMSON, 1983. Sexual dimorphism in zooplankton (Copepoda, Cladocera, and Rotifera). Annual Review of Ecology and Systematics, **14**: 1-33. DOI:10.1146/annurev.es.14.110183.000245.

GISMERVIK, I., 1997. Stoichiometry of some marine planktonic crustaceans. Journal of Plankton Research, **12**: 279-285. DOI:10.1093/plankt/19.2.279.

GRASSHOFF, K., M. EHRHARDT & K. KREMLING (eds.), 1983. Methods of seawater analysis: 1-419. (Verlag Chemie, Weinheim, NY).

HANNIDES, C. C. S., M. R. LANDRY, C. R. BENITEZ-NELSON, R. M. STYLES, J. P. MONTOYA & D. M. KARL, 2009. Export stoichiometry and migrant-mediated flux of phosphorus in the North Pacific Subtropical Gyre. Deep-Sea Research (Part I, Oceanographic Research Papers), **56**: 73-88. DOI:10.1016/j.dsr.2008.08.003.

HELCOM, 1988. Guidelines for the Baltic monitoring programme for the third stage. Baltic Sea Environment Proceedings, **27**: 1-193.

HELCOM, 2014. Eutrophication status of the Baltic Sea 2007-2011 — a concise thematic assessment. Baltic Sea Environment Proceedings, **143**: 7-36.

HOGFORS, H., N. H. MOTWANI, S. HAJDU, R. EL-SHEHAWY, T. HOLMBORN, A. VEHMAA, J. ENGSTRÖM-ÖST, A. BRUTEMARK & E. GOROKHOVA, 2014. Bloom-forming cyanobacteria support copepod reproduction and development in the Baltic Sea. PLoS ONE, **9**(11): e112692. DOI:10.1371/journal.pone.0112692.

HOLLILAND, P. B., B. I. AHLBECK, E. WESTLUND & S. HANSSON, 2012. Ontogenetic and seasonal changes in diel vertical migration amplitude of the calanoid copepods *Eurytemora affinis* and *Acartia* spp. in a coastal area of the northern Baltic proper. Journal of Plankton Research, **34**: 298-307. DOI:10.1093/plankt/fbs001.

HUNTLEY, M. & E. R. BROOKS, 1982. Effects of age and food availability on diel vertical migration of *Calanus pacificus*. Marine Biology, **71**: 23-31. DOI:10.1007/BF00396989.

JÓNASDÓTTIR, S. H., 2019. Fatty acid profiles and production in marine phytoplankton. Marine Drugs, **17**: 151. DOI:10.3390/md17030151.

JURGENSONE, I., J. CARSTENSEN, A. IKAUNIECE & B. KALVEKA, 2011. Long-term changes and controlling factors of phytoplankton community in the Gulf of Riga (Baltic Sea). Estuaries and Coasts, **34**: 1205-1219. DOI:10.1007/s12237-011-9402-x.

KIØRBOE, T., 1989. Phytoplankton growth rate and nitrogen content: implications for feeding and fecundity in a herbivorous copepod. Marine Ecology Progress Series, **55**: 229-234.

KIØRBOE, T., 1997. Population regulation and role of mesozooplankton in shaping marine pelagic food webs. Hydrobiologia, **363**: 13-27. DOI:10.1023/A:1003173721751.

KLAUSMEIER, C. A., E. LITCHMAN, T. DAUFRESNE & S. A. LEVIN, 2004. Optimal nitrogen-to-phosphorus stoichiometry of phytoplankton. Nature, **429**: 171-174. DOI:10.1038/nature02454.

KOSKI, M., 1999. Carbon: nitrogen ratios of Baltic Sea copepods — indication of mineral limitation? Journal of Plankton Research, **21**: 1565-1573. DOI:10.1093/plankt/21.8.1565.

KOSKI, M., J. ENGSTRÖM & M. VIITASALO, 1999. Reproduction and survival of the calanoid copepod *Eurytemora affinis* fed with toxic and non-toxic cyanobacteria. Marine Ecology Progress Series, **186**: 187-197. DOI:10.3354/meps186187.

KOTTA, J., V. LAURINGSON, G. MARTIN, M. SIMM, I. KOTTA, K. HERKÜL & H. OJAVEER, 2008. Gulf of Riga and Pärnu Bay. In: U. SCHIEWER (ed.), Ecology of Baltic coastal waters. Ecological Studies, **197**: 217-243. (Springer-Verlag, Berlin-Heidelberg).

LAMPERT, W., 1989. The adaptive significance of diel vertical migration of zooplankton. Functional Ecology, **3**: 21-27. DOI:10.2307/2389671.

LEE, C. E. & B. W. FROST, 2002. Morphological stasis in the *Eurytemora affinis* species complex (Copepoda: Temoridae). Hydrobiologia, **480**: 111-128. DOI:10.1023/A:1021293203512.

LIVDĀNE, L., I. PUTNIS, G. RUBENE, D. ELFERTS & A. IKAUNIECE, 2016. Baltic herring prey selectively on older copepodites of *Eurytemora affinis* and *Limnocalanus macrurus* in the Gulf of Riga. Oceanologia, **58**: 46-53. DOI:10.1016/j.oceano.2015.09.001.

MAUCHLINE, J., 1998. The biology of calanoid copepods. Advances in Marine Biology, **33**: i-x: 1-710. (Academic Press Inc., San Diego, London).

MEUNIER, C. L., M. BOERSMA, K. H. WILTSHIRE & A. M. MALZAHN, 2016. Zooplankton eat what they need: copepod selective feeding and potential consequences for marine systems. Oikos, **125**: 50-58. DOI:10.1111/oik.02072.

MOTWANI, N. H., J. DUBERG, J. B. SVEDÉN & E. GOROKHOVA, 2017. Grazing on cyanobacteria and transfer of diazotrophic nitrogen to zooplankton in the Baltic Sea. Limnology and Oceanography, **63**: 672-686. DOI:10.1002/lno.10659.

PERTOLA, S., M. KOSKI & M. VIITASALO, 2002. Stoichiometry of mesozooplankton in N- and P-limited areas of the Baltic Sea. Marine Biology, **140**: 425-434. DOI:10.1007/s00227-001-0723-3.

R CORE TEAM, 2018. R: a language and environment for statistical computing. (R Foundation for Statistical Computing, Vienna, Austria). https://www.R-project.org/.

SKUDRA, M. & U. LIPS, 2017. Characteristics and inter-annual changes in temperature, salinity and density distribution in the Gulf of Riga. Oceanologia, **59**: 37-48. DOI:10.1016/j.oceano.2016.07.001.

SNOEIJS-LEIJONMALM, P. & E. ANDRÉN, 2017. Why is Baltic Sea so special to live in?. In: P. SNOEIJS-LEIJONMALM, H. SCHUBERT & T. RADZIEJEWSKA (eds.), Biological oceanography of the Baltic Sea: 23-84. (Springer, Dordrecht, Netherlands).

STERNER, R. W. & J. J. ELSER, 2002. Ecological stoichiometry: the biology of elements from molecules to the biosphere: 1-440. (Princeton University Press, Princeton, NJ).

STERNER, R. W. & D. O. HESSEN, 1994. Algal nutrient limitation and the nutrition of aquatic herbivores. Annual Review of Ecology and Systematics, **25**: 1-29. DOI:10.1146/annurev.es.25.110194.000245.

SUKHIKH, N., A. SOUISSI, S. SOUISSI, G. WINKLER, V. CASTRIC, A. C. HOLL & V. ALEKSEEV, 2016. Genetic and morphological heterogeneity among populations of *Eurytemora affinis* (Crustacea: Copepoda: Temoridae) in European waters. Comptes Rendus Biologies, **339**: 197-206. DOI:10.1023/A:1021293203512.

VILLAR-ARGAIZ, M., J. M. MEDINA-SÁNCHEZ & P. CARRILLO, 2002. Linking life history strategies and ontogeny in crustacean zooplankton: implications for homeostasis. Ecology, **83**: 1899-1914. DOI:10.1890/0012-9658(2002)083[1899:LLHSAO]2.0.CO;2.

VREDE, T., T. ANDERSEN & D. O. HESSEN, 1999. Phosphorus distribution in three crustacean zooplankton species. Limnology and Oceanography, **44**: 225-229. DOI:10.4319/lo.1999.44.1.0225.

VREDE, T., D. R. DOBBERFUHL, S. KOOIJMAN & J. J. ELSER, 2004. Fundamental connections among organism C : N: P stoichiometry, macromolecular composition, and growth. Ecology, **85**: 1217-1229. DOI:10.1890/02-0249.

VUORINEN, I., 1987. Vertical migration of *Eurytemora* (Crustacea, Copepoda): a compromise between the risks of predation and decreased fecundity. Journal of Plankton Research, **9**: 1037-1046. DOI:10.1093/plankt/9.6.1037.

WALVE, J. & U. LARSSON, 1999. Carbon, nitrogen and phosphorus stoichiometry of crustacean zooplankton in the Baltic Sea: implications for nutrient recycling. Journal Plankton Research, **21**: 2309-2321. DOI:10.1093/plankt/21.12.2309.

WICKHAM, H., 2009. ggplot2: elegant graphics for data analysis. (Springer-Verlag, New York, NY).

YURKOVSKIS, A., E. KOSTRICHINA & A. IKAUNIECE, 1999. Seasonal succession and growth in the plankton communities of the Gulf of Riga in relation to long-term nutrient dynamics. Hydrobiologia, **393**: 83-94. DOI:10.1007/978-94-017-0912-5_9.

First received 30 August 2019.
Final version accepted 9 December 2019.

[When citing this volume, please refer to *Crustaceana* 93 (2020), 241-547]

*Proceedings of the* Eurytemora *conference, St. Petersburg, 2019*

# COMPARISON OF RESPIRATION RATE AND ELECTRON TRANSPORT SYSTEM (ETS) ENZYME-MEDIATED REDUCTION ASSAY OF THE INVASIVE COPEPOD *EURYTEMORA CAROLLEEAE* ALEKSEEV & SOUISSI, 2011 (CALANOIDA, TEMORIDAE) IN GREEN BAY, WI, U.S.A.

BY

ALEXANDER W. TIMPE and BART T. DE STASIO[1])
Biology Department, Lawrence University, 711 East Boldt Way, Appleton, WI 54911, U.S.A.

## ABSTRACT

Anthropogenic activities can cause important changes in aquatic ecosystems, such as warming due to climate change, nutrient loading from agricultural runoff and urban areas, and decreased concentrations of oxygen in bottom waters. These changes may lead to impacts on both organism performance and ecosystem functionality. Studying planktonic species that form an aquatic ecosystem's foundation is an important step towards understanding the entire food web and predicting how it may respond to a changing environment. One important planktonic species in the Laurentian Great Lakes is the invasive calanoid copepod *Eurytemora carolleeae* (formerly considered part of the *Eurytemora affinis* species complex). This study analyzes the metabolic activity of *E. carolleeae* from Green Bay, Lake Michigan, U.S.A. using two different methods, over a range of temperatures from 9 to 26°C. Total oxygen consumption was measured directly using a micropulse oxygen probe system, and the activity of aerobic metabolic enzymes in the electron transport system (ETS) was quantified using in vitro reduction of iodonitrotetrazolium chloride (INT). Respiration rate of *E. carolleeae* increases approximately linearly from 9 to 26°C. Measurements of ETS activity indicate that the copepod's metabolic enzymes have an Arrhenius activation energy of $46.5 \pm 15.6$ kJ/mol with a thermal maximum between 22 and 26°C. Overall, *E. carolleeae* ETS rates increased by approximately 7% per °C over the range 9 to 22°C. This thermal limit has implications for future performance of this species, as the combination of higher temperatures and disappearance of oxygenated colder-water refuges may limit *E. carolleeae*'s success in the Green Bay system following warmer climate and increased nutrient conditions.

## RÉSUMÉ

Les activités anthropiques peuvent causer des changements importants dans les écosystèmes aquatiques, comme un réchauffement dû au changement climatique, un enrichissement en nutriments par le ruissellement agricole et les zones urbaines, et une diminution de la concentration en oxygène des eaux profondes. Ces changements peuvent entrainer des impacts sur les performances des organismes et le fonctionnement des écosystèmes. Etudier les espèces planctoniques qui forment la

[1]) Corresponding author; e-mail: bart.t.destasio@lawrence.edu

 | DOI:10.1163/9789004465565_015

base de l'écosystème aquatique est une étape importante vers la compréhension du réseau trophique, et prédire comment il pourrait répondre à un changement d'environnement. Une espèce planctonique importante des grands lacs laurentiens est le copépode calanoïde invasif *Eurytemora carolleeae* (considéré antérieurement comme faisant partie de l'espèce *Eurytemora affinis*). Cette étude analyse l'activité métabolique de *E. carolleeae* de Green Bay, Lac Michigan, U.S.A., en utilisant deux méthodes, sur une échelle de températures de 9° à 26°C. La consommation d'oxygène a été mesurée directement avec une sonde à oxygène, et l'activité des enzymes du métabolisme aérobie dans le système de transport des électrons (ETS) a été quantifiée par la réduction in vitro du chlorure d'iodonitrotetrazolium (INT). Le taux respiratoire de *E. carolleeae* augmente linéairement de 9° à 26°C. Les mesures de l'activité ETS indiquent que les enzymes métaboliques du copépode ont une énergie d'activation d'Arrhenius de 46,5 ± 15,6 kJ/mol avec un maximum thermique entre 22° et 26°C. Globalement les taux ETS de *E. carolleeae* ont augmenté d'approximativement 7% par °C sur l'intervalle de 9° à 22°C. Cette limite thermale a des implications pour les performances futures de cette espèce, la combinaison de températures plus hautes et la disparition des eaux plus froides oxygénées comme refuge pourrait limiter le succès de *E. carolleeae* dans le système de Green Bay à la suite d'un réchauffement climatique et d'une augmentation des conditions de nutrition.

## INTRODUCTION

Invasions by crustacean zooplankton present important opportunities to understand how organisms and ecosystems may respond to changes over time. When invasive species become established in new environments and begin interacting with native species, it is possible to gain insight into mechanisms controlling both structure and function of these systems. A good example is the recent invasion of the Laurentian Great Lakes of North America by *Eurytemora carolleeae* Alekseev & Souissi, 2011 (formerly as *E. affinis* (Poppe, 1880): Alekseev & Souissi, 2011). *E. carolleeae* has been in the Great Lakes since at least 1958 (Anderson & Clayton, 1959; Lee, 1999), and has been recorded since 1966 in the largest freshwater estuary of the Great Lakes, Green Bay of Lake Michigan (Gannon, 1974). The expansion into freshwater habitats required changes in physiology, especially related to osmoregulation (Lee et al., 2003, 2011; Gerber et al., 2016). Work by Lee et al. (2013) has shown that the ability of this species to cope with freshwater conditions depends on high food availability, which makes eutrophic embayments like Green Bay prime locations for successful establishment. Further work has demonstrated evolution of various enzyme systems related to osmoregulation, with increased activity of V-type $H^+$-ATPase and reduction of $Na^+/K^+$-ATPase enzymes in the freshwater *E. carolleeae* populations relative to ancestral populations from saline environments (Lee et al., 2011; Gerber et al., 2016). It is unclear whether these shifts in enzyme systems during invasion of freshwater habitats have resulted in shifts in overall metabolic activity, but the increased survival with higher food concentrations may suggest increased metabolic demands in freshwater conditions. Further studies of the metabolic activity of *E. carolleeae* are needed to understand how this species survives in the invaded habitats, performs compared to potential competitors, and how future changes in environmental conditions may affect these ecosystems.

Previous studies of *E. carolleeae* in Green Bay demonstrate that it can grow and reproduce on local food resources. Richman et al. (1980) demonstrated that it is more selective for large food particles and has higher filtering rates than native calanoid copepods (*Skistodiaptomus oregonensis* (Lilljeborg, 1889), *Leptodiaptomus siciloides* (Lilljeborg in Guerne & Richard, 1889), and *L. ashlandi* (Marsh, 1893)). Engström-Öst et al. (2017) showed that ingestion rates are negatively affected by toxins released by the cyanobacterium *Microcystis* but that there are no significant effects of toxins on survivorship or egg production when animals are fed a mixed diet of the cyanobacterium *Microcystis* and the green alga *Scenedesmus*. These results are similar to studies of the ancestral species, *E. affinis*, in eutrophied areas of the Baltic Sea (Adrian et al., 1999). Animals in eutrophic regions demonstrated higher feeding rates and higher growth efficiencies (e.g., feeding adjusted for metabolic demands) than other calanoid copepods (*Acartia* spp.). In the Green Bay system, nothing is known about the metabolic costs incurred by *E. carolleeae*, or how these might compare to those of potential competitors like native copepods. To further understand the current food web interactions and potential future changes in this ecosystem there is a clear need for studies of metabolic activity and respiration rates of *E. carolleeae* and native copepods.

Here we report results of metabolic activity over a range of temperatures for *E. carolleeae* from Green Bay, Lake Michigan, U.S.A. We determined respiration rates as a direct measure of metabolic demands and compare them to the standard Electron Transport System (ETS) enzyme-mediated reduction assay often used to measure copepod metabolic potential (King & Packard, 1975; Hernández-Léon, 2000). ETS has often been used as a rapid assessment tool of potential metabolic activity of animals but there are a number of factors that affect the relationship between ETS activity and respiration rates (Hernández-Léon & Gómez, 1996). As a result, there is a need to determine these relationships for specific situations, especially following a recent invasion where animals are functioning under relatively novel conditions.

## MATERIAL AND METHODS

### Collection and maintenance of *Eurytemora carolleeae*

Experiments were conducted on the calanoid copepod *Eurytemora carolleeae*. Animals were collected between 25 July and 19 October 2018 in their natural assemblage from Little Sturgeon Bay (LSB) area of Green Bay, Lake Michigan, U.S.A. (44°50′42.72″N 87°33′30.24″W), using a 0.5 m-diameter plankton net with 250 $\mu$m mesh. Samples were collected with duplicate net tows in nearshore environments and rinsed into buckets containing approximately 15 litres of water from LSB. Juvenile fish and large clumps of macrophytes were removed from samples prior to transport to the laboratory. Live adult *E. carolleeae* in samples were identified and sorted manually using wide-mouthed pipettes at 10-40× magnification within 48 hours of transport back to the lab, and incubated in containers containing aged tap water at a range of temperatures. We used aged tap water to limit the potential effects of bacterial growth on respiration rate measurements due to high dissolved organic matter concentrations in LSB water. The source of tap water was Lake Winnebago, WI, U.S.A., the main tributary to Green Bay, previously shown to provide a suitable substitute for LSB water in experiments with *E. carolleeae* (cf. Engström-Öst et al., 2017). *Eurytemora carolleeae* were identified according to Balcer et. al. (1984), assuming that all *Eurytemora* were *E. carolleeae*, as has been determined previously for the Laurentian Great Lakes (Vasquez et al., 2016).

## Measurement of respiration

Copepods used in respiration experiments were incubated in the dark for at least 48 hours at a given experimental temperature prior to respiration measurement (table I). Field samples were placed initially in an incubator at 18°C and then sorted to isolate *E. carolleeae* for experiments. The monospecific collections of *E. carolleeae* were placed in incubators or at room temperature (22°C) and reached target temperatures within 2-4 hours. During the subsequent 48 hour acclimation period animals were fed ad libitum with the green alga *Scenedesmus quadricauda* (Turpin) Brébisson, 1835, shown previously to provide good survivorship and reproduction for *E. carolleeae* from this location (Engström-Öst et al., 2017). After acclimation, an average of 23 healthy, adult *E. carolleeae* were gently rinsed by transfer into filtered, aged tap water [oxygen-saturated, Whatman GF/C filters] to remove attached algae (table I). Animals were then placed in 55 ml glass chambers containing filtered aged tap water and sealed with an oxygen probe to avoid trapped air bubbles. Multiple computer-controlled Clarke-type micropulse sensors were employed, with measurements taken at 15 min intervals (Endeco). Respiration experiments included two or three experimental bottles and two or one control bottles run concurrently. Control bottles contained filtered, aged tap water. To reduce temperature fluctuations during experiments all bottles were submerged in a 4 litre bucket containing water at the incubation temperature and held in an incubator in the dark. Experiments were conducted for 12-24 hours and were

TABLE I

Details of metabolic rate experiments on *Eurytemora carolleeae* Alekseev & Souissi, 2011, indicating temperatures tested, number of replicate runs at each temperature, and mean number of copepods tested in each replicate run

| Parameter | Temperature (°C) | Number of replicates | Mean number of copepods tested per replicate |
|---|---|---|---|
| Respiration rate | 9 | 4 | 30 |
| | 14 | 6 | 23 |
| | 18 | 7 | 17 |
| | 22 | 3 | 19 |
| | 26 | 3 | 28 |
| ETS Activity | 9 | 4 | 30 |
| | 14 | 6 | 23 |
| | 18 | 7 | 17 |
| | 22 | 10 | 31 |
| | 26 | 3 | 28 |

terminated when oxygen concentration in one experimental bottle reached 80% of the starting value. Oxygen concentrations often varied during the initial period of experiments (first 2-4 hours), so respiration rates were determined following this time period when concentrations were decreasing linearly. After each experiment, respiration chambers were dried and cleaned with 95% ethanol to prevent bacterial growth.

Changes in oxygen concentration in chambers were determined using a least-squares linear best fit regression over time using Microsoft Excel. Respiration rates were calculated based on 40-80 data points obtained for each bottle. The average $r^2$ for all regressions was >0.99. Respiration rates were calculated by subtracting the mean slope for control bottles from the slope for experimental bottles and transforming appropriately to obtain final units as $\mu$l $O_2$/copepod per h. Effect of temperature on respiration rates was tested using one-way ANOVA and described by ordinary least squares linear regression. All statistical tests were performed using the Paleontological Statistics (PAST) software package (Hammer et al., 2001).

## Determination of ETS activity

Potential metabolic activity was determined for *E. carolleeae* using the Electron Transport System (ETS) procedure following Hernández-Léon (2000) and Simčič & Brandcelj (2004), but with centrifugation performed at room temperature. Additionally, dilution of the homogenate was not necessary to maintain activity within the sensitivity level of the experiment, as recommended by Hernández-Léon (2000).

Groups of active, heathy-looking *E. carolleeae* (table I; mean $n = 25.8$ per replicate) were rinsed in filtered, aged tap water and placed into Teflon glass grinders containing homogenization buffer (0.1 M sodium phosphate buffer pH 8.4, 75 mM $MgSO_4$, 0.15% (w/v) polyvinyl pyrrolidone, 0.2% (v/v) Triton-X-100). Organisms were manually crushed for 2 min, then transferred to a 4 ml screw-top plastic centrifuge tube and placed on ice. Homogenate was sonicated using a Branson ultrasonic tissue disrupter for 20 s at 40 W (20% amplitude) and centrifuged for 4 min at 10 000 rpm.

Each sample was prepared in triplicate or quadruplicate as technical replicates and incubated for 40 to 60 min at the experimental temperature along with a substrate blank and triplicate reagent blanks. Each sample combined 0.5 ml supernatant, 1.5 ml substrate solution, and 0.5 ml from a frozen stock of INT solution (2.5 mM 2-*p*-iodo-phenyl 3-*p*-nitrophenyl 5-phenyl tetrazolium chloride (Cayman, Ann Arbor, MI, U.S.A.) in milli-Q water). Substrate solution was prepared daily by combining stock buffer (0.1 M sodium phosphate buffer pH 8.4, 0.2% (v/v) Triton-X-100; kept frozen at −20°C) with NADH and NADPH (Cayman) to achieve molar concentrations of 1.7 mM and 0.25 mM, respectively. Substrate blanks (0.5 ml supernatant and 1.5 ml homogenization buffer) were prepared for each sample, and triplicate reagent blanks (0.5 ml homogenization buffer, 1.5 ml substrate daily substrate buffer, and 0.5 ml INT) were prepared for each incubation. Each incubation contained between 2 and 4 samples. After the incubation period, 0.5 ml quench solution (equal parts formalin (36%, aqueous) and concentrated $H_3PO_4$) was added to each sample and reagent blank to cease enzymatic activity and stop the abiotic reduction of INT.

Amount of INT-formazan was determined by spectrophotometry (Cary 50 UV-Vis scanning spectrophotometer, Varian, Palo Alto, CA, U.S.A.) immediately after the addition of the quench solution. Absorbance was measured at 450 nm, with absorbance at 750 nm subtracted as a turbidity blank. ETS activity ($\mu$l $O_2$/animal per h) was calculated following Hernández-Léon (2000), using the standard conversion factor of INT to volume of oxygen of 1.42.

### Determination of ETS : Respiration ratios

Comparisons of ETS and respiration measurements were obtained by performing ETS assays on the same individuals that were used in respiration measurement experiments. At the end of each experiment all *E. carolleeae* were re-counted and washed with fresh, filtered aged tap water. Any *E. carolleeae* that had died in the respiration experiment or those that looked sickly were excluded from the subsequent ETS analysis. Excess water was removed by placing animals on filter paper (Whatman No. 1 qualitative filter), and total wet weight of each batch of copepods was measured using a microbalance (Cahn C-35 ultra-microbalance, Thermo

Electron, Beverly, MA, U.S.A.). Animals were stored on pieces of aluminum foil at −20°C immediately after weighing. ETS activity of those animals was determined as described above except they were not re-rinsed with aged tap water. Instead, the frozen mass of *E. carolleeae* was removed from the foil and added directly to the homogenization buffer.

### Determination of Arrhenius activation energy ($E_A$) and $Q_{10}$ values

Results of the ETS : Respiration ratio experiments were used to determine the Arrhenius activation energy ($E_A$; expressed in kcal/mol) according to the calculations in Packard et al. (1975). The $E_A$ is determined as $E_A = -RS$, where $R$ is the gas constant (1.987 kcal/mol) and $S$ is the slope of a linear least-squares regression. An Arrhenius plot with a linear least-squares regression was created in Microsoft Excel using the natural log (ln) of ETS activity against the reciprocal of absolute temperature (in K). The ETS value used for each temperature was the mean of replicate ETS activities used to determine the ETS : Resp ratio at that temperature. The 95% confidence intervals of the slope (and by consequence 95% confidence intervals for $E_A$) were determined using a bivariate ordinary least-squares regression in PAST (Hammer et al., 2001). Values of $Q_{10}$ as defined by the Van't Hoff rule were calculated as $Q_{10} = (k_1/k_2)^{10/(t_1-t_2)}$ where $k_1$ and $k_2$ are ETS rates corresponding to temperatures $t_1$ and $t_2$ (Hernández-Léon & Ikeda, 2005). Also, a measure of the percent change in ETS rate per °C was calculated using the log-linear regression technique employed by Heine et al. (2019).

### Statistical analyses

Statistical analyses were performed with the Paleontological Statistics (PAST) software package (Hammer et al., 2001). Data distributions were examined for deviations from normality using the Shapiro-Wilks test. Data that deviate from normality were tested using non-parametric statistical analyses (Kruskal-Wallis & Mann-Whitney tests) because transformations did not eliminate non-normality issues.

## RESULTS

### Respiration rates

Oxygen use by *Eurytemora carolleeae* responded directly to temperature of incubation (fig. 1). Respiration rates increased approximately ten-fold over the temperature range examined, from 0.025 ± 0.011 $\mu$l $O_2$/animal per h at 9°C (mean ± 95% CI, $n = 4$) to 0.238 ± 0.088 $\mu$l/animal per h at 26°C ($n = 3$).

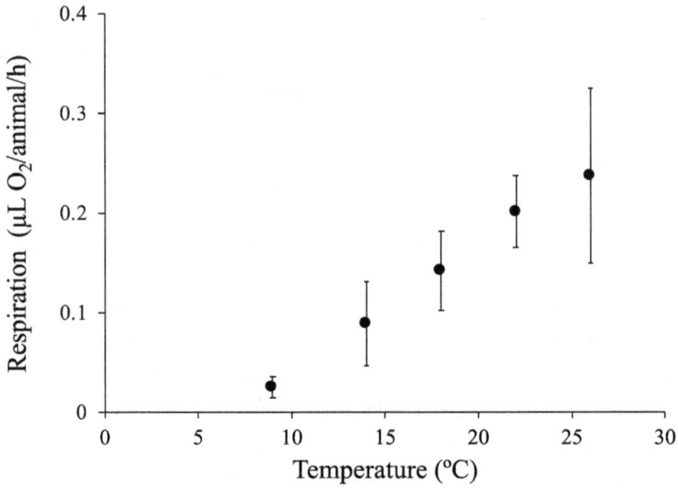


Fig. 1. Mean respiration rate of *Eurytemora carolleeae* Alekseev & Souissi, 2011 at different temperatures. Individuals were acclimated to experimental temperatures for at least 48 hours prior to experiments. Error bars represent 95% confidence intervals.

There was a significant effect of temperature on respiration rates ($F_{4,18} = 10.35$; $p = 0.00016$). The data can be best modeled by the linear least-squares regression $R = 0.0128 *$ Temperature $(°C) - 0.0897$ ($r^2 = 0.996$; $p = 0.007$).

## Effect of respiration experiments on ETS activity

Using *E. carolleeae* in respiration experiments did not significantly affect ETS activity levels compared to those not used in experiments (fig. 2). Mean ETS rates at 18°C were approximately 0.063 $\mu$l $O_2$/copepod per h, whereas ETS rates at 22°C were 0.071 $\mu$l $O_2$/copepod per h. There was no significant effect on ETS activity of copepods being used or not used in respiration experiments (2-way ANOVA: $F_{1,31} = 0.1076$; $p = 0.745$) or between temperatures ($F_{1,31} = 0.591$; $p = 0.448$).

## ETS activity of *E. carolleeae*

The calculated in situ ETS activity of *E. carolleeae* increased with increasing temperature until reaching a peak at 22°C (fig. 3). ETS was lowest at 9°C with a value of $0.031 \pm 0.014$ $\mu$l $O_2$/animal per h (mean $\pm$ 95% CI, $n = 4$) and increased to its maximum of $0.076 \pm 0.007$ $\mu$l $O_2$/animal per h at 22°C ($n = 4$). Mean ETS activity subsequently decreased to $0.053 \pm 0.014$ $\mu$l $O_2$/animal per h at 26°C ($n = 3$). There was a significant effect of temperature on ETS activity (ANOVA: $F_{4,25} = 7.358$; $p = 0.0005$) with significant differences between rates at 22°C compared to both 14 and 9°C (Tukey's pairwise tests: $p < 0.002$ for both). The

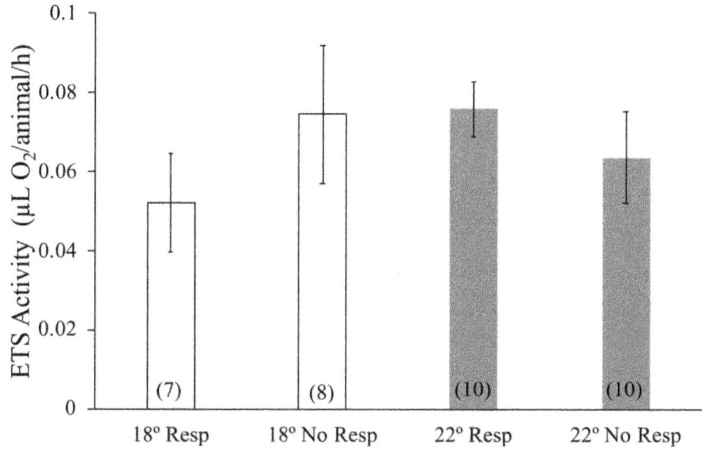


Fig. 2. ETS activity in *Eurytemora carolleeae* Alekseev & Souissi, 2011 used in respiration experiments (Resp) compared to those not used in respiration experiments (No Resp) at 18 and 22°C. Number of sample runs for each category shown in parentheses (with $n = 23$ copepods per sample). Error bars indicate 95% confidence intervals.

relationship between mean ETS activity and temperature from 9 to 22° can be best explained by the exponential least-squares regression:

$$\mathrm{ETS} = 0.0162 * e^{0.0675 * \mathrm{Temp}(^\circ C)} \quad \left(r^2 > 0.97\right).$$

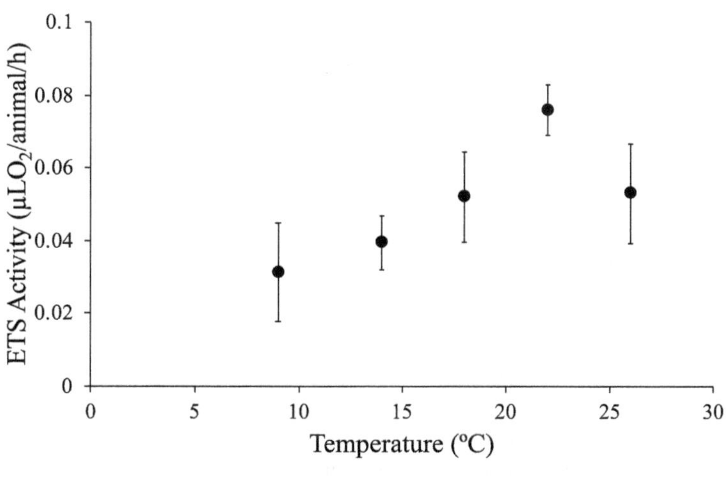


Fig. 3. Mean ETS activity of *Eurytemora carolleeae* Alekseev & Souissi, 2011 as a function of temperature. Error bars indicate 95% confidence intervals.

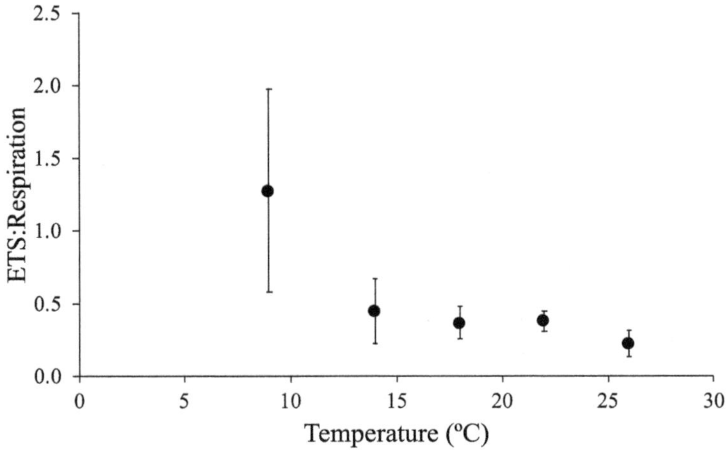


Fig. 4. ETS : Respiration ratio (mean ± 95% CI) for *Eurytemora carolleeae* Alekseev & Souissi, 2011 at 9, 14, 18, 22 and 26°C. ETS activity was measured directly after each respiration experiment.

## ETS : Resp ratios

The ETS : Resp ratio for *E. carolleeae* acclimated to the respiration temperatures decreased significantly with increasing temperature (fig. 4; Kruskal-Wallis test: $H = 10.41$; $p = 0.034$). The calculated ratio was highest at 9°C with a value of $1.28 \pm 0.70$ ($n = 4$; mean ± 95% CI) and was significantly greater than rates at the other temperatures (Mann-Whitney tests: $p \leqslant 0.05$ for all comparisons). Mean ratio values for all other temperatures were less than 0.5.

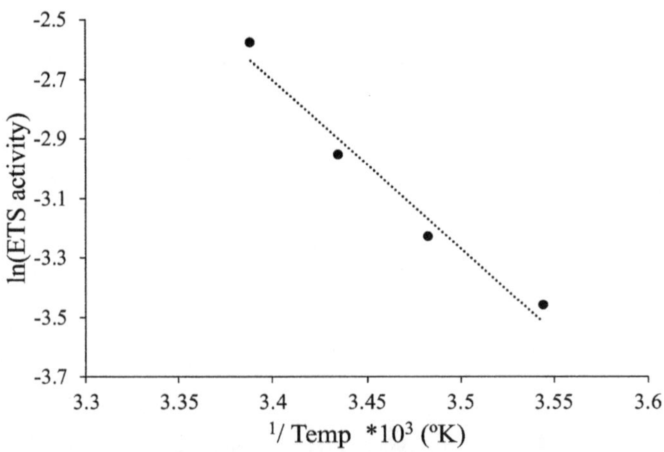


Fig. 5. Arrhenius plot showing mean ln (ETS activity) against reciprocal of absolute temperature for *Eurytemora carolleeae* Alekseev & Souissi, 2011. All ETS assays were incubated at acclimation temperatures (9, 14, 18 and 22°C). Dashed line indicates least-squares linear regression ($r^2 = 0.97$).

### Arrhenius activation energy ($E_A$) and $Q_{10}$ values

There was a strong negative linear relationship between the natural log of ETS activity and the inverse of the absolute temperature ($r^2 = 0.97$; fig. 5). The slope of the linear regression indicates an Arrhenius activation energy ($E_A$) for *E. carolleeae* of 46.5 ± 15.6 kJ/mol (mean ± 95% CI). ETS data for 26°C were excluded as they were past the temperature of maximum enzyme activity (fig. 3) and therefore do not accurately reflect the dependence of enzyme rates on temperature. Using ETS data between 9 and 22°C the calculated $Q_{10}$ value for *E. carolleeae* was 1.97. The percent change in ETS rate was 6.99 ± 0.79 per °C (mean ± SE) as calculated with the log-linear technique of Heine et al. (2019).

## DISCUSSION

This study's primary focus is metabolic activity of the invasive calanoid copepod *Eurytemora carolleeae*. Many aspects are consistent with those determined previously for marine calanoid copepods, but some features differ from general relationships reported in the literature. Metabolic activity quantified using traditional respirometry indicates that oxygen consumption increases as a function of temperature between 9 and 26°C (fig. 1), while the Electron Transport System (ETS) assay demonstrated that potential metabolic activity increased exponentially between 9 and 22°C then decreased again at 26°C (fig. 2). However, the ETS rates we observed were lower than those generally reported for marine copepods. Other measures of changes in metabolic activity with temperature (Arrhenius activation energy ($E_A$), $Q_{10}$, and percent change of rates per degree change in temperature) were consistent with published values for other copepods. The ratio of ETS activity and respiration rate varied inversely with temperature, casting doubt on the use of ETS as a general proxy for respiration rate estimates. These findings provide a first step in assessing the potential responses of this invasive copepod to anthropogenic perturbations in the Green Bay system and how changing environmental conditions may affect its success relative to its native competitors.

### Respiration rates and competitive ability

Respiration rates for *E. carolleeae* in this experiment compare well with the range of few previously published values for *Eurytemora*. In their 1999 study, Adrian et al. measured respiration rates of *E. affinis* in the Baltic Sea, showing that rates ranged from 0.04 $\mu$l $O_2$/animal per h at 6°C to 0.3 $\mu$l $O_2$/animal per h at 18°C, very similar to our range for *E. carolleeae* (fig. 1). Other studies on *Eurytemora* have found mean rates similar to these with values in the 0.1 to

0.25 $\mu$l $O_2$/animal per h range at 15-20°C (Raymont, 1959; Barthel, 1983; Ikeda et al., 2001; Paffenhöfer, 2006). A summary by Lampert (1984) of respiration data for various freshwater calanoid copepod species shows respiration rates between 0.05 and 0.1 $\mu$l $O_2$/animal per h at temperatures ranging from 15 to 20°C. All of these rates represent what is typically termed "routine respiration" rates because animals are swimming freely during the experiments, and some portion of the metabolic activity is due to costs of swimming (Hernández-Léon & Ikeda, 2005). This component of metabolism is difficult to control in experiments, but should not be an important factor contributing to the difference between published values and our estimates because of the similarity of experimental conditions employed. Some studies also focus on separate components of metabolism by determining basal ("standard") versus "routine respiration" rate (e.g., Kiørboe et al., 1985). While this distinction is important, it was not the focus of our work. When comparing respiration rate with the ETS enzyme system, it may be more appropriate to use "routine respiration" rates. The ETS enzyme system integrates total potential oxic production of ATP, which includes investment in foraging, as well as any cellular processes and other aspects of feeding (i.e., specific dynamic action) including mechanical ingestion and digestion, assimilation, growth, and somatic tissue maintenance (Hernández-Léon & Ikeda, 2005). The maintained level of ETS enzymes in a copepod should reflect its optimal long-term activity requirements and consequently should be less affected by temporary stressors such as short bouts of activity. Therefore, the relationship between ETS activity and routine metabolic rate is a comparison of the potential ATP "machinery" and its actual usage under normal conditions.

The respiration rates of *Eurytemora* species presented in this study and those reported by Adrian et al. (1999) under low food availability, are higher than other rates determined for *Eurytemora*, especially at higher temperatures. Rates one order of magnitude lower than our rates have been reported for animals of comparable sizes (Gyllenberg & Lundqvist, 1978; Boak & Goulder, 1983; Roddie et al., 1984). For *E. carolleeae* this may be explained in part by the recent expansion into fresh water from its native brackish habitat (Mills et al., 1993; Kipp et al., 2013). Lee et al. (2013) found that food abundance was an important determinant of the survival rate of North American *Eurytemora* coping with low-salinity environments. They found that survival rate was significantly higher when there was a large supply of available food, as the copepod could offset metabolic costs of maintaining homeostasis in a low-ion environment. Other studies of this same species demonstrated that evolutionary shifts in ion regulation following invasion of freshwater habitats may contribute to these increased metabolic demands (Lee et al., 2011; Gerber et al., 2016). The authors also propose that this phenomenon will occur until the established population adapts over evolutionary time. The high respiration rates observed in our study are consistent with this hypothesis given the invasion of *E. carolleeae* into the Great Lakes is recent on an evolutionary timescale. Our data for *E. carolleeae* raise the question of whether *Eurytemora* generally exhibits higher metabolic activity than other copepods or if the higher rates we measured are due to the recent invasion of freshwater habitats. Overall, very few studies have focused on *Eurytemora* metabolic activity and clearly there is a need for further investigations of this topic.

We were able to compare respiration rates of *E. carolleeae* to those of the most likely competitor in Green Bay, the native calanoid copepod *Skistodiaptomus oregonensis*. Respiration rates of *E. carolleeae* were significantly higher than those of *S. oregonensis* at 22°C but not at 14°C (A.W. Timpe & B.T. De Stasio, unpubl. data). This could indicate that *E. carolleeae* would be at a competitive disadvantage at higher temperatures; however, this does not consider possible differences in feeding abilities. Feeding rates for these two species in Little Sturgeon Bay have been determined in previous studies, indicating that *E. carolleeae* is a more selective feeder and exhibits lower filtering rates than *S. oregonensis* at 22°C (Richman et al., 1980). Previous researchers have used the ratio of (filtering rate: respiration rate) as a means of combining feeding and metabolic demands into a single measure of zooplankton performance (Richman & Dodson, 1983; Adrian et al., 1999). Combining the previously published filtering rate data and our respiration rates for these two species at 22°C indicates that *S. oregonensis* filtering rates are 8.5 times as high as respiration rates, whereas the ratio for *E. carolleeae* is only 1.1. This means that respiration specific filtering rates for the native copepod *S. oregonensis* are 7.9 times as high as for the invasive *E. carolleeae*. This indication of a relative competitive advantage for *S. oregonensis* is consistent with population data for these two species in Green Bay. *S. oregonensis* abundances are consistently greater than those of *E. carolleeae* (cf. Gannon, 1974; Torke, 2001; B.T. De Stasio, unpubl. data). While these rough estimates are suggestive, additional studies combining feeding and respiration rates at varying temperatures and food abundances are needed to further our understanding of the competitive interactions between the invasive and native copepods in Green Bay, the rest of the Laurentian Great Lakes (e.g., Winkler et al., 2008), as well as other areas such as the Baltic Sea (e.g., Sukhikh et al., 2019).

### ETS activity and ETS : Respiration ratios

Metabolic rates we measured as ETS activity were generally lower than those reported for other copepods (fig. 3), with rates for *E. carolleeae* being half to a third of other published values (e.g., Ikeda, 1977; Devol & Packard, 1978; Båmstedt, 1980; Cammen, 1990). As a result, the values we measured for ETS : Resp are considerably lower than many reported values for both marine and freshwater zooplankton (fig. 4). Owens & King (1975) found a ratio of 2.02 for the predatory copepod *Calanus pacificus* Brodsky, 1948, while Packard (1985) and Rai (2002) cite 2.04 and 1.96, respectively, as more typical ETS : Resp ratios for zooplankton. Simčič & Brancelj (2004) found ETS : Respiration ratios of approximately 2 in several *Daphnia* hybrids. However, an extensive survey of the literature for marine zooplankton species indicated a range from 0.4 to over 5 (Hernández-Léon & Gómez, 1996), with clear effects of animal activity, temperature, and food limitation on the relationship. Our findings do show that ETS : Respiration ratios were negatively related to temperature (fig. 4). To examine the effects that starvation during respiration had on the measured ETS, we directly compared the enzymatic activities of animals after being acclimated while feeding and those used in respiration experiments (e.g., animals starved during measurements). We found no significant difference between groups (fig. 2), indicating that starvation was not likely a cause for low ratios. Rates for ETS theoretically should always be equal to or greater than respiration rates (Packard, 1985) because ETS is thought to measure maximum potential enzymatic activity. However, the recurring observation of ETS : Respiration ratios less than 1, as shown in our study, suggests other factors may be occurring to alter the relationship between ETS and overall respiration rates in some groups. In the case of *E. carolleeae* this could be related to the recent invasion of freshwater habitats and subsequent evolution of enzyme systems determining osmoregulation (e.g., Lee et al., 2011).

### Arrhenius activation energy ($E_A$) and $Q_{10}$

Despite the low measured ETS values of *E. carolleeae*, the Arrhenius plot presented in fig. 5 closely mirrors the relationship between activity and temperature as described by Packard et al. (1975). The data in this study display the characteristic exponential increase in enzymatic reaction rates followed by a sharp decrease after a temperature maximum (Packard et al., 1975; Simčič & Brancelj, 2004). The thermal maximum observed in *E. carolleeae* around 22°C was lower than the reported maximum temperatures near 40°C for marine copepod assemblages as reported by Packard (1975) but were close to the 25°C inflection point determined by Simčič & Brancelj (2004) for freshwater cladocerans. This is also consistent with observations from natural assemblages in the Great Lakes, where populations disappear in the middle of the summer when water temperatures exceed 24°C. Lab experiments have shown that at temperatures above 22°C, growth is maximized but the number and quality of offspring produced drops dramatically, suggesting that temperatures below 20°C are optimal for *E. carolleeae* (cf. Lloyd et al., 2013). The differences in thermal maxima illustrate the importance of species identity on enzymatic kinetics and potential changes following invasion of freshwater habitats.

The calculated Arrhenius activation energy $E_A$ of 46.5 kJ/mol is also comparable to $E_A$ values obtained by other investigators for planktonic organisms or assemblages. Packard & King (1975) determined an $E_A$ of 63.6 kJ/mol for a marine copepod assemblage and reported values between 48.9 and 91.7 kJ/mol for other bulk plankton assemblages, with a general $E_A$ of 62.8 kJ/mol used in subsequent experiments (King & Packard, 1975; Devol & Packard, 1978). For freshwater species, Simčič & Brancelj (2004) recorded $E_A$ values between 46.22 and 74.36 for hybrid *Daphnia* populations and Pascal & Chong (2016) reported an $E_A$ of 34.8 to 76.2 for the copepod species *Pseudodiaptomus annandalei* Sewell, 1919. Generally, *E. carolleeae* has a lower $E_A$ and enzymatic thermal maximum than other species. But, given *E. carolleeae* is native to colder waters (Balcer et al., 1984) and is most common during colder parts of the year in the Great Lakes system (Torke, 2001), that result perhaps is not surprising. Enzymes with lower activation energies, while unstable at higher temperatures, are more efficient at the lower temperatures under which *E. carolleeae* likely evolved.

The $Q_{10}$ value obtained for *E. carolleeae* based on ETS activity is consistent with other published information. Summary data compiled for marine copepods show $Q_{10}$ values of 2.03 for epipelagic species and 3.0 for mesopelagic groups (Hernández-Léon & Ikeda, 2005). More detailed work for epipelagic species produced values of 1.8-2.0 (Ikeda et al., 2001). Our value of 1.97 for *E. carolleeae* is consistent with that of the epipelagic groups, and is also similar to the general value of 2.0 typically reported for freshwater zooplankton (Lampert, 1984). Using the higher resolution technique of Heine et al. (2019) our calculated percent change of respiration based on our ETS values was 6.99% per °C, very similar to a mean value of 6.92% reported by Heine et al. (2019) for marine calanoid copepods. The changes in metabolic activity across different temperatures for *E. carolleeae* indicate it will likely respond to future temperature changes in a way similar to other copepods.

## CONCLUSIONS AND IMPLICATIONS

This study has several implications for both the utility of the ETS assay as a proxy measurement of respiration rate and for the future success of *E. carolleeae* in Green Bay, WI, U.S.A.. The respiration rate of *E. carolleeae* was higher than that of its native competitor *Skistodiaptomus oregonensis* at temperatures above 14°C, and all respiration rates are within the ranges presented in other studies. In combination with filtering rates, the data support the competitive advantage of *S. oregonensis* in the Green Bay system. The Arrhenius activation energy also compares favorably to other zooplankton. ETS data suggest that the thermal limit of *E. carolleeae* is between 22 and 26°C, which supports previous findings. Because *E. carolleeae* is native to and most common in colder, brackish environments, the warming trends associated with climate change may make the copepod less competitive in warm, fresh water due to higher energetic demands. However, due to the copepod's ability to survive and reproduce in the presence of toxic cyanobacteria that are associated with eutrophication (Engström-Öst et al., 2017), and increasing concentrations of cyanobacteria in Green Bay over the past few decades (De Stasio et al., 2008, 2014), those deficits and extra metabolic expenditures may be overcome as this important aquatic ecosystem changes in the future.

## ACKNOWLEDGEMENTS

This research was supported by the Dennis and Charlot Nelson Singleton Professorship in the Biological Sciences, the Lawrence University Excellence in Science Fund, and the Sherman Fairchild Foundation. We thank Sydney De Mets for field and lab assistance, and Wayne Kreuger and JoAnn Stamm for technical support. The manuscript benefitted from comments on earlier drafts by Molly Doruska, Mark Jenike and Peter Peregrine.

## REFERENCES

ADRIAN, R., S. HANSSON, B. SANDIN, B. DE STASIO & U. LARSSON, 1999. Effects of food availability and predation on a marine zooplankton community — a study on copepods in the Baltic Sea. International Review of Hydrobiology, **84**: 609-626. DOI:10.1002/iroh.199900055.

ALEKSEEV, V. R. & A. SOUISSI, 2011. A new species within the *Eurytemora affinis* complex (Copepoda: Calanoida) from the Atlantic coast of U.S.A., with observations on eight morphologically different European populations. Zootaxa, **2767**: 41-56.

ANDERSON, D. V. & D. CLAYTON, 1959. Plankton in Lake Ontario: 1-7. (Ontario Department of Lands and Forests, Maple, Ontario).

BALCER, M. D., N. L. KORDA & S. I. DODSON, 1984. Zooplankton of the Great Lakes: 1-175. (The University of Wisconsin Press, Madison, WI).

BÅMSTEDT, U., 1980. ETS activity as an estimator of respiratory rate of zooplankton populations. The significance of variations in environmental factors. Journal of Experimental Marine Biology and Ecology, **42**: 267-283. DOI:10.1016/0022-0981(80)90181-1.

BARTHEL, K. G., 1983. Food uptake and growth efficiency of *Eurytemora affinis* (Copepoda: Calanoida). Marine Biology, **74**: 269-274. DOI:10.1007/BF00403450.

BOAK, A. C. & R. GOULDER, 1983. Bacterioplankton in the diet of the calanoid copepod *Eurytemora* sp. in the Humber estuary. Marine Biology, **73**: 139-149. DOI:10.1007/BF00406881.

CAMMEN, L. M., S. CORWIN & J. P. CHRISTENSEN, 1990. Electron transport system (ETS) activity as a measure of benthic macrofaunal metabolism. Marine Ecology Progress Series, **65**: 171-182. DOI:10.3354/meps065171.

DE STASIO, B. T., M. B. SCHRIMPF, A. E. BERANEK & W. C. DANIELS, 2008. Increased chlorophyll *a*, phytoplankton abundance, and cyanobacteria occurrence following invasion of Green Bay, Lake Michigan by dreissenid mussels. Aquatic Invasions, **3**: 21-27. DOI:10.3391/ai.2008.3.1.5.

DE STASIO, B. T., M. SCHRIMPF & B. CORNWELL, 2014. Phytoplankton communities in Green Bay, Lake Michigan after invasion by dreissenid mussels: Increased dominance by cyanobacteria. Diversity, **6**: 681-704. DOI:10.3390/d6040681.

DEVOL, A. H. & T. T. PACKARD, 1978. Seasonal changes in respiratory enzyme activity and productivity in Lake Washington microplankton. Limnology and Oceanography, **23**: 104-111. DOI:10.4319/lo.1978.23.1.0104.

ENGSTRÖM-ÖST, J., N. BARRETT, A. BRUTEMARK, A. VEHMAA, A. DWYER, A.-K. ALMEN & B. T. DE STASIO, 2017. Feeding, survival, and reproduction of two populations of *Eurytemora* (Copepoda) exposed to local toxic cyanobacteria. Journal of Great Lakes Research, **43**: 1091-1100. DOI:10.1016/j.jglr.2017.09.009.

GANNON, J. E., 1974. The crustacean zooplankton of Green Bay, Lake Michigan: 28-51. (17th Conference of Great Lakes Research; Internat. Assoc. Great Lakes Res., Hamilton, Ontario).

GERBER, L., C. E. LEE, E. GROUSSET, E. BLONDEAU-BIDET, N. B. BOUCHEKER, C. LORIN-NEBEL, M. CHARMANTIER-DAURES & G. CHARMANTIER, 2016. The legs have it: in situ expression of ion transporters V-type H+-ATPase and Na+/K+-ATPase in the osmoregulatory leg organs of the invading copepod *Eurytemora affinis*. Physiological and Biochemical Zoology, **89**: 233-250. DOI:10.1086/686323.

GYLLENBERG, G. & G. LUNDQVIST, 1978. Oxygen consumption of *Eurytemora hirundoides* nauplii and adults as a function of salinity. Annales Zoologici Fennici, **15**: 328-330.

HAMMER, Ø., D. A. T. HARPER & P. D. RYAN, 2001. PAST: paleontological statistics software package for education and data analysis. Palaeontologia Electronica, **4**: 1-9.

HEINE, K. B., A. ABEBE, A. E. WILSON & W. R. HOOD, 2019. Copepod respiration increases by 7% per °C increase in temperature: a meta-analysis. Limnology and Oceanography Letters, **4**: 53-61. DOI:10.1002/lol2.10106.

HERNÁNDEZ-LÉON, S., 2000. Enzymatic method — electron transfer system. In: R. HARRIS, P. WIEBE, J. LENZ, H. R. SKJOLDAL & M. HUNTLEY (eds.), ICES zooplankton methodology manual: 506-532. (Academic Press, San Diego, CA).

HERNANDEZ-LEÓN, S. & M. GÓMEZ, 1996. Factors affecting the respiration/ETS ratio in marine zooplankton. Journal of Plankton Research, **18**: 239-255. DOI:10.1093/plankt/18.2.239.

HERNÁNDEZ-LÉON, S. & T. IKEDA, 2005. Zooplankton respiration. In: P. A. DEL GIORGIO & P. J. L. B. WILLIAMS (eds.), Respiration in aquatic ecosystems: 57-82. (Oxford University Press, New York, NY).

IKEDA, T., 1977. The effect of laboratory conditions on the extrapolation of experimental measurements to the ecology of marine zooplankton. IV. Changes in respiration and excretion rates of boreal zooplankton species maintained under fed and starved conditions. Marine Biology, **252**: 241-252. DOI:10.1007/BF00394910.

IKEDA, T., Y. KANNO, K. OZAKI & A. SHINADA, 2001. Metabolic rates of epipelagic marine copepods as a function of body mass and temperature. Marine Biology, **139**: 587-596. DOI:10.1007/s002270100608.

KING, F. D. & T. T. PACKARD, 1975. Respiration and the activity of the respiratory electron transport system in marine zooplankton. Limnology and Oceanography, **20**: 849-854. DOI:10.4319/lo.1975.20.5.0849.

KIØRBOE, T., F. MØHLENBERG & K. HAMBURGER, 1985. Bioenergetics of the planktonic copepod *Acartia tonsa*: relation between feeding, egg production and respiration, and composition of specific dynamic action. Marine Ecology Progress Series, **26**: 85-97. DOI:10.3354/meps026085.

KIPP, R. M., A. J. BENSON, J. LARSON, T. H. MAKLED & A. FUSARO, 2013. Eurytemora affinis Poppe, 1880. In: Nonindigenous aquatic species database. (U.S. Geological Survey, Gainesville, FL). Retrieved 17 May 2019, from: https://nas.er.usgs.gov/queries/FactSheet.aspx?SpeciesID=178.

LAMPERT, W., 1984. The measurement of respiration. In: J. A. DOWNING & F. H. RIGLER (eds.), A manual on methods for the assessment of secondary productivity in fresh waters: 413-468. (Blackwell Scientific Publications, Oxford, U.K.).

LEE, C. E., 1999. Rapid and repeated invasions of fresh water by the copepod *Eurytemora affinis*. Evolution, **53**: 1423-1434. DOI:10.1111/j.1558-5646.1999.tb05407.x.

LEE, C. E., M. KIERGAARD, G. W. GELEMBIUK, B. D. EADS & M. POSAVI, 2011. Pumping ions: rapid parallel evolution of ionic regulation following habitat invasions. Evolution, **65**: 2229-2244. DOI:10.1111/j.1558-5646.2011.01308.x.

LEE, C. E., W. E. MOSS, N. OLSON, K. F. CHAU, Y.-M. CHANG & K. E. JOHNSON, 2013. Feasting in fresh water: impacts of food concentration on freshwater tolerance and the evolution of food × salinity response during the expansion from saline into fresh water habitats. Evolutionary Applications, **6**: 673-689. DOI:10.1111/eva.12054.

LEE, C. E., J. L. REMFERT & G. W. GELEMBIUK, 2003. Evolution of physiological tolerance and performance during freshwater invasions. Integrative and Comparative Biology, **43**: 439-449. DOI:10.1093/icb/43.3.439.

LLOYD, S. S., D. T. ELLIOTT & M. R. ROMAN, 2013. Egg production by the copepod, *Eurytemora affinis*, in Chesapeake Bay turbidity maximum regions. Journal of Plankton Research, **35**: 299-308. DOI:10.1093/plankt/fbt003.

MILLS, E. L., J. H. LEACH, J. T. CARLTON & C. L. SECOR, 1993. Exotic species in the Great Lakes: a history of biotic crises and anthropogenic introductions. Journal of Great Lakes Research, **19**: 1-54. DOI:10.1016/S0380-1330(93)71197-1.

OWENS, T. G. & F. D. KING, 1975. The measurement of respiratory electron-transport-system activity in marine zooplankton. Marine Biology, **30**: 27-36. DOI:10.1007/BF00393750.

PACKARD, T. T., 1985. Measurement of electron transport activity of marine microplankton. Advances in Aquatic Microbiology, **3**: 208-235.

PACKARD, T. T., A. H. DEVOL & F. D. KING, 1975. The effect of temperature on the respiratory electron transport system in marine plankton. Deep-Sea Research and Oceanographic Abstracts, **22**: 237-249. DOI:10.1016/0011-7471(75)90029-7.

PAFFENHÖFER, G.-A., 2006. Oxygen consumption in relation to motion of marine planktonic copepods. Marine Ecology Progress Series, **317**: 187-192.

PASCAL, L. & V. C. CHONG, 2016. Does developmental temperature modulate copepods respiratory activity through adult life? Journal of Plankton Research, **38**: 1215-1224. DOI:10.1093/plankt/fbw058.

RAI, H., 2002. Electron transport system (ETS activity) as a measure of respiratory activity of microorganisms. In: D. V. SUBBA RAO & A. A. LISSE (eds.), Pelagic ecology methodology: 169-175. (Balkema Publishers, Rotterdam).

RAYMONT, J. E. G., 1959. The respiration of some planktonic copepods. Limnology and Oceanography, **4**: 479-491. DOI:10.4319/lo.1959.4.4.0479.

RICHMAN, S., S. A. BOHON & S. E. ROBBINS, 1980. Grazing interactions among freshwater calanoid copepods. In: W. C. KERFOOT (ed.), Evolution and ecology of zooplankton communities: 219-233. (The University Press of New England, Lebanon, NH).

RICHMAN, S. & S. I. DODSON, 1983. The effect of food quality on feeding and respiration by *Daphnia* and *Diaptomus*. Limnology and Oceanography, **28**: 948-956. DOI:10.4319/lo.1983.28.5.0948.

RODDIE, B. D., R. J. G. LEAKEY & A. J. BERRY, 1984. Salinity-temperature tolerance and osmoregulation in *Eurytemora affinis* (poppe) (Copepoda, Calanoida) in relation to its distribution in the zooplankton of the upper reaches of the Forth Estuary. Journal of Experimental Marine Biology and Ecology, **79**: 191-211. DOI:10.1016/0022-0981(84)90219-3.

SIMČIČ, T. & A. BRANCELJ, 2004. Respiratory electron transport system (ETS) activity as an estimator of the thermal tolerance of two *Daphnia* hybrids. Journal of Plankton Research, **26**: 525-534. DOI:10.1093/plankt/fbh056.

SUKHIKH, N., A. SOUISSI, S. SOUISSI, A.-C. HOLL, N. V. SCHIZAS & V. ALEKSEEV, 2019. Life in sympatry: coexistence of native *Eurytemora affinis* and invasive *Eurytemora carolleeae* in the gulf of Finland (Baltic Sea). Oceanologia, **61**: 227-238. DOI:10.1016/j.oceano.2018.11.002.

TORKE, B., 2001. The distribution of calanoid copepods in the plankton of Wisconsin lakes. Hydrobiologia, **453**: 351-365. DOI:10.1023/A:1013185916287.

VASQUEZ, A. A., P. L. HUDSON, M. FUJIMOTO, K. KEELER, P. M. ARMENIO & J. L. RAM, 2016. *Eurytemora carolleeae* in the Laurentian Great Lakes revealed by phylogenetic and morphological analysis. Journal of Great Lakes Research, **42**: 802-811. DOI:10.1016/j.jglr.2016.04.001.

WINKLER, G., J. J. DODSON & C. E. LEE, 2008. Heterogeneity within the native range: population genetic analyses of sympatric invasive and noninvasive clades of the freshwater invading copepod *Eurytemora affinis*. Molecular Ecology, **17**: 415-430. DOI:10.1111/j.1365-294X.2007.03480.x.

First received 1 September 2019.
Final version accepted 16 November 2019.

[When citing this volume, please refer to *Crustaceana* 93 (2020), 241-547]

*Proceedings of the* Eurytemora *conference, St. Petersburg, 2019*

# EFFECTS OF THE CHINA COASTAL CURRENT ON THE COMMUNITY STRUCTURE OF PLANKTONIC COPEPODS IN EARLY SPRING, WITH NOTES ON *EURYTEMORA PACIFICA* SATO, 1913 IN THE WESTERN TAIWAN STRAIT

BY

YAN-GUO WANG[1,2]), LI-CHUN TSENG[1]), ROU-XIN SUN[2]), ZHI-YONG LIU[3]), MAO LIN[2,4,6]) and JIANG-SHIOU HWANG[1,3,5,7])

[1]) Institute of Marine Biology, National Taiwan Ocean University, Keelung 20224, Taiwan

[2]) Third Institute of Oceanography, Ministry of Natural Resources, Xiamen 361005, Fujian, P.R. China

[3]) Xiamen Marine Environment Monitoring Center Station of the SOA, Xiamen 361008, Fujian, P.R. China

[4]) Collaborative Innovation Center of Deep Sea Biology, Hangzhou 310012, P.R. China

[5]) Center of Excellence for the Oceans, National Taiwan Ocean University, Keelung, Taiwan

## ABSTRACT

The Taiwan Strait, located between Taiwan Island and the southeast of the mainland of China, is the main passageway connecting the East China Sea and the South China Sea. The particular coastline of the mainland created several semi-enclosed embayments along the west coast of the Taiwan Strait. Runoffs from land bring large amounts of nutrients into the bays, which made these bays important natural spawning and breeding grounds for several economically important marine organisms. In order to reveal the effects of the China Coastal Current (CCC) on the zooplankton communities in Fuqing Bay in northeastern Fujian, zooplankton samples were collected at 12 stations in early March 2014. The average surface seawater temperature was 12.53 ± 0.14°C, and salinity was 28.33 ± 0.21 PSU in the investigation area during the research period. In total, 23 identified copepod species and in addition several unidentified benthic harpacticoid copepods were recorded with an average abundance of 77.44 ± 60.07 ind. $m^{-3}$. In the present study, the most dominant group consisted of juveniles (copepodites) with an average density of 59.97 ± 51.49 ind. $m^{-3}$, which was followed by *Calanus sinicus* Brodsky, 1965 with an average density of 5.04 ± 4.95 ind. $m^{-3}$. The occurrence rate of *Calanus sinicus* was 91.67% in our study, which indicates that the research area was controlled by the CCC water mass. So, we concluded that the CCC played an important role in transporting cold water copepod species from the Bohai Sea and the Yellow Sea to the western Taiwan Strait. A noteworthy discovery in our samples was *Eurytemora pacifica* Sato, 1913, which was recorded for the first time in waters of the western Taiwan Strait with an occurrence rate of 33.33% and an average density of 0.81 ± 1.91 ind. $m^{-3}$. *Eurytemora pacifica* was first reported at

[6]) Co-corresponding author; e-mail: linmao@tio.org.cn

[7]) Corresponding author; e-mail: jshwang@mail.ntou.edu.tw; Fax: +886-2-24629464

 | DOI:10.1163/9789004465565_016

Yantai harbour and is widely distributed in waters of the northern Pacific Ocean. The co-occurrence of *Calanus sinicus* and *Eurytemora pacifica* in our research area possibly indicates that *E. pacifica* could be used as a bioindicator for the directional movement of the CCC. The high occurrence of this species in our research area indicated that the CCC affects the community structure of copepods in the western Taiwan Strait in early spring.

Key words. — Copepod, *Eurytemora pacifica*, Taiwan Strait, China Coastal Current, bioindicator

## RÉSUMÉ

Le détroit de Taïwan, situé entre l'île de Taïwan et le sud-est de la Chine continentale, est la voie de passage reliant la mer de Chine orientale et la mer de Chine méridionale. La côte continentale particulière a créé des échancrures semi-fermées le long de la côte ouest du détroit de Taïwan. Les ruissellements provenant des terres apportent de grandes quantités de nutriments dans les baies, qui en font des zones importantes de frayères et de reproduction pour plusieurs organismes marins d'importance économique. Afin de montrer les effets du Courant Côtier de Chine (China Coastal Current, CCC) sur les communautés de zooplancton de la baie de Fuqing dans le nord-est de la province de Fujian, des échantillons de zooplancton ont été collectés sur 12 stations au début de mars 2014. La température de surface de l'eau de mer était de 12,53 ± 0,14°C, et la salinité de 28,33 ± 0,21 PSU dans la zone d'investigation au cours de la période de recherche. Au total, 23 espèces de copépodes et plusieurs espèces de copépodes harpacticoïdes non identifiées ont été rapportées avec une abondance moyenne de 77,44 ± 60,07 inds. $m^{-3}$. Dans cette étude, le groupe dominant était constitué de juvéniles (copépodites) avec une densité moyenne de 59,97 ± 51,49 inds. $m^{-3}$, suivi par l'espèce *Calanus sinicus* Brodsky, 1965 avec une densité moyenne de 5,04 ± 4,95 inds. $m^{-3}$. Le taux de fréquence de *Calanus sinicus* était de 91,67% dans notre étude, ce qui indique que la zone de recherche était contrôlee par la masse d'eau du CCC. Ainsi, nous avons conclu que le CCC jouait un rôle important en transportant des espèces de copépodes d'eau froide de la mer de Bohai et de la mer Jaune jusqu'au détroit de Taïwan occidental. Une découverte remarquable dans nos échantillons est la présence de *Eurytemora pacifica* Sato, 1913, qui a été identifiée pour la première fois dans le détroit de Taïwan occidental avec une fréquence de 33,33% et une densité moyenne de 0,81 ± 1,91 inds. $m^{-3}$. *Eurytemora pacifica* a été mentionnée pour la première fois du port de Yantai et est largement répartie dans les eaux de l'Océan Pacifique septentrional. La co-occurrence de *Calanus sinicus* et de *Eurytemora pacifica* dans notre aire de recherche indique vraisemblablement que *E. pacifica* pourrait être utilisée comme bioindicateur du mouvement directionnel du CCC. La forte occurrence de cette espèce dans notre zone de recherche indique que le CCC a une influence sur la structure des communautés de copépodes dans le détroit de Taïwan occidental au début du printemps.

Mots clés. — Copepoda, *Eurytemora pacifica*, détroit de Taiwan, China Coastal Current, courant côtier de Chine, bioindicateur

## INTRODUCTION

The Taiwan Strait, connecting the South China Sea and the East China Sea, is an important channel for seawater transport and biological dispersal between these two major marginal seas (Chang et al., 2013; Lan et al., 2014). According to the meteorological data, winter is understood here as the period from December to February when the northeasterly wind prevails in the Taiwan Strait. In contrast, during the summer period from June to August, the southwesterly and southerly winds dominate in the area of the Taiwan Strait. Therefore, the southwestward currents prevail in winter and the northeastward currents dominate in summer (Jan et al., 2002). Except the wind driven current in the surface water, the counter-winder current extends through the whole western Taiwan Strait and occupies the entire water column during the winter monsoon relaxation (Shen et al., 2017).

According to the interaction of different currents, the physical oceanography is particularly complex in the whole Taiwan Strait. The ocean currents and the surface runoff waters supply nutrients to promote the primary production in this area. Previous reports showed that the waters of the Taiwan Strait might contribute a sizeable nutrient budget to the East China Sea, especially for the P-limiting conditions in the huge coastal plume of the East China Sea (Liu, 2000; Chung et al., 2001). High nutrient supplies increase the primary production and transfer more energy to higher trophic levels. Abundant food supply, therefore, makes the Taiwan Strait a productive marine fishing ground and a place for mariculture (Chang et al., 2013).

Zooplankton is a key link for energy transfer from primary production to higher trophic levels. It is dominating the abundance and biomass of multicellular animals in pelagic marine systems. Ruhl & Smith (2004) noted that zooplankton not only supports higher trophic levels in aquatic ecosystems but also sustains benthic and microbial communities. However, several results have pointed out that the functional groups of zooplankton are very sensitive to changes in physico-chemical variables in marine systems (Sun et al., 2010; Shi et al., 2015). Usually, copepods represent the dominant zooplankton taxon comprising 70% of the total abundance and 30% of the total biomass among pelagic communities in the open tropical ocean (McKinnon & Duggan, 2014).

In the present investigation, the calanoid copepod *Eurytemora pacifica* Sato, 1913 was identified from the samples, which was the first record in the middle of the western Taiwan Strait. The genus *Eurytemora*, belonging to the family Temoridae, order Calanoida, comprises ca. 23 species globally. Most species in this genus are freshwater species and some of them are brackish. Among them, *E. pacifica* is a species distributed in the northern Pacific Ocean (Razouls et al., 2005-2019). The centre of its population is around the Bering Strait, covering both Siberia (Brodsky, 1950) and Alaska (Heron, 1964). They spread southward along the western Pacific rim. *E. pacifica* was recorded in samples from Japan (Takuo, 1952; Uye et al., 2000), Korea (Park et al., 2002; Moon et al., 2006, 2012, 2016; Yoo et al., 2010), and China (Chen & Zhang, 1965; Sun et al., 2013; Yin, 2017). Ohtsuka & Nishida (2017) reported *E. pacifica* as a cold-water species along the Japanese coastlines affected by the Oyashio waters. Previous records identified this species collected in waters of Yantai harbour (latitude 37.55°N), located at the coast of the eastern Bohai Sea (Chen & Zhang, 1965). So far, the southernmost boundary at the Chinese mainland coast of this species was recorded in Xiangshan Bay (latitude 29.66°N), located at the northern coast of Zhejiang province (Zhu et al., 2012; Du et al., 2013). At the eastern Pacific coast, this species is limited southward to the Strait of Georgia, British Columbia, and Canada (DiBacco et al., 2012). In the northern Atlantic, *E. pacifica* was first recorded as a non-indigenous species in the Charente estuary in 1985 (Goulletquer et al., 2002).

Fuqing Bay, which was an important area for fishing and aquaculture in Fujian province, is located in the northwest of the Taiwan Strait. This area is affected by waters of the CCC and the South China Sea. The influence of CCC on copepod communities in the western Taiwan Strait is an as yet poorly studied region with regard to rare biological survey opportunities. This study was aiming to understand the abundance, diversity, and community characteristics of the planktonic copepods in this area during the advent of the CCC in early spring. The objectives of the present study were to understand: (1) the community structure of zooplankton when the CCC occupies the coastal area, and (2) aspects of the CCC transporting specific species of copepods living in the Bohai Sea and the Yellow Sea to the study area.

## MATERIAL AND METHODS

### Field collections

Samples were collected from 12 stations in Fuqing Bay, on the northwest coast of the Taiwan Strait (fig. 1); the location and depth of the sampling stations is shown in table I. The collection of samples was carried out during daytime on 1 March 2014 by vertical towing from the near bottom in 25 m depth to the surface, using a type I zooplankton net (mouth diameter 50 cm, net length 145 cm, and mesh size 505 $\mu$m) in early spring. A Hydrobios flow meter was mounted in the centre of the net opening. The filtered water volume was calculated by the flow meter reading multiplied by the coefficient and the area of the net mouth. All samples were stored in 5% formalin seawater buffer solution, immediately after being rinsed from the net. Seawater temperature and salinity were also synchronously measured by a portable CTD [Conductivity, Temperature, Depth meter]. The images of the Group for High Resolution Sea Surface Temperature Pilot Project (GHRSST-PP)/Operational SST and Sea Ice Analysis (OSTIA) were obtained from the Fisheries Research Institute, Council of Agriculture, Taiwan. The images of the sea surface temperature were used for analysing the relationship between water mass transport and the occurrence of *E. pacifica*.

### Copepod identification and abundance estimation

Species identification and their abundance estimations are the basis of any community analysis. In the laboratory, the sample was poured into a beaker to allow thorough mixing until the organisms were distributed randomly before taking

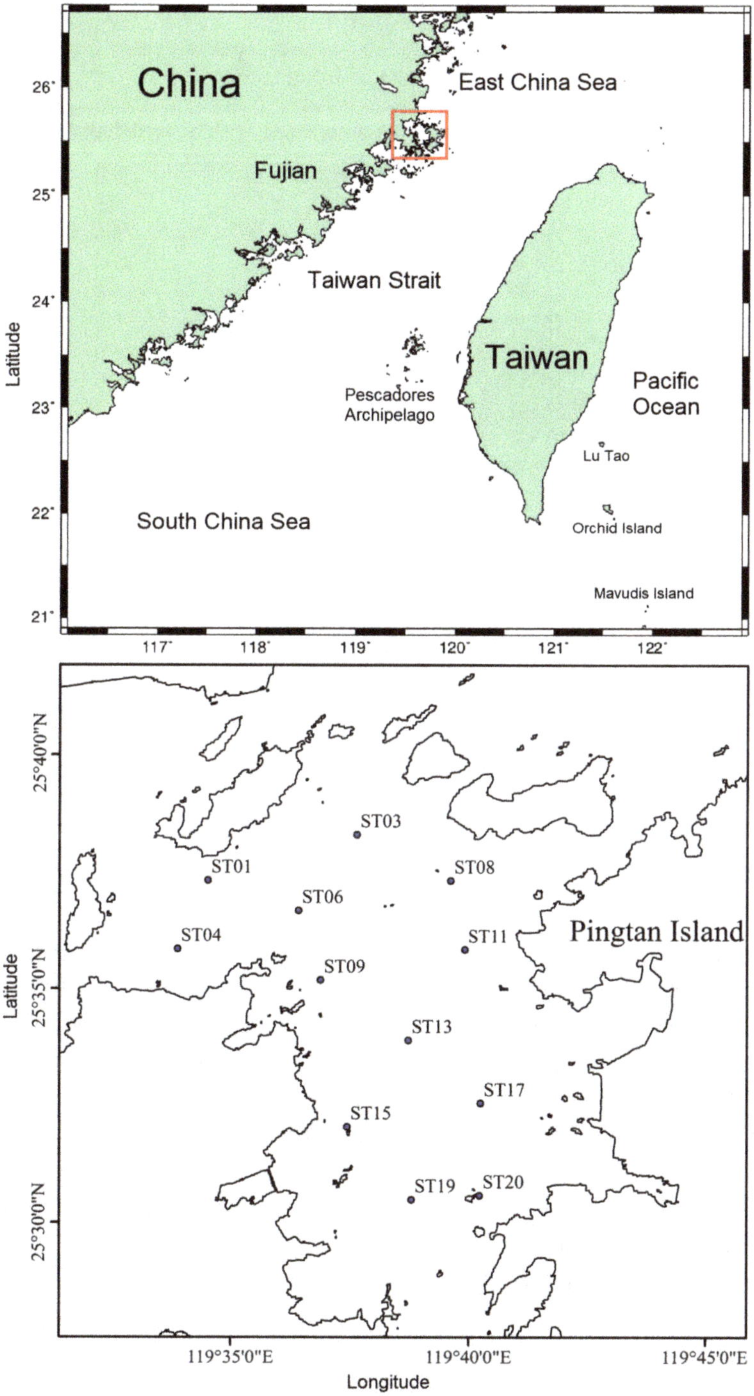


Fig. 1. Top, map of the study area; and, bottom, location of the sampling stations in Fuqing Bay, western Taiwan Strait.

TABLE I
Sampling stations, their location, and seabed depth in Fuqing Bay on the western part of Taiwan Strait

| Station | Longitude (N) | Latitude (E) | Seabed depth (m) |
|---|---|---|---|
| ST01 | 25°37.32′ | 119°34.51′ | 5.0 |
| ST03 | 25°38.29′ | 119°37.63′ | 5.5 |
| ST04 | 25°35.85′ | 119°33.86′ | 5.5 |
| ST06 | 25°36.67′ | 119°36.42′ | 15.0 |
| ST08 | 25°37.31′ | 119°39.61′ | 15.0 |
| ST09 | 25°35.18′ | 119°36.88′ | 5.0 |
| ST11 | 25°35.83′ | 119°39.91′ | 5.0 |
| ST13 | 25°33.90′ | 119°38.72′ | 8.0 |
| ST15 | 25°32.04′ | 119°37.44′ | 2.8 |
| ST17 | 25°32.55′ | 119°40.25′ | 4.0 |
| ST19 | 25°30.49′ | 119°38.78′ | 14.0 |
| ST20 | 25°30.58′ | 119°40.23′ | 12.0 |

an aliquot. A subsample was moved into a counting chamber and counted under the stereomicroscope. Adult copepods were identified to species level; juvenile individuals were recorded as an additional "taxon" [or group] of copepodites. Species identification for adults was using the following keys: Sars (1903), Rose (1933), Chen & Zhang (1965), and Chihara & Murano (1997). Abundance (as density in the seawater) was calculated by the number of individuals divided by the filtered water volume.

## Statistical analyses

The distribution pattern of samples was calculated based on the similarity analysis of species composition and abundance using the PAST software (Hamner et al., 2001). Before clustering, a data matrix of copepod abundance was transformed by $\log(x + 1)$. To avoid bias in the results, the abundance of all copepod species, except the taxa calanoid copepodites and benthic harpacticoids, in each sample was used to calculate similarities before clustering. Significance levels of Bray-Curtis similarity among stations of the copepod community were calculated. SIMPER analysis was also used to describe the average similarity of stations in the same group. Indicator species of copepods among each cluster group were calculated using the value index (IndVal) (Dufrêne & Legendre, 1997). The Shannon-Wiener diversity index was used to evaluate species diversity, while the Margalef richness index and Pielou's evenness were used to estimate the composition of the copepod community in each sample that was calculated by the PAST software. In addition, to identify the differences among the groups of cluster results, the present study adopted one-way ANOVA with post hoc Tukey's honestly significant difference (HSD) tests.

## RESULTS

### Hydrological structure

The study area is controlled by the China Coastal Current during winter and early spring (fig. 2). The surface seawater temperature varied in a narrow range between 12.4°C and 12.9°C, with an average temperature of 12.53 ± 0.14°C (fig. 2A). The low seawater temperature indicated that the China Coastal Current

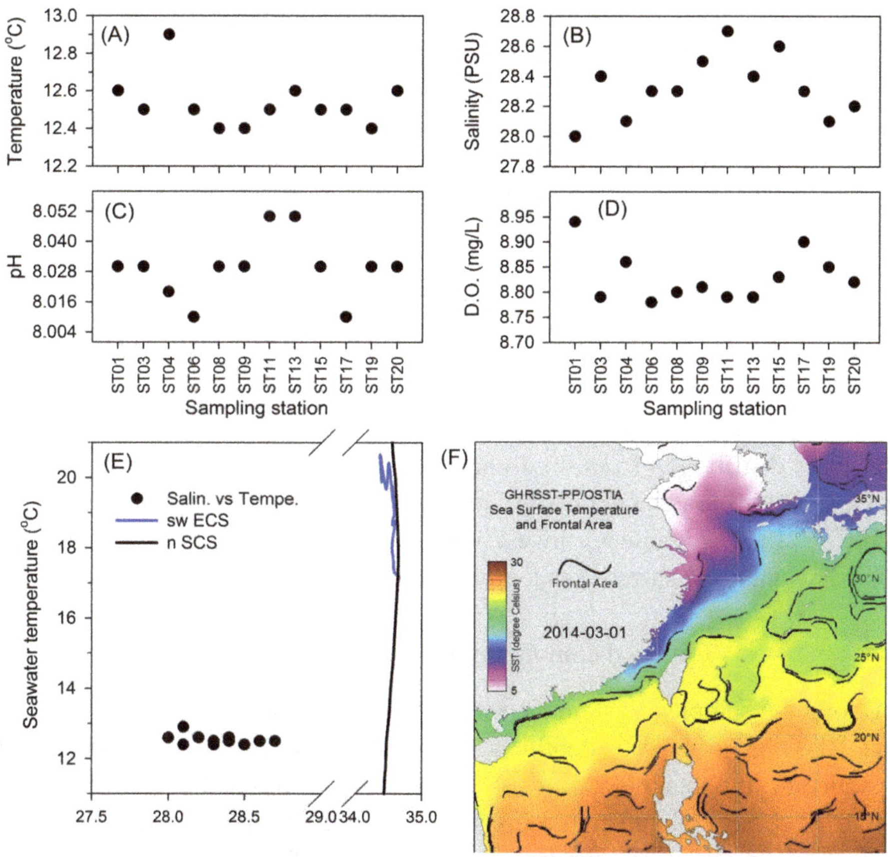


Fig. 2. Physico-chemical parameters at the various sampling stations: A, seawater temperature; B, salinity; C, pH; and, D, dissolved oxygen; at each of the 12 sampling stations considered in the present study. In addition: E, the temperature vs. salinity diagram shows the distribution of values of these 12 sampling stations during the study period; furthermore, two reference temperature-salinity curves of the southwestern East China Sea (26.25°N 122.5°E; abbreviated ECS) and northern South China Sea (21°N 118°E; abbreviated SCS) in the period March-May, were provided by the Ocean Data Bank of the Ministry of Science and Technology, Republic of China. F, The sea surface water temperature image on 1 March 2014, derived from a satellite image of GHRSST-PP/OSTIA [see text for explanation of abbreviation], demonstrates the water mass distribution in the East China Sea, the Taiwan Strait, and the South China Sea.

was strongly affecting the study area. The salinity of the surface seawater ranged from 28.00 to 28.70 PSU with an average value of 28.33 ± 0.21 PSU between sampling stations, revealing that the area was characterized by brackish water throughout our study (fig. 2B). In addition, the study area was a semi-enclosed coastal bay; the terrestrial fresh water carried great amounts of nutrients into the surface layer of Fuqing Bay. The pH of the surface seawater ranged from 8.01 to 8.05 with an average value of 8.03 ± 0.01 (fig. 2C). The dissolved oxygen ranged from 8.78 to 8.94 mg/l with an average of 8.83 ± 0.05 mg/l (fig. 2D). The surface seawater in the research area was neither acidified nor eutrophicated.

The relationship between seawater temperature and salinity of each station in the sampling period is shown in fig. 2E. The present study applied two reference temperature-salinity (TS) curves of the southwestern East China Sea (26.25°N 122.5°E) (depth above 90 m) and the northern South China Sea (21°N 118°E) (depth above 350 m) in a period March-May (provided by the Ocean Data Bank of the Ministry of Science and Technology, Republic of China). The water mass properties showed that salinity and temperature of the surface seawaters in the study area were lower than in the southwestern East China Sea (fig. 2E). The surface seawater temperature image provided by GHRSST-PP/OSTIA [Group for High Resolution Sea Surface Temperature/Operational Sea Surface Temperature and Sea Ice Analysis; Donlon et al., 2012] on 1 March 2014 demonstrated the cold water mass distribution from the Bohai Sea, the Yellow Sea, South Korea, Japan, along the eastern coast of mainland China, and extending southwards to the Taiwan Strait (fig. 2F). The whole northwest of the Taiwan Strait was then occupied by the CCC cold water; the surface sea temperature was about 15°C (fig. 2F).

## Copepod community structure and abundance

Among the samples, systematic investigation provided 22 identified species of Copepoda, one not further identified *Acartia* sp., and several unidentified benthic harpacticoids, the total belonging to 4 orders, 14 families and 15 genera (table II). Overall, the most abundant group was calanoid copepodites (59.97 ± 51.49 ind. $m^{-3}$, with a relative abundance, RA: 77.44%), followed by *Calanus sinicus* Brodsky, 1965 (5.04 ± 4.95 ind. $m^{-3}$, RA: 6.51%), and benthic harpacticoids (2.53 ± 6.05 ind. $m^{-3}$, RA: 3.27%), respectively. However, the highest occurrence rate was recorded for calanoid copepodites with 100% occurrence, followed by *Calanus sinicus* with 91.7% occurrence (table II). Integrating the assemblage data, maximum copepod abundance was recorded at station 13 (220.0 ind. $m^{-3}$), followed by station 11 (154.29 ind. $m^{-3}$), whereas the lowest abundance was recorded at station 9 (22.0 ind. $m^{-3}$) (fig. 3A), with a mean abundance of 77.44 ± 60.07 ind. $m^{-3}$. The number of species per sample ranged from 3 to 8 (fig. 3B).

TABLE II

Abundance (ind. $m^{-3}$, mean ± standard deviation), occurrence ratio (OR, %), and relative abundance (RA, %) of copepods recorded from the western Taiwan Strait on 1 March 2014

| Systematic classification | Abundance | OR | RA |
|---|---|---|---|
| Order CALANOIDA | | | |
| ACARTIIDAE | | | |
| *Acartia (Odontacartia) pacifica* Steuer, 1915* | 1.90 ± 5.76 | 16.67 | 2.46 |
| *Acartia* sp. | 0.15 ± 0.42 | 16.67 | 0.19 |
| *Acartia (Acanthacartia) tsuensis* Ito, 1956 | 0.37 ± 0.95 | 33.33 | 0.48 |
| CALANIDAE | | | |
| *Calanus sinicus* Brodsky, 1965* | 5.04 ± 4.95 | 91.67 | 6.51 |
| *Cosmocalanus darwini* (Lubbock, 1860) | 0.12 ± 0.41 | 8.33 | 0.15 |
| CENTROPAGIDAE | | | |
| *Centropages abdominalis* Sato, 1913 | 0.13 ± 0.44 | 8.33 | 0.17 |
| *Centropages tenuiremis* Thompson & Scott, 1903 | 0.03 ± 0.11 | 8.33 | 0.04 |
| EUCHAETIDAE | | | |
| *Euchaeta concinna* Dana, 1849 | 0.14 ± 0.37 | 16.67 | 0.19 |
| *Euchaeta plana* Mori, 1937 | 0.28 ± 0.96 | 8.33 | 0.36 |
| PARACALANIDAE | | | |
| *Paracalanus aculeatus* Giesbrecht, 1888 | 0.32 ± 0.53 | 33.33 | 0.41 |
| *Paracalanus parvus* (Claus, 1863) | 0.69 ± 1.75 | 16.67 | 0.89 |
| PONTELLIDAE | | | |
| *Labidocera euchaeta* Giesbrecht, 1889 | 0.13 ± 0.25 | 25.00 | 0.16 |
| PSEUDODIAPTOMIDAE | | | |
| *Pseudodiaptomus inopinus* Burckhardt, 1913 | 0.31 ± 0.96 | 16.67 | 0.40 |
| *Pseudodiaptomus marinus* Sato, 1913 | 0.50 ± 1.03 | 25.00 | 0.65 |
| *Pseudodiaptomus poplesia* (Shen, 1955) | 0.06 ± 0.22 | 8.33 | 0.08 |
| SCOLECITRICHIDAE | | | |
| *Scolecithricella nicobarica* (Sewell, 1929) | 0.03 ± 0.11 | 8.33 | 0.04 |
| TEMORIDAE | | | |
| *Eurytemora pacifica* Sato, 1913* | 0.81 ± 1.91 | 33.33 | 1.05 |
| TORTANIDAE | | | |
| *Tortanus (Eutortanus) derjugini* Smirnov, 1935* | 2.71 ± 7.60 | 41.67 | 3.50 |
| Calanoid copepodites* | 59.97 ± 51.49 | 100.0 | 77.44 |
| Order CYCLOPOIDA | | | |
| OITHONIDAE | | | |
| *Dioithona rigida* (Giesbrecht, 1896) | 0.60 ± 1.06 | 33.33 | 0.77 |
| *Oithona similis* Claus, 1866 | 0.20 ± 0.58 | 16.67 | 0.26 |
| Order HARPACTICOIDA | | | |
| ECTINOSOMATIDAE | | | |
| *Microsetella norvegica* (Boeck, 1864) | 0.27 ± 0.65 | 16.67 | 0.35 |
| TISBIDAE | | | |
| *Tisbe tenera* (Sars, 1905) | 0.10 ± 0.36 | 8.33 | 0.13 |
| Benthic harpacticoid spp.* | 2.53 ± 6.05 | 58.33 | 3.27 |
| Order POECILOSTOMATOIDA | | | |
| CORYCAEIDAE | | | |
| *Corycaeus (Ditrichocorycaeus) affinis* McMurrich, 1916 | 0.03 ± 0.12 | 8.33 | 0.04 |
| Total | 77.44 ± 60.07 | – | 100 |

*Copepods with relative abundance values exceeding 1.0%.

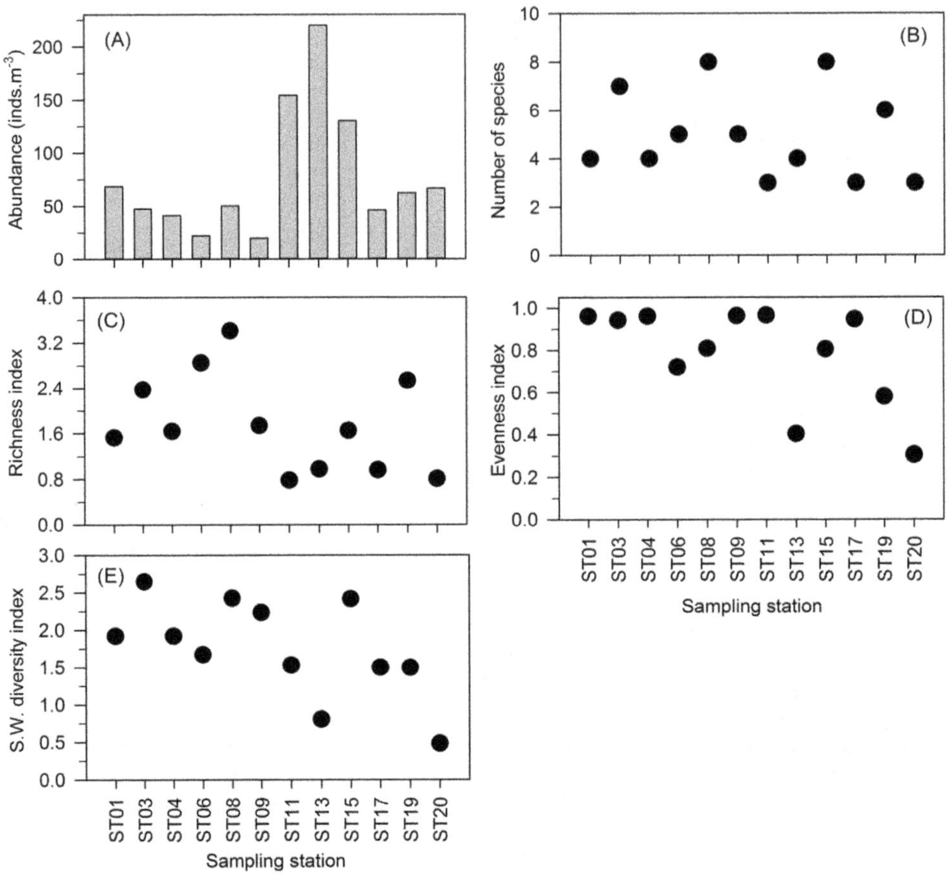


Fig. 3. Variation in values of the plankton copepods at each sampling station in Fuqing Bay: A, abundance; B, number of species; and in the indices of, C, richness; D, evenness; and E, Shannon-Wiener diversity.

The richness index of each sample ranged from 0.78 (station 11) to 3.41 (station 8), and the average value was 1.77 ± 0.86 (fig. 3C). The evenness index of each sample ranged between 0.30 (station 20) and 0.97 (station 11), and the average value was 0.78 ± 0.23 (fig. 3D). The Shannon-Wiener diversity index ranged from 0.48 (station 20) to 2.65 (station 3), with a mean value of 1.75 ± 0.65 (fig. 3E).

The varied patterns of the proportions in the five most dominant copepod species at each sampling station were clear (fig. 4). The dynamic proportion of each sampling station indicated a clear geographic difference. The group calanoid copepodites showed a relatively high proportion by ranging from 42.86% (station 9) to 89.58% (station 1). The species *C. sinicus* had a relatively high proportion at station 20 (16.43%) and station 6 (11.67%). The species *Tortanus* (*Eutortanus*) *derjugini* Smirnov, 1935 had a relatively high proportion at station 15 (20.51%) and station 9 (14.29%); *Acartia* (*Odontacartia*) *pacifica* Steuer, 1915 had a high

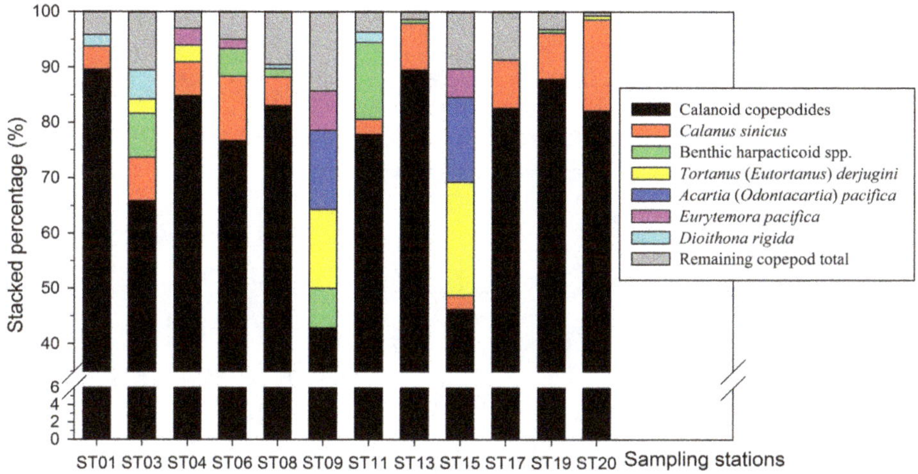


Fig. 4. Relative abundance of 5 dominant copepod groups/taxa/species at each of the sampling stations in Fuqing Bay during the study.

proportion at station 15 (15.38%) and station 9 (14.29%). In addition, the species *Eurytemora pacifica* showed a high proportion at station 9 (7.14%) and station 15 (5.13%) (fig. 4).

## Hierarchical classification of copepod assemblages

The species association of copepods was evaluated by Bray-Curtis similarity distances. Samples with similar composition of copepod species formed a cluster. The results of cluster analysis by the stations showed a tendency to form three groups. (fig. 5A). SIMPER analysis results showed that the average similarity in Group A stations was 61.41%. The average similarity was 51.62% in station Group B1 and 42.53% in the Group B2 stations. Index values (IndVal) for each dominant copepod species with a relative abundance > 5%, and features of samples of the associated cluster grouping results for each are given in table III. The first assemblage (Group A) displayed its principal association in samples collected from 3 stations located in the east of Fuqing Bay (fig. 5B), and this group was characterized by a higher abundance (28.75 ± 35.77 ind. $m^{-3}$) and a higher number of species (5.67 ± 2.08). The major three copepod species of group A were: *T.* (*E.*) *derjugini* (IndVal: 35.68%), *A.* (*O.*) *pacifica* (17.68%) and *E. pacifica* (10.84%) (table III). The second group (Group B1) contained 4 stations located in the north of Fuqing Bay, and was characterized by a lower abundance (10.07 ± 3.03 ind. $m^{-3}$) and a higher number of species (5.50 ± 2.38). The major three copepod species of group B1 were: *Calanus sinicus* (IndVal: 33.48%), *Dioithona rigida* (Giesbrecht, 1896) (17.78%), and *Paracalanus parvus* (Claus, 1863) (10.84%). The third group (Group B2) included 5 stations located in the south of Fuqing Bay,

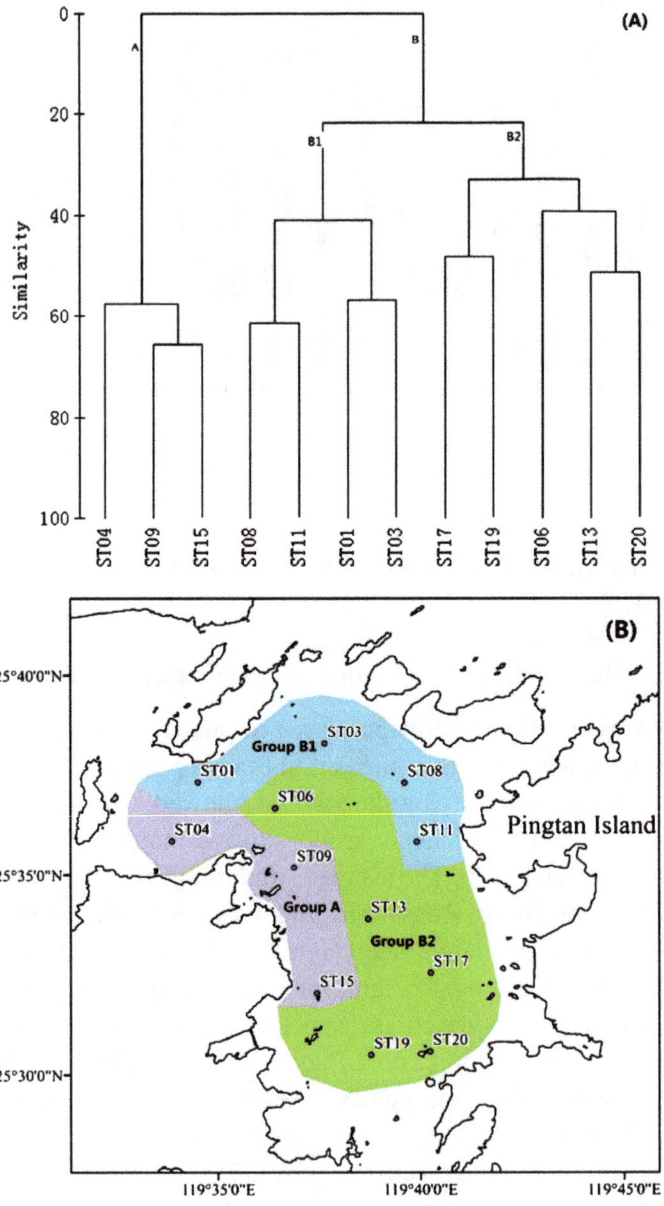


Fig. 5. Pattern and structure of the distribution of copepod zooplankton described through: A, a hierarchical clustering dendrogram using Bray-Curtis cluster analysis of all copepod species found; and, B, the geographical distribution of that cluster grouping over the 12 sampling stations in Fuqing Bay, March 2014.

TABLE III

The Indicator Values (IndVal) of the dominant copepod species exceeding 5%, and features of samples (mean ± standard deviation) of the associated cluster results shown in fig. 5. Various superscripts indicate the significant differences between the cluster groups as analysed by one-way ANOVA. For authorities with species names, see table II

| Species | Group A | Group B1 | Group B2 | $F$-value | $p$-value |
|---|---|---|---|---|---|
| *Acartia (Odontacartia) pacifica* | 17.67 | | | | |
| *Calanus sinicus* | | 33.48 | 78.17 | | |
| *Eurytemora pacifica* | 10.84 | | | | |
| *Dioithona rigida* | | 17.77 | | | |
| *Paracalanus aculeatus* | | 6.37 | | | |
| *Paracalanus parvus* | | 10.31 | | | |
| *Pseudodiaptomus marinus* | 6.97 | | | | |
| *Tortanus (Eutortanus) derjugini* | 35.68 | | | | |
| Features of samples | | | | | |
| Number of species | 5.67 ± 2.08 | 5.50 ± 2.38 | 4.20 ± 1.30 | 0.77 | 0.49 |
| Abundance (ind. $m^{-3}$) | 28.75 ± 35.77 | 10.07 ± 3.03 | 10.54 ± 6.75 | 1.35 | 0.33 |
| Richness index | 1.67 ± 0.06 | 2.02 ± 1.13 | 1.63 ± 0.98 | 0.23 | 0.80 |
| Evenness index | 0.91 ± 0.09 | 0.92 ± 0.07 | 0.59 ± 0.26 | 4.67 | 0.041 |
| Diversity index | $2.19 \pm 0.25^{a}$ | $2.13 \pm 0.50^{a}$ | $1.19 \pm 0.52^{b}$ | 6.26 | 0.02* |

and this was characterized by a lower abundance (10.54 ± 6.75 ind. $m^{-3}$), and a lower number of species (4.20 ± 1.30). The only major copepod species in group B2 was *C. sinicus* (IndVal: 78.17%). Obviously, *C. sinicus* was the most dominant species in groups B1 and B2.

In addition, one-way ANOVA was used to compare differences between groups. The results showed that the average density of *E. pacifica* in group A was significantly higher than in Group B1 ($p = 0.049$) and in Group B2 ($p = 0.045$). In Group B1, the average density of *D. rigida* was significantly higher than in Group A ($p = 0.014$) and in Group B2 ($p = 0.007$). The value of the Shannon-Wiener diversity index was significantly higher in group A than in groups B1 ($p = 0.04$) and B2 ($p = 0.036$). The results of community cluster grouping indicated that the distribution of copepod assemblages shows geographical difference during the period the CCC water mass comes to Fuqing Bay, i.e., in early spring (fig. 5B).

## DISCUSSION

### Diversity and environmental characters

The species diversity of zooplankton is governed, according to the literature, mostly by temperature and other environmental parameters. The highest diversity records were found in tropical and subtropical regions; whereas the lowest records were in extreme environments, such as low temperature or brackish water areas (Harris et al., 2000). The present study collected zooplankton in Fuqing Bay, i.e., in a subtropical area. During the investigation period, this area was controlled by the northeast monsoon and the seawater temperature was relatively low. Hardly any studies report on the zooplankton communities in the waters of Taiwan Strait, and most of these deal with differences in sampling tactics in relation to sampling gears and spatio-temporal scales. Compared with those reports, we found that the abundance of copepods in the present study was lower than the existing records from the open water area of the Taiwan Strait, with an abundance of 231.96 ind. $m^{-3}$ during mid-spring (Wang et al., 2012). In contrast, our recorded abundances were higher than records at Sansha Bay (Lin et al., 1997), and the Nanhai Island coastal area (Sun et al., 2014) with 45.46 ind. $m^{-3}$ and 55 ind. $m^{-3}$, respectively. The present study thus adds basic information on the zooplankton, that contributes to the knowledge on the waters of Taiwan Strait. However, continuously sampling and analysing zooplankton in this region will help to understand the biodiversity as well as the succession of communities of copepods in the coastal area of mainland China.

A noteworthy result is also, that the value of the Shannon-Wiener diversity index was higher than at the coast of Hainan Island, with an average of 1.17 (Sun et al., 2014). In our research, the study area was affected by surface runoffs, the interplay waters of CCC, and local seawater, which might have contributed to a higher diversity of copepods (Tseng et al., 2015). Interactions of different water masses brought rich nutrients that enhanced the primary productivity. The concentration of chlorophyll *a* in the seawater ranged from 0.75 $mg.m^{-3}$ (station 17) to 2.55 $mg.m^{-3}$ (station 01), with an average about 1.15 $mg.m^{-3}$ during the present study period (unpubl. data). A higher primary productivity could support high rates of reproduction and growth of copepods, and it might increase the diversity of copepods as well.

## Role of the China Coastal Current

The present study found that *Calanus sinicus* was dominant, and distributed in the area east of Fuqing Bay. *C. sinicus* prefers low seawater temperatures and is one of the dominant species in the continental shelf waters of both the Yellow Sea and the East China Sea (Uye, 2000; Hwang & Wong, 2005). Its distribution extends to the south of the northern part of the South China Sea, as far as Hainan Island, and Vietnam (Chen & Zhang, 1965; Hulsemann, 1994; Uye, 2000; Hwang & Wong, 2005; Yin et al., 2017). Temperature is an important factor regulating the distribution of *C. sinicus* in waters of the shelf ecosystem in the northwestern Pacific Ocean (Uye, 2000). *Calanus sinicus* was an important indicator species of

CCC in the southern East China Sea in northern Taiwan (Hwang et al., 2004, 2006; Chou et al., 2012; Molinero et al., 2016, 2018). A previous study reported that the upper limit of seawater temperature for *C. sinicus* was about 26-27°C (Wang et al., 2003). The population of *C. sinicus* was not endemic in waters of the China shelf in the south of Fujian Province; the CCC transported it to the Taiwan Strait during the northeast monsoon period in the winter (Chen, 1992). This report pointed out that the distribution of *C. sinicus* was reaching Hong Kong and Guangdong Province along the mainland China coast, associated with the movement of the cold water CCC (Hwang & Wong, 2005). Monsoons and coastal currents played an important role in shaping the population of *C. sinicus* at different spatio-temporal scales in the northwest continental shelf of the South China Sea (Yin et al., 2011). A previous study showed that *C. sinicus* was abundant (198.85 ind. $m^{-3}$) and dominant (RA: 92.66%) in open waters of the northern Taiwan Strait during mid-spring (Wang et al., 2012). The present results revealed that *C. sinicus* was the dominant species next to the group of calanoid copepodites, with a high occurrence ratio (91.67%) in Fuqing Bay. Our results confirmed that the CCC is an important conveyor, transporting *C. sinicus* southwards into the coastal waters of the China mainland.

## A note on *Eurytemora pacifica*

Notably, the calanoid copepod *E. pacifica* was found in the samples collected in the west of Fuqing Bay. This was the first record from the western Taiwan Strait. The historical records revealed that *E. pacifica* prefers low salinity waters, as a coastal species (Chen & Zhang, 1965), and was distributed in waters of the northern Pacific Ocean (Chen & Zhang, 1965; Uye et al., 2000; Sun et al., 2013; Jeong et al., 2014). In the coastal waters of China, the first record of this species was made at Yantai harbour (Chen & Zhang, 1965). In waters of Japan, *E. pacifica* was known to occur in low temperature environments ($\leqslant$ 9.38°C; Uye et al., 2000). In southern Brittany, France, *E. pacifica* was found in waters of a temperature range between 18.5-19.5°C, and a salinity range of 35.0-35.5 PSU (Brylinski et al., 2016). The present study recorded *E. pacifica* in samples with seawater temperatures ranging from 12.4 to 12.9°C and salinities from 28.10 to 28.60 PSU. Previous reports confirmed that *E. pacifica* belongs to a group of marine and brackish water species (Walter & Boxshall, 2019). In waters of southern Korea, *E. pacifica* was not abundant ($3 \pm 1$ ind. $m^{-3}$, RA: 0.05%) and not often recorded (OR: 19.13%) (Jeong et al., 2014).

A review of *E. pacifica* along the coast of the Chinese mainland showed a high variation in abundance as well as in the ratio of occurrence (table IV). The mesh size of the zooplankton net applied in the sampling appeared to be an important factor, causing a bias in the abundance estimations (Tseng at al., 2011). Examples

TABLE IV

Review of recorded locations, sampling period, mesh size, and abundance of *Eurytemora pacifica* Sato, 1913 along the China coastal area

| Location | Period | Mesh size ($\mu$m) | Abundance (ind. $m^{-3}$) | Reference |
|---|---|---|---|---|
| Bohai Sea | – | – | – | Chen & Zhang (1965) |
| Xiangshan Bay | March | 505 | 0.06 | Zhu et al. (2012) |
| Xiangshan Bay | March | 160 | 1.02 | Zhu et al. (2012) |
| Xiaoshi Island | March | 160 | $15 \pm 31$ | Sun et al. (2013) |
| Laizhou Bay | – | 505 | 21.17 | Jiang et al. (2015) |
| Zhangzi Island | November | 500 | – | Yin et al. (2017) |
| Bohai Sea | May, July, October | 505 | – | Dai et al. (2018) |
| Fuqing Bay | March | 505 | $0.81 \pm 1.91$ | Present study |

– = not mentioned.

of the samples collected by using a 160 $\mu$m mesh size net: the abundance of *E. pacifica* was counted as 1.02 ind. $m^{-3}$ in Xiangshan Bay (Zhu et al., 2012), and was measured as $15 \pm 31$ ind. $m^{-3}$ in the Yellow Sea (Sun et al., 2013). When changing to the use of a 505 $\mu$m mesh size net, the abundance of *E. pacifica*, as apparent from the data, became 0.06 ind. $m^{-3}$ in Xiangshan Bay (Zhu et al., 2012), and 21.17 ind. $m^{-3}$ in Laizhou Bay (Jiang et al., 2015), and the species was found to be dominant in the artificial reef area of Tianjin seashore (Dai et al., 2018). In addition, previous investigations found that both *C. sinicus* and *E. pacifica* in the Bohai Sea were closely associated to the origin of the CCC (Dai et al., 2018). The present results suggest from the records of *Eurytemora pacifica*, that this species could be applied as an bioindicator species of the CCC movement, similar to the copepod *Calanus sinicus*.

## CONCLUSIONS

In conclusion, the present study discovered: (1) the community structure of copepods shows a clear geographical distribution pattern in the waters of western Taiwan Strait during the CCC (China Coastal Current) intruding period in early spring; (2) the CCC played an important role in the transport of the copepods *Calanus sinicus* and *Eurytemora pacifica* from the Bohai Sea and the Yellow Sea to the western Taiwan Strait; and (3) the hydrological environment of *E. pacifica* suggested it as a bioindicator species for the directional movement of the CCC. In addition, previous records demonstrated that *E. pacifica* was distributed in the waters of the Bohai Sea and the Yellow Sea, and the present study reported it in the bay environments of the western Taiwan Strait.

Future studies on the phylogenetic position of *E. pacifica* within the genus *Eurytemora* and of that genus in the family Temoridae, should provide a better understanding of its historical and recent dispersal routes as well as of its detailed geographic distribution.

## ACKNOWLEDGEMENTS

The first two authors, i.e., YAN-GUO WANG and LI-CHUN TSENG, both contributed equally to the present contribution. This study was funded by the Third Institute of Oceanography, Ministry of Natural Resources, grant No. 2017010, as well as by the Bilateral Cooperation of Maritime Affairs, Global Climate Change and Ocean Atmosphere Interaction Research, grant No. GASI-01-02-04 to Y.-G. Wang; Ministry of Science and Technology (MOST) of Taiwan, through grants nos. MOST 107-2811-M-019-004 and MOST 108-2811-M-019-504 to L.-C. Tseng; and grants nos. MOST 104-2611-M-019-004, MOST 105-2621-M-019-001, MOST 105-2918-I-019-001, MOST 106-2621-M-019- 001, MOST 107-2621-M-019-001 and MOST 108-2621-M-019-003 to J.-S. Hwang. We would like to thank Yan-Shuang Xie and Guo-Mei Wei of the College of Ocean and Earth Science, Xiamen University, for their reviewing of the physical oceanography around the research area. The authors acknowledge support from the Fisheries Research Institute of Taiwan for providing satellite sea surface temperature images.

Conflict of interest.— No potential conflict of interest was reported by the authors.

## ETHICAL APPROVAL

All applicable international, national, and/or institutional guidelines for the care and use of animals were followed.

## REFERENCES

BRODSKY, K. A., 1950. Vesnologie rachki Calanoida dal'Nevostochnykh morei SSSR i Polyarnogo Basseina [Calanoida of the far-eastern seas of the USSR and the Polar Basin]. Opred. Faune SSSR, **35**: 1-440. [In Russian; English translation 1967 by Israel Program for Scientific Translations, Jerusalem, under the title: Calanoida of the far-eastern seas and Polar Basin of the USSR.]

BRYLINSKI, J. M., L. COURCOT, V. DAVID & B. SAUTOUR, 2016. Expansion of the north Pacific copepod *Eurytemora pacifica* Sato, 1913 (Copepoda Calanoida Temoridae) along the Atlantic coast of France. BioInvasions Records, **5**(4): 245-250.

CHANG, Y., M. A. LEE, K. T. LEE & K. T. SHAO, 2013. Adaptation of fisheries and mariculture management to extreme oceanic environmental changes and climate variability in Taiwan. Marine Policy, **38**: 476-482.

CHEN, Q. C., 1992. Zooplankton of China seas, **1**: 1-87. (Science Press, Beijing).

CHEN, Q.-C. & S.-Z. ZHANG, 1965. The planktonic copepods of the Yellow Sea and the East China Sea. I. Calanoida. Studia Marina Sinica, **7**: 20-131. [In Chinese, with English summary.]

CHIHARA, M. & M. MURANO, 1997. An illustrated guide to marine plankton in Japan: 1-1574. (Tokai Univ. Press, Tokyo). [In Japanese.]

CHOU, C., L.-C. TSENG, C.-H. OU, Q.-C. CHEN & J.-S. HWANG, 2012. Seasonal succession of planktonic copepods in bight environments of northeastern Taiwan. Zoological Studies, **51**(8): 1380-1396.

CHUNG, S. W., S. JAN & K. K. LIU, 2001. Nutrient fluxes through the Taiwan Strait in spring and summer 1999. Journal of Oceanography, **57**(1): 47-53.

DAI, Y.-Y., C.-Q. HOU, S. YANG, B.-L. ZHANG, Y. YU, S. WANG, W. CHEN, E.-J. FANG, Y. GAO, B. GUO, W.-Y. YANG & J.-W. ZHANG, 2018. Community structure of zooplankton in artificial reef area of Tianjin seashore and its relationship with environmental factors. Transactions of Oceanology and Limnology, **164**: 165-172. [In Chinese, with English abstract.]

DIBACCO, C., D. B. HUMPHREY, L. E. NASMITH & C. D. LEVINGS, 2012. Ballast water transport of non-indigenous zooplankton to Canadian ports. ICES Journal of Marine Science/Journal du Conseil, **69**(3): 483-491.

DONLON, C. J., M. MARTIN, J. D. STARK, J. ROBERTS-JONES, E. FIEDLER & W. WIMMER, 2011. The Operational Sea Surface Temperature and Sea Ice Analysis (OSTIA). Remote Sensing of the Environment, **116**: 140-158. DOI:10.1016/j.rse.2010.10.017.

DU, X.-N., Y. L. WANG, T. CHEN & Y. LIAO, 2013. Seasonal variation of zooplankton copepod community in Xiangshan Bay of east China. Chinese Journal of Ecology, **32**(10): 2756-2763. [In Chinese, with English abstract.]

DUFRÊNE, M. & P. LEGENDRE, 1997. Species assemblages and indicator species: the need for a flexible asymmetrical approach. Ecological Monographs, **67**: 345-366.

GOULLETQUER, P., G. BACHELET, P. G. SAURIAU & P. NOEL, 2002. Open Atlantic coast of Europe – a century of introduced species into French waters. In: E. LEPPÄKOSKI, S. GOLLASCH & S. OLENIN (eds.), Invasive aquatic species of Europe. Distribution, impacts and management: 276-290. (Springer Verlag, Dordrecht).

HAMMER, Ø., D. A. T. HARPER & P. D. RYAN, 2001. PAST: paleontological statistics software package for education and data analysis. Palaeontol. Electron., **4**: 4.

HARRIS, R., P. WIEBE, J. LENZ, H. R. SKJOLDAL & M. E. HUNTLEY, 2000. ICES zooplankton methodology manual: 1-684. (Academic Press, London).

HERON, G. A., 1964. Seven species of *Eurytemora* (Copepoda) from northwestern North America. Crustaceana, **7**: 199-211.

HULSEMANN, K., 1994. *Calanus sinicus* Brodsky and *C. jashnovi*, nom. nov. (Copepoda: Calanoida) of the north-west Pacific Ocean: a comparison, with notes on the integumental pore pattern in *Calanus* s. str. Invert. Taxon., **8**: 1461-1482.

HWANG, J.-S., S. SOUISSI, L.-C. TSENG, L. SEURONT, F. G. SCHMITT, L.-S. FANG, S.-H. PENG, C.-H. WU, S.-H. HSIAO, W.-H. TWAN, T.-P. WEI, R. KUMAR, T.-H. FANG, Q.-C. CHEN & C.-K. WONG, 2006. A 5-year study of the influence of the northeast and southwest monsoons on copepod assemblages in the boundary coastal waters between the East China Sea and the Taiwan Strait. Journal of Plankton Research, **28**(10): 943-958.

HWANG, J.-S., Y.-Y. TU, L.-C. TSENG, L.-S. FANG, S. SOUISSI, T.-H. FANG, W.-T. LO, W.-H. TWAN, S.-H. HSAIO, C.-H. WU, S.-H. PENG, T.-P. WEI & Q.-C. CHEN, 2004. Taxonomic composition and seasonal distribution of copepod assemblages from waters adjacent to Nuclear Power Plant I and II in northern Taiwan. Journal of Marine Science and Technology, **12**(5): 380-391.

HWANG, J.-S. & C.-K. WONG, 2005. The China Coastal Current as a driving force for transporting *Calanus sinicus* (Copepoda: Calanoida) from its population centers to waters off Taiwan and Hong Kong during the winter northeast monsoon period. Journal of Plankton Research, **27**(2): 205-210.

JAN, S., J. WANG, C.-S. CHERN & S.-Y. CHAO, 2002. Seasonal variation of the circulation in the Taiwan Strait. Journal of Marine Systems, **35**(3): 249-268.

JEONG, H.-G., H.-L. SUH, W. LEE & H.-Y. SOH, 2014. Seasonal variation of the neustonic zooplankton community in southern waters of Korea. Ocean Science Journal, **49**(3): 167-181.

JIANG, H.-C., H.-G. CHEN, X.-K. SONG, N. LIU, J.-L. HE, L. CHENG & Y.-X. WANG, 2015. Zooplankton community structure in Jincheng area of Laizhou Bay and its relationship with environmental factors. Acta Ecologica Sinica, **35**(22): 7308-7319. [In Chinese with English abstract.]

LAN, K.-W., M.-A. LEE, C.-I. ZHANG, P.-Y. WANG, L.-J. WU & K.-T. LEE, 2014. Effects of climate variability and climate change on the fishing conditions for grey mullet (*Mugil cephalus* L.) in the Taiwan Strait. Climatic Change, **126**(1-2): 189-202.

LIN, J.-H., X.-P. WANG & R.-X. CHEN, 1997. Distribution of planktonic copepods in Sansha Bay, Fujian. Marine Science Bulletin, **16**(6): 13-19. [In Chinese with English Abstract.]

LIU, K.-K., 2000. Cross-shelf and along-shelf nutrient fluxes derived from flow fields and chemical hydrography observed in the southern East China Sea off northern Taiwan. Continental Shelf Research, **20**(4): 493-523.

MCKINNON, A. D. & S. DUGGAN, 2014. Community ecology of pelagic copepods in tropical coastal waters. In: L. SEURONT (ed.), Copepods: diversity, habitat and behavior: 25-49. (Nova Science Publishers, New York, NY).

MOLINERO, J. C., L.-C. TSENG, C. LÓPEZ-ABBATE, E. RAMIREZ-ROMERO & J.-S. HWANG, 2018. Interannual changes in zooplankton echo subtropical and high latitude climate effects in the southern East China Sea. PLoS One, **13**(5): e0197382.

MOLINERO, J. C., L.-C. TSENG, L. LÓPEZ-LÓPEZ, U. SOMMER, S. SOUISSI & J.-S. HWANG, 2016. Climate driven winter variations of *Calanus sinicus* abundance in the East China Sea. Fisheries Oceanography, **25**(6): 555-564.

MOON, S. Y., M. H. SEO, Y. S. SHIN & H. Y. SOH, 2012. Seasonal variation of mesozooplankton communities in the semi-enclosed Muan Bay, Korea. Ocean Polar Res., **34**: 1-18.

MOON, S. Y., H. S. YOON, H. Y. SOH & S. D. CHOI, 2006. Environmental factors and variation characteristics of zooplankton communities in Gamak Bay. Ocean Polar Res., **8**: 79-94.

MOON, S. Y., J. H. CHOI, J. N. KIM & K. K. SUN, 2016. New extension range and complementary description of *Eurytemora americana* (Calanoida: Temoridae) in northern Korea. Marine Biodiversity Records, **9**: 76.

OHTSUKA, S. & S. NISHIDA, 2017. Copepod biodiversity in Japan: recent advances in Japanese copepodology. In: M. MOTOKAWA & H. KAJIHARA (eds.), Species diversity of animals in Japan: i-xiii, 1-721. (Springer Japan, Minato).

PARK, C., P. G. LEE & S. R. YANG, 2002. Variation of zooplankton distribution in the Seomjin River estuary with respect to season and salinity gradients. J. Korean Soc. Oceanogr., **7**: 51-59.

RAZOULS, C., F. DE BOVÉE, J. KOUWENBERG & N. DESREUMAUX, 2005-2019. Diversity and geographic distribution of marine planktonic copepods. (Sorbonne University, CNRS, Paris). Available at: http://copepodes.obs-banyuls.fr/en (accessed 26 May 2019).

ROSE, M., 1933. Copépodes pélagiques. Faune de France, **26**: i-iv, 1-374. (Librairie de la Faculté des Sciences, Paris). [Reprinted 1970 by Kraus Reprint, Nendeln, Liechtenstein.]

RUHL, H. A. & K. L. SMITH, 2004. Shifts in deep-sea community structure linked to climate and food supply. Science, **305**(5683): 513-515.

SARS, G. O., 1903. Copepoda, Calanoida. In: An account of the Crustacea of Norway with short descriptions and figures of all the species, **IV**: 1-171. (Bergen Museum, Bergen).

SHEN, J.-Q., Y. QIU, X.-G. GUO, A.-J. PAN & X.-W. ZHOU, 2017. The spatio-temporal variation of wintertime subtidal currents in the western Taiwan Strait. Acta Oceanologica Sinica, **36**(11): 4-13.

SHI, Y.-Q., S. SUN, G.-T. ZHANG, S.-W. WANG & C.-L. LI, 2015. Distribution pattern of zooplankton functional groups in the Yellow Sea in June: a possible cause for geographical separation of giant jellyfish species. Hydrobiologia, **754**(1): 43-58.

SUN, R.-X., Y.-G. WANG, G.-S. LIAN & M. LIN, 2014. Distribution and community characteristics of planktonic copepods in the northwest coastal waters off Hainan Island. Biodiversity Science, **22**(3): 320-328. [In Chinese with English abstract.]

SUN, S., Y. HUO & B. YANG, 2010. Zooplankton functional groups on the continental shelf of the Yellow Sea. Deep-Sea Research II, **57**(11): 1006-1016.

SUN, X.-H., Z.-L. LIANG, J.-X. ZOU & L.-X. WANG, 2013. Seasonal variation in community structure and body length of dominant copepods around artificial reefs in Xiaoshi Island, China. Chinese Journal of Oceanology and Limnology, **31**(2): 282-289.

TAKUO, C., 1952. Untersuchungen über das Plankton im Hafen von Shimonoseki-I. über den Copepoden "*Eurytemora pacifica* Sato". Bulletin of the Japanese Society of Scientific Fisheries, **18**(2): 63-66.

TSENG, L.-C., H.-U. DAHMS, J.-J. HUNG, Q.-C. CHEN & J.-S. HWANG, 2011. Can different mesh sizes affect the results of copepod community studies? Journal of Experimental Marine Biology and Ecology, **398**: 47-55.

TSENG, L.-C., J.-J. HUNG, J. C. MOLINERO, Q.-C. CHEN & J.-S. HWANG, 2015. Indicator species and seasonal succession of planktonic copepod assemblages driven by the interplay of subtropical and temperate waters in the southern East China Sea. Crustaceana, **88**(1): 96-112.

UYE, S., T. SHIMAZU, M. YAMAMURO, Y. ISHITOBI & H. KAMIYA, 2000. Geographical and seasonal variations in mesozooplankton abundance and biomass in relation to environmental parameters in Lake Shinji-Ohashi River-Lake Nakaumi brackish-water system, Japan. Journal of Marine Systems, **26**(2): 193-207.

WALTER, T. C. & G. BOXSHALL, 2019. World of copepods database. *Eurytemora pacifica* Sato, 1913. World Register of Marine Species. Available at: http://www.marinespecies.org/aphia.php?p=taxdetails&id=232028 [accessed 14 June 2019].

WANG, R., T. ZUO & K. WANG, 2003. The Yellow Sea cold bottom water — an oversummering site for *Calanus sinicus* (Copepoda, Crustacea). Journal of Plankton Research, **25**(2): 169-183.

WANG, Y.-G., J.-H. LIN, C.-G. WANG & M. LIN, 2012. Species composition and distribution characteristics of pelagic copepods in the northern sea of Fujian during withdraw of Zhe-Min coastal current. Environmental Science, **33**(6): 1839-1845. [In Chinese with English abstract.]

YIN, J.-H., G.-T. ZHANG, C.-L. LI, S.-W. WANG, Z.-X. ZHAO & A.-Y. WAN, 2017. Community composition, abundance and biomass of zooplankton in Zhangzi Island waters, northern Yellow Sea. Chinese Journal of Oceanology and Limnology, **35**(5): 1144-1151. [In Chinese, with English summary.]

YIN, J.-Q., L.-M. HUANG, K.-Z. LI, S.-M. LIAN, C.-L. LI & Q. LIN, 2011. Abundance distribution and seasonal variations of *Calanus sinicus* (Copepoda: Calanoida) in the northwest continental shelf of South China Sea. Continental Shelf Research, **31**(14): 1447-1456.

YOO, J. K., C. S. MYUNG, J. K. CHOI, H. P. HONG & E. S. KIM, 2010. Spatial and temporal variation of mesozooplankton community in Lake Shihwa, Korea. Ocean and Polar Res., **32**(3): 187-201.

ZHU, Y.-F., Y. WANG, X. LIN, Y.-P. WU, J.-L. ZHAO & X.-J. YAN, 2012. A comparison of zooplankton communities collected by two types of nets with different mesh sizes in Xiangshan Bay of Zhejiang, east China. Chinese Journal of Applied Ecology, **33**(8): 2277-2286. [In Chinese, with English summary.]

First received 24 June 2019.
Final version accepted 2 January 2020.

[When citing this volume, please refer to *Crustaceana* 93 (2020), 241-547]

*Proceedings of the* Eurytemora *conference, St. Petersburg, 2019*

# HOW MANY GENERATIONS DOES *EPISCHURA BAIKALENSIS* (COPEPODA, CALANOIDA) HAVE IN LAKE BAIKAL?

BY

ELENA YU. NAUMOVA[1,3]) and VICTOR R. ALEKSEEV[2])
[1]) Limnological Institute of the Siberian Branch of the Russian Academy of Sciences, Ulan-Batorskaja Str., 3, Irkutsk, Russia
[2]) Zoological Institute of the Russian Academy of Sciences, University Emb., 1, St. Petersburg, Russia

## ABSTRACT

*Epischura baikalensis* Sars is a dominant pelagic species of the Lake Baikal zooplankton. We empirically determined the duration of the development of *Epischura* from field data, as it is not possible to determine this experimentally. Based on our data, we conclude that the central pelagic food web of metazoan species in Lake Baikal has not two, as was previously accepted, but three generations during a year and each lasts not 6 but instead 3 to 4 months. The life cycle of the species also possibly includes a 3-months long winter embryonic diapause, starting in November and ending in late February with the start of the under-ice algae bloom. The high nauplii density in spring is not the result of high adult density, suggesting the presence of resting eggs in the *E. baikalensis* life cycle. We also found a large morphological difference between the pelagic eggs of the species in the winter and in the summertime. The possible participation of diatom algal toxins in the regulation of the seasonal dynamics of their main consumer in Lake Baikal, similar to those known in marine ecosystems, is discussed. The population dynamics in a year with relatively low diatom algae density show four peaks of reproduction, i.e., in February, June, September, and December. In years with significant (more than 15 000 cells/l) diatom blooming we found mass mortality of *E. baikalensis* juveniles in the plankton, with only two clear peaks of nauplii density. A negative correlation was found between the number of population density peaks for *E. baikalensis* and the spring-blooming of diatom algae.

## RÉSUMÉ

*Epischura baikalensis* Sars est une espèce pélagique dominante du zooplancton du lac Baïkal. Nous avons déterminé empiriquement la durée du développement d'*Epischura* à partir des données de terrain car il n'est pas possible de la déterminer expérimentalement. À partir de nos données, nous avons conclu que le réseau alimentaire pélagique central des espèces de métazoaires dans le lac Baïkal avait non deux, comme auparavant accepté, mais trois générations par an, chacune durant non 6 mois, mais 3 ou 4 mois. Le cycle de vie de l'espèce inclut aussi probablement une longue diapause

[3]) Corresponding author; e-mail: elena.baikalensis@gmail.com

 | DOI:10.1163/9789004465565_017

embryonnaire hivernale de trois mois, commençant en novembre et se terminant à la fin de février avec le départ du bloom algal sous la glace. La densité élevée de nauplies au printemps n'est pas le résultat d'une densité élevée des adultes, et suggère la présence d'œufs de diapause dans le cycle de vie de *E. baikalensis*. Nous avons aussi trouvé une différence morphologique importante entre les œufs pélagiques de l'espèce en hiver et ceux de l'été. La participation possible des toxines algales de diatomées dans la régulation des dynamiques saisonnières de leur principal consommateur dans le lac Baïkal, similaire à celles connues dans les écosystèmes marins, est discutée. La dynamique de population sur une année avec une densité de diatomées relativement faible montre quatre pics de reproduction, soit en février, juin, septembre et décembre. Dans les années avec un bloom significatif de diatomées (plus de 15 000 cell./l) nous avons trouvé une mortalité massive de juvéniles de *E. baikalensis* dans le plancton, avec seulement deux pics clairs de densité des nauplies. Une corrélation négative a été trouvée entre le nombre de pics de densité de population de *E. baikalensis* et le bloom de printemps des algues diatomées.

## INTRODUCTION

*Epischura baikalensis* Sars, 1900, a species endemic to Lake Baikal, makes up to 90% of the zooplankton in the lake (Kozhov, 1963; Afanasieva, 1977, 1998). *E. baikalensis*, as well as many marine species of calanoids, is characterized by complex relationships with algae. The interannual dynamics of the abundance of *Epischura* are largely related to the rhythms of development of diatoms of the genus *Aulacoseira*, on which it feeds (Sheveleva et al., 1992; Naumova et al., 2015). During years of high abundance of diatoms of this genus and in places of their accumulation, the numbers of *Epischura* sharply decrease (Melnik, 1995; Hampton et al., 2014).

A feature of the seasonal succession of the pelagic community of Lake Baikal is the presence of two complexes of planktonic algae separated in time: "spring large-cell", having cell sizes of 20 $\mu$m or more, with intensive development in April-May, and "summer small-cell", picoplankton, 1.5 $\mu$m and smaller, reaching a seasonal maximum in August-September. The fundamental difference between these two complexes lies in the composition of the species, the size of the algal cells, and the nature of seasonal and interannual dynamics. If the productivity of the summer complex, represented by the picoplanktonic forms of green algae and cyanobacteria, is approximately comparable from year to year (Bondarenko & Guselnikova, 1989; Bondarenko, 1997; Belykh & Sorokovikova, 2003), then the spring complex demonstrates pronounced bursts of production that is larger by two orders of magnitude and higher. The homothermic period, which divides the two complexes in time, usually occurs in Lake Baikal in June. At this time, spring algae, including clusters of species of the genus *Aulacoseira*, slowly sink from the surface 50-meter layer into the deep layers of the lake.

The reasons for this depression in the population of *Epischura* are not well understood. There are several explanations for this phenomenon, for example, mechanical breaking of the filtration process in copepods, avoiding layers of increased concentrations of diatoms by *Epischura* (Melnik, 1978), or the toxic effect of algal metabolites (Ianora & Poulet, 1993; Ban et al., 1997; Miralto et al., 1999; Harðardóttir et al., 2018). Recently, it has been suggested that limiting the reproductive fitness of the diatom consumer can explain the "algae-grazer" interactions observed in the marine pelagic zone (Poulet et al., 1994; Ianora et al., 2003; Caldwell et al., 2004). The life cycle parameters of *E. baikalensis*, including feeding and reproduction, are still poorly studied and it is difficult to conclude if the reproductive fitness of this species is affected by algal metabolites. Also, it is not clear why the smallest species of this genus has the lowest productivity among congeners and only two generations per year (Afanasieva, 1977).

Freshwater calanoid copepods are able to include a diapause in their life cycle, and the photoperiod, the food concentration, and the temperature can change the duration of the developmental stages (Pasternak & Arashkevich, 1999; Seebens et al., 2007). Copepods of the family Temoridae are well known for the production of diapausing eggs (Castellani & Lucas, 2003) and species of the genus *Epischura* can disappear from the plankton in the wintertime (Chow-Freser & Maly, 1988; Folt & Byron, 1989; Folt & Schulze, 1993; Chen & Folt, 1996; Barabanshchikov, 2004).

In this study, we examined the seasonal dynamics of *E. baikalensis* in the years with different diatom algae density and analysed published data covering more than 20 years. Our primary objective was to estimate the duration of the *E. baikalensis* generation in different seasons and test our hypothesis on the diapause presence in the life cycle of this species. Answering these questions could significantly change the evaluation of the energy flow in the pelagic ecosystem of Lake Baikal.

## MATERIAL AND METHODS

Plankton samples were taken in 1998 and 2000 with a monthly periodicity in the open waters of South Baikal (5 km from Berezovy cape, with a site depth of 1440 m) from a ship (May-December) or from ice (February-April). Sampling was carried out using a plankton net of 37-cm diameter and with a mesh size of 88 $\mu$m from three depth intervals, 0-50 m, 50-150 m, 150-250 m, and/or covering the entire range of 0-250 m. The samples were preserved with 4% formalin. At the same site, we measured the water temperature with an SBE-25 CTD probe (SeaBirdElectronics, Bellevue, WA, U.S.A.) and we collected algae with a 10 l bathometer. In the laboratory, all stages of development of *E. baikalensis* were identified and calculated separately, which allowed us to reconstruct the rate of the species' metamorphosis and the shift in maximal density at the different stages of development. For the studies using a scanning electron microscope (SEM) QUANTA 200 FEI, the samples were washed in distilled water. The *Epischura* eggs were then placed on the filter and next processed with alcohol of the following percentages: 30, 50, 70 and 96%. Alcohol residues were removed by using hexamethyldisilazane (Laforsch & Tollrian, 2000), and then the eggs were covered with gold. Quantitative samples of phytoplankton were fixed with Lugol's iodine solution (Utermöhl, 1958; Kuzmin, 1975) and allowed to settle; thereafter, algal cell counts in 0.1-ml drops were taken by the Gensen method (Kiselev, 1969), in two to three replications, under the compound light microscope.

For statistical analyses, one-way ANOVA analysis (on the basis of Statistica-6), we used the published data from previous publications on diatom algae blooming (Evstafiev & Bondarenko, 2002) and *E. baikalensis* population densities in Lake Baikal for about a 20-year period (Kozhova & Izmestieva, 1998). A detailed field study of the *E. baikalensis* life cycle was conducted by Afanasieva (1977), but there was no linear periodicity in the sampling. We used these data, too, but recalculated the population density in different developmental stages per month.

## RESULTS

### The annual dynamics of diatoms and *Epischura baikalensis*

Diatom dynamics.— According to our data, in 1998 and 2000 the abundance of the summer algae complex was practically the same in both years (fig. 1). However, the spring-complex differed sharply in both composition and abundance. In 1998, the maximum abundance of algae (with a minimal contribution from the diatoms of the genus *Aulacoseira* of 200 cells/l) occurred in May and reached 300 000 cells/l (fig. 1); small forms of Cryptophyta and Chlorophyta prevailed and diatoms of the species *Synedra acus* Kützing, 1844. In 2000, in contrast, there was an increase in the number of diatoms, predominantly consisting of species of the genus *Aulacoseira*, beginning in March. During the period of intensive vegetation growth (March-May) their average number exceeded that of 1998 by four orders of magnitude (375 000 thousand cells/l); the maximum abundance of diatoms of this genus (470 000 cells/l) was recorded in March-April.

*Epischura* dynamics.— In the population dynamics of *E. baikalensis* in those years, only two clear maxima of nauplii density were evident, indicating the breeding periods of two generations, one in spring and one in autumn (fig. 2). Other developmental stages also produced two periods of high density, approximately at the same time. However, the males in the population appeared three times per year, i.e., in May, July (low density), and October.

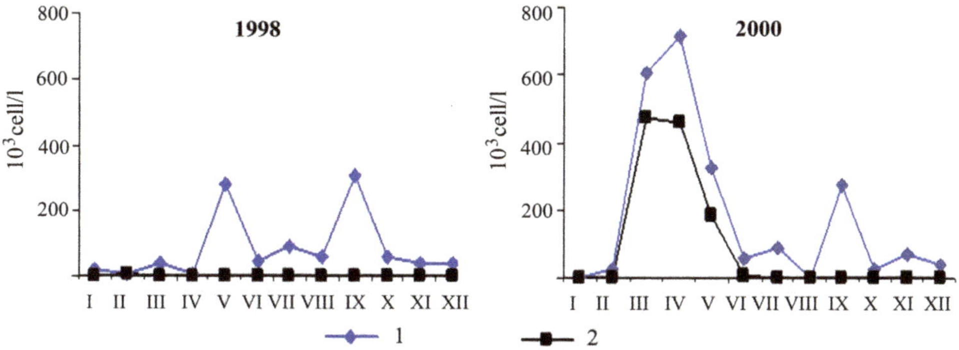


Fig. 1. Seasonal dynamics of pelagic phytoplankton in South Baikal: 1, All phytoplankton; 2, *Aulacoseira baicalensis* (K. Meyer) Simonsen, 1979.

A similar trend in population dynamics with a high density of nauplii in spring and a decline of that density in the summer, was obtained after recalculation of the data on monthly averages (Afanasieva, 1977), collected in 1962 (fig. 1). According to Evstafiev & Bondarenko (2002), significant diatom algae blooming occurred in South Baikal in 1962 as well.

In 1998, there was no significant diatom blooming, and in the *E. baikalensis* population dynamics for this year four density maxima of nauplii can clearly be seen (fig. 2). The total population density also produced four density maxima in this year, as did total *Epischura* biomass.

To determine if the number of maxima in the *E. baikalensis* population dynamics corresponds with the presence of spring diatom blooming in the lake, we analysed published data for both parameters, obtained over a long-term period in South

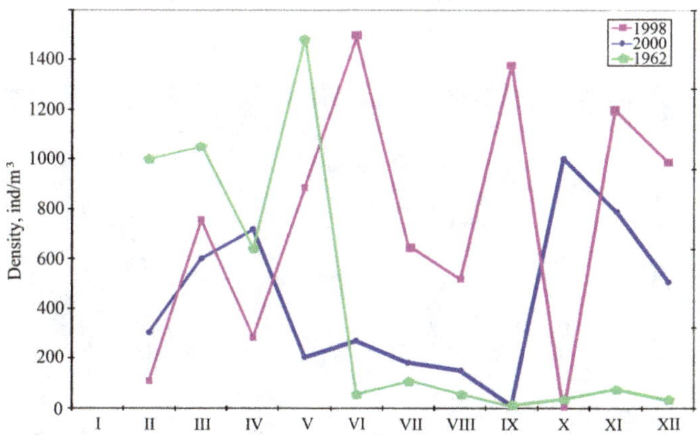


Fig. 2. Seasonal dynamics of *Epischura baikalensis* Sars, 1900 nauplii in South Baikal in 1962, 1998 and 2000. The data of monthly averages from Afanasieva (1977), collected in 1962, were recalculated.

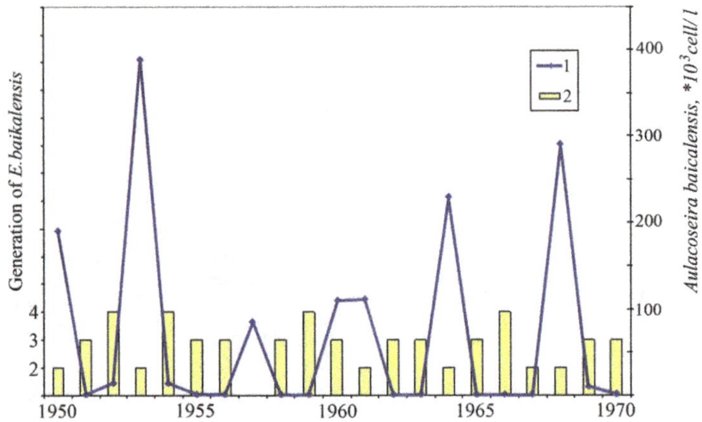


Fig. 3. 1, Spring diatom algae blooming; and, 2, the number of successive generations in *Epischura baikalensis* Sars, 1900, in South Baikal. The diagram was created by combining the data from Evstafiev & Bondarenko (2002) for algae, and Kozhova & Izmestieva (1998) for *Epischura*.

Baikal, with one-way ANOVA (fig. 3). This analysis confirmed that the number of generations in *E. baikalensis* strongly depends on diatom algae blooming in the lake ($F = 20.5655$; $p = 0.00027$). A negative range Spearman correlation coefficient between the number of annual biomass maxima of *E. baikalensis* and diatom algae density (in cells/l) in Baikal in the spring, was also statistically significant ($R = -0.545$; $p<0.05$).

## Egg morphology

Two types of *E. baikalensis* eggs were found in the plankton in 2000. One type of eggs, with a smooth shell, was found throughout most of the period (fig. 4A). The other type of egg, with a more complicated casing structure, was discovered

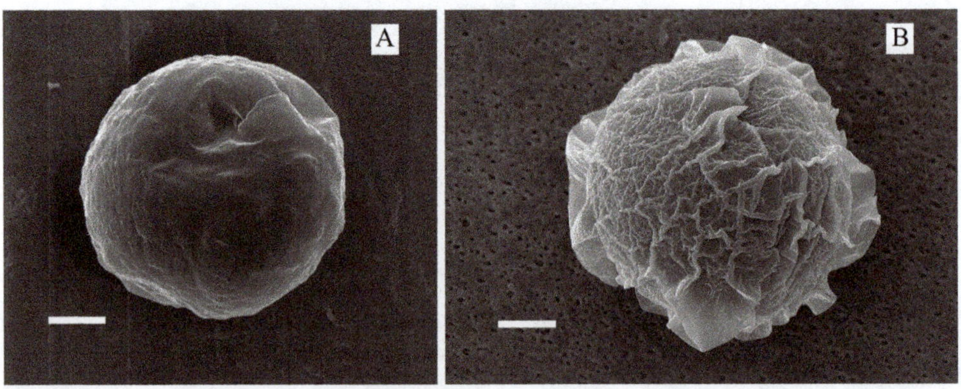


Fig. 4. SEM photographs of two types of eggs in *Epischura baikalensis* Sars, 1900: A, summer egg; B, winter (diapause) egg. Scale bar = 10 $\mu$m.

only during the last generation time in November-December (fig. 4B). These differences in the structure of the egg shells are similar to those found for active and resting eggs in another brackish water calanoid, *Temora longicornis* (O. F. Müller, 1785) (cf. Castellani & Lucas, 2003).

## DISCUSSION

*Epischura baikalensis*, as other congeners, lays eggs directly in the water column. For this reason it is impossible to trace breeding periods among a generation and its effect on the dynamics of fecundity. According to E. L. Afanasieva (1977), adult females of *E. baikalensis* can live for more than a year after the breeding period ends, which adds additional problems to the life cycle analyses of female density changes in population dynamics. To avoid this obstacle in this study, we used the density of other developmental stages with an obviously shorter developmental time, such as nauplii of stages one to two, or males, as a marker of the reproduction maximum for each generation. The development of other stages was also traced by the shifting in their maximum density over time. These data were also used to estimate the duration of a generation.

The maxima for the developmental stages, despite the crude assessment via monthly samplings, suggest that the duration of the development of one generation of *Epischura* is closer to 3 rather than to 6 months (fig. 2).

For the majority of the copepods studied in this regard, the duration of development of each stage is approximately the same, with variations in the stages of orthonauplius and copepodites of the fourth stage of 8-10% of the entire duration of the metamorphosis (Ivanova, 1978). For *Epischura* this duration is, at a temperature of 5°C, about 6 days. Since *Epischura* passes through the 12-stage metamorphosis development, the approximate period of the complete metamorphosis of *Epischura* at this temperature should be considered to be 72 days or about 2.5 months. Given the need for young females to mature and copulate, and for the subsequent embryonic egg development, the duration of the "from nauplius to nauplius" cycle is about 3 months, which is consistent with our observations.

We used the duration of the development of *Epischura*, established by field data, to calculate the rate of weight growth and compared the Baikal species with other crustaceans of similar ecology and size. With a metamorphosis period of 3 months, the Baikal species completely fits into the framework known for freshwater crustaceans in general, and is closest to the cold-water species of copepods (Ivanova, 1978). At the same time, the developmental period previously stated for the Baikal *Epischura* (6 months) clearly does not fit into the general pattern. An experimental study of the growth parameters of this species could bring additional clarity to the solution of this problem. The cultivation of *Epischura* in full has not yet been carried out, due to the difficulty of maintaining the necessary conditions for a long time. To the knowledge of the authors, there has been only one attempt to cultivate and grow the species (Kondratieva, 1980): metamorphosis from the fourth naupliar stage to the sixth reliably took place during a 30-day period at a temperature of 4-7°C, but the metamorphosis did not continue past this stage. The copepod spent 30 days (maximum) in three stages. If the stage took 10 days, then the whole metamorphosis was about 120 days of development at 6°C, which at 10°C would be 90 days (including about 10 days for embryogenesis and the maturation of young females).

Instead of the previously indicated two cycles of *Epischura* development in the pelagic zone of Lake Baikal, we thus distinguish three complete cycles with a total period of 8 months and, possibly, a dormancy stage lasting about 4 months. The process of metamorphosis in the first two generations lasts 3 months, and in the autumn (with decreasing day length and maximum heating of the epilimnion) it lasts 2 months.

The rate of change in the morphological characteristics of *E. baikalensis* during the year (Ermakov & Rusanovskaya, 2015) which is most likely associated with genetic parameters (Zaidykov et al., 2015) also suggests the presence of more than two generations.

The under-ice period of reactivation of resting eggs possibly starts at the end of January, but as collecting material under ice has long been impossible, it can only be seen in samples since February. During this time, the population density of the adults was very low. Therefore, we suspect that these females are not able to produce enough eggs to establish the first maximum of nauplii (fig. 2). In the plankton, we discovered *Epischura* eggs of various structures. At least some of them were similar in morphology to diapausing eggs in other copepods. This also does not contradict our assumption about the participation of diapausing eggs in the formation of the first peak in the number of nauplii. Therefore, we conclude that at least part of the population of this species undergoes a winter egg diapause lasting from November to February.

In the first maximum of *E. baikalensis* during algae blooming, a high proportion of dead juveniles was found, contributing to up to 80% of the total density (Naumova et al., 2008).

The spring period of breeding is the first real maximum of reproduction in the year. It was observed at the end of May or the beginning of June. For this reason, the first generation can be presumed to take 3.5-4 months at an average temperature 5-6°C in the layer of 0-50 m deep, where most of the juveniles were staying within this period.

The summer maximum of nauplii density was only seen in 1998 and disappeared in 2000, when diatom algae had bloomed. In 1998, this maximum occurred 3.5 months after the previous one and the average surface temperature during this period was 7-8°C.

The autumn maximum of nauplii was observed 2.5 months after the summer peak in 1998, in November, which corresponded with the warmest surface water temperature usually found in Baikal in this period (10-12°C). The nauplii produced by these females became adult in October-November. In November (late fall) these females laid two types of eggs, normal and diapausing ones (fig. 4). The resting eggs sank and became active the following year. The chance of survival for the eggs that activated immediately and produced this last maximum of nauplii was not high. Due to a lack of food in wintertime, most of these nauplii died and only a small part could possibly have survived until under-ice algae blooming commenced and participated in population reproduction in February. This is supported by the declining biomass of *E. baikalensis* in wintertime, observed every year (Kozhov, 1967; Melnik, 1995).

Thus, in the annual cycle of *E. baikalensis*, we distinguished three generations that lasted 2.5-4 months in accordance with the surface (0-50 m) water temperature in Lake Baikal, instead of two generations of 6 months each, as reported in previous studies (Afanasieva, 1977). This conclusion changes our understanding about the rate of productivity of *E. baikalensis* in Lake Baikal, especially in summer. The impact of this species on the annual secondary production of Lake Baikal should be considered significantly higher than before. Therefore, it now comes close to the rate of productivity of cold-water living copepods of a similar size. For more detailed answers about the effect of this change on the energy budget of the pelagic system of Baikal, a model assessment needs to be carried out.

The fact that there are different numbers of generations in years with and without diatom algae blooming, led to the question regarding the interaction between consumer *Epischura baikalensis* and its prey, species of the genus *Aulacoseira*, in particular *A. baicalensis* (K. Meyer) Simonsen, 1979 in Baikal. By taking into consideration the role of the toxic metabolites of marine diatoms in limiting the grazer's reproductive fitness via affecting embryonic development in its offspring (Poulet et al., 1994; Caldwell et al., 2004; Ianora et al., 2004), we can predict a similar kind of interaction between *Epischura* and *Aulacoseira* in Lake Baikal as well. In our case, during the "melosira" years, as possibly caused by exposure to the toxic metabolites of the diatoms in June-September, there are depressions observed in the population of *E. baikalensis* after the diatom blooming, and nauplii maxima are only observed in early spring and autumn.

## ACKNOWLEDGEMENTS

This work is devoted to the memory of Dr. Natalia G. Melnik (1951-2009), who first inspired this study. Many thanks to Nina A. Bondarenko for help in processing the phytoplankton samples. This study was supported with RFBR grant 20-04-00035, by Government Program No. AAAA-A19-119020690091-0, by the Russian Academy of Sciences topics 65.4 and 65.5 and the Federal Zoological Collection, field research by State project No. 0345-2019-0008 (AAAA-A16-116122110065-4).

## REFERENCES

AFANASIEVA, E. L., 1977. Biologija baikalskoj epishury (The biology of *Epischura* in Lake Baikal): 1-144. (Nauka, Novosibirsk). [In Russian.]

AFANASIEVA, E. L., 1998. Life cycle of *Epischura baicalensis* Sars (Copepoda, Calanoida) in Lake Baikal. J. of Mar. Syst., **15**: 351.

BAN, S., C. BURNS, J. CASTEL, Y. CHAUDRON, E. CHRISTOU, R. ESCRIBANO, S. FONDA UMANI, S. GASPARINI, F. GUERRO RUIZ, M. HOFFMEYER, A. IANORA, H.-K. KANG, M. LAABIR, A. LACOSTE, A. MIRALTO, X. NING, S. POULET, V. RODRIGUEZ, J. RUNGE, J. SHI, M. STARR, S. UYE & Y. WANG, 1997. The paradox of diatom-copepod interactions. Mar. Ecol. Progr. Ser., **157**: 287-293.

BARABANSHCHIKOV, E. I., 2004. Zooplankton ozera Khanka (Zooplankton of Lake Khanka). (Ph.D. Thesis, St. Petersburg). [In Russian.]

BELYKH, O. I. & E. G. SOROKOVIKOVA, 2003. Autotrophic picoplankton in Lake Baikal: abundance, dynamics, and distribution. Aquat. Ecosyst. Health Manag., **6**(3): 251-261.

BONDARENKO, N. A., 1997. Struktura i produktsionnyye kharakteristiki fitoplanktona ozera Baykal (The structure and production characteristics of phytoplankton of Lake Baikal). (Ph.D. Thesis, IBIW RAS, Borok, Russia). [In Russian.]

BONDARENKO, N. A. & N. E. GUSELNIKOVA, 1989. Znacheniye vodorosley piko- i nanoplanktona v produktsionnykh protsessakh v ozere Baykal (The importance of algae pico- and nanoplankton in production processes in Lake Baikal). Biologicheskiye Nauki, **12**: 34-36. [In Russian.]

CALDWELL, G. S., S. B. WATSON & M. G. BENTLEY, 2004. How to assess toxin ingestion and post-ingestion partitioning in zooplankton? J. Plankton Res., **26**: 1369-1377.

CASTELLANI, C. & I. A. N. LUCAS, 2003. Seasonal variation in egg morphology and hatching success in the calanoid copepods *Temora longicornis*, *Acartia clausi* and *Centropages hamatus*. J. Plankton Res., **25**: 527-537.

CHEN, C. Y. & C. L. FOLT, 1996. Consequences of fall warming for zooplankton overwintering success. Limnol. and Oceanogr., **5**: 1077-1086.

CHOW-FRESER, P. & E. J. MALY, 1988. Aspects of mating, reproduction, and co-occurrence in three freshwater calanoid copepods. Freshwater Biology, **19**: 95-108.

ERMAKOV, E. L. & O. O. RUSANOVSKAYA, 2015. Seasonal dynamics of the phenotypic structure of a natural *Epischura baicalensis* Sars population in terms of quantitative morphological traits. Contemp. Probs. Ecol., **2**: 193-201. DOI:10.15372/sej20150207.

EVSTAFIEV, V. K. & N. A. BONDARENKO, 2002. Priroda javlenia "melozitnih let" v ozere Baikal (About the origin of "melozira years" in Lake Baikal). Hydrobiol. J., **38**: 3-12. [In Russian.]

FOLT, C. L. & E. R. BYRON, 1989. A comparison of the effects of prey and non-prey neighbours on foraging rates of *E. nevadensis* (Copepoda: Calanoida). Freshwater Biology, **21**: 283-293.

FOLT, C. L. & P. C. SCHULZE, 1993. Spatial patchiness, individual performance and predator impacts. Oikos, **68**: 560-566.

HAMPTON, S. E., D. K. GRAY, L. R. IZMESTIEVA, M. V. MOORE & T. OZERSKY, 2014. The rise and fall of plankton: long-term changes in the vertical distribution of algae and grazers in Lake Baikal, Siberia. PLoS ONE, **9**(2): e88920. DOI:10.1371/journal.pone.0088920.

HARÐARDÓTTIR, S., B. KROCK, S. WOHLRAB, U. JOHN, T. G. NIELSEN & N. LUNDHOLM, 2018. Can domoic acid affect escape response in copepods? Harmful Algae, **79**: 50-52. DOI:10.1016/j.hal.2018.08.009.

IANORA, A., A. MIRALTO, S. A. POULET, Y. CAROTENUTO, I. BUTTINO, G. ROMANO, R. CASOTTI, G. POHNERT, T. WICHARD, L. COLUCCI-D'AMATO, G. TERRAZZANO & V. SMETACEK, 2004. Aldehyde suppression of copepod recruitment in blooms of a ubiquitous planktonic diatom. Nature, **429**: 403-407.

IANORA, A. & S. A. POULET, 1993. Egg viability in the copepod *Temora stylifera*. Limnol. Oceanogr., **38**: 1615-1626.

IANORA, A., S. A. POULET & A. MIRALTO, 2003. The effects of diatoms on copepod reproduction: a review. Phycologia, **42**: 351-363.

IVANOVA, M. B., 1978. Produktsiya planktonnykh rakoobraznykh v presnykh vodakh (Production of planktonic crustaceans in fresh waters). Hydrobiol. J., **14**(6): 17-26. [In Russian.]

KISELEV, I. A., 1969. Plankton morei i kontinental'nykh vodoemov (Plankton of seas and continental waterbodies): 1-439. (Nauka, Leningrad). [In Russian.]

KONDRATIEVA, I. A., 1980. Ekologo-toksikologicheskiy kontrol' kachestva ochishchennykh stochnykh vod sul'fat-tsellyuloznogo proizvodstva (Ecological and toxicological quality control of treated wastewater from sulphate-cellulose production). (Ph.D. Thesis, State University, Petrozavodsk, Russia).

KOZHOV, M. M., 1963. Lake Baikal and its life: 1-344. (Dr. W. Junk, The Hague).

KOZHOVA, O. M. & L. R. IZMESTIEVA, 1998. Lake Baikal: 1-447. (Backhuys Publishers, Leiden).

KUZMIN, G. V., 1975. Phytoplankton: species composition and abundance. In: F. D. MORDUKHAY-BOLTOVSKY (ed.), Metodika izucheniya biogeocenozov vnutrennikh vodoemov (Methods of studies on biogeocenoses of inland water bodies): 73-87. (Nauka, Moscow).

LAFORSCH, C. & R. TOLLRIAN, 2000. A new preparation technique of daphnids for scanning electron microscopy using hexamethyldisilazane. Arch Hydrobiol., **149**: 587-596.

MELNIK, N. G., 1978. O vlijanii na sezonnuju dinamiku chislennosti epishury bioticheskih factorov (Effects of biotic factors on seasonal density dynamics in *Epischura*). Matematicheskoe modelirovanie vodnuh ecologichaskih system: 133-144. (Irkutsk). [In Russian.]

MELNIK, N. G., 1995. Itogi issledovanija zooplanktona (Results of zooplankton studies). In: O. A. TIMOSHKIN (ed.), Guide and key to pelagic animals of Baikal (with ecological notes): 34-84. (Nauka, Novosibirsk). [In Russian.]

MELNIK, N. G., E. V. DZYUBA, V. A. DEGTYAREV, M. I. LAZAREV, E. YU. NAUMOVA & E. P. TEREZA, 2008. Life strategy of pelagic animals in Lake Baikal. Verh. Internat. Verein. Limnol., **30**(2): 291-294.

MIRALTO, A., G. BARON, S. A. POULET, A. IANORA, G. L. RUSSO, I. BUTTINO, G. MAZZARELLA, M. CABRINI & M. G. GIACOBBE, 1999. The insidious effect of diatoms on copepod reproduction. Nature, **402**: 173-176.

NAUMOVA, E. YU., E. P. TEREZA, G. A. FEDOROVA, N. A. BONDARENKO & N. G. MELNIK, 2008. The reasons of mortality of *Epischura baicalensis* (Copepoda, Calanoida) in Lake Baikal. In: $10^{th}$ Intern. Conference on Copepoda (13-19 July 2008, Pattaya, Thailand): 73.

NAUMOVA, E. YU., I. YU. ZAIDYKOV, V. L. TAUSON & YE. V. LIKHOSHWAY, 2015. Features of the fine structure and Si content of the mandibular gnathobase of four freshwater species of *Epischura* (Copepoda: Calanoida). J. Crust. Biol., **35**(6): 741-746.

PASTERNAK, A. F. & E. G. ARASHKEVICH, 1999. Resting stages in the life cycle of *Eudiaptomus graciloides* (Lill.) (Copepoda: Calanoida) in Lake Glubokoe. J. Plankton Res., **21**(2): 309-325.

POULET, S. A., A. IANORA, A. MIRALTO & D. L. MEIJER, 1994. Do diatoms arrest embryonic development in copepods? Mar. Ecol. Prog. Ser., **111**: 79-86.

SEEBENS, H., D. STRAILE, R. HOEGG, H.-B. STICH & U. EINSLE, 2007. Population dynamics of a freshwater calanoid copepod: complex responses to changes in trophic status and climate variability. Limnol. Oceanogr., **52**(6): 2364-2372.

SHEVELEVA, N. G., N. F. LOGACHEVA, A. E. BALAYAN & L. D. ZUBAREVA, 1992. Quantitative aspects of feeding of Baikal *Epischura*. Hydrobiol. J., **28**(6): 19-23.

UTERMÖHL, H., 1958. Vervollkommung der quantitativen Phytoplankton-Methodik (Perfecting the quantitative phytoplankton methodology). Mitt. Int. Verein. Theor. Angew. Limnol., **9**: 1-39.

ZAIDIKOV, I. YU., T. YU. MAIOR, L. V. SUKHANOVA, S. V. KIRILCHIK & E. YU. NAUMOVA, 2015. MtDNA polymorphism of Lake Baikal *Epischura* — an endemic key species of the plankton community. Russ. J. Genet., **51**(9): 935-937.

First received 19 September 2019.
Final version accepted 13 January 2020.

[When citing this volume, please refer to *Crustaceana* 93 (2020), 241-547]

*Proceedings of the Eurytemora conference, St. Petersburg, 2019*

# A MULTI-YEAR INVESTIGATION OF THE TEMORIDAE (COPEPODA, CALANOIDA) ASSEMBLAGE SUCCESSION WITHIN THE INTERPLAY WATERS OF THE NORTHERN SOUTH CHINA SEA

BY

LI-CHUN TSENG[1,2]), YAN-GUO WANG[1,3]), GUANG-SHAN LIAN[3]) and JIANG-SHIOU HWANG[1,2,4])

[1]) Institute of Marine Biology, National Taiwan Ocean University, Keelung, Taiwan
[2]) Center of Excellence for the Oceans, National Taiwan Ocean University, Keelung, Taiwan
[3]) Third Institute of Oceanography, Ministry of Natural Resources, Xiamen 361005, Fujian, P.R. China

## ABSTRACT

The South China Sea is the largest marginal sea in the world. The northern South China Sea has a complex hydrographic system affected by interplay waters of the Kuroshio Branch Current and the South China Sea waters. To understand the changes in the community structure of the planktonic calanoid copepod family Temoridae in relation to the interplay waters, we collected zooplankton samples in the northern South China Sea from surface waters (0 to 2 m) from November 2001 to January 2007 during 24 research cruises using a standard North Pacific zooplankton net (with a mesh size of 200 $\mu$m, a length of 1.8 m, and a mouth diameter of 0.45 m). Among a total of 253 samples, 5 species belonging to 3 genera of Temoridae were identified: *Eurytemora pacifica* Sato, 1913, *Temora discaudata* Giesbrecht, 1889, *Temora stylifera* (Dana, 1849), *Temora turbinata* (Dana, 1849), and *Temoropia mayumbaensis* Scott T., 1894. The 3 most abundant species were: *T. turbinata* (relative abundance, RA: 53.28%; mean density $\pm$ standard deviation, MD: 29.33 $\pm$ 89.53 ind. $m^{-3}$), *T. stylifera* (RA: 36.02%, MD: 19.83 $\pm$ 145.22 ind. $m^{-3}$), and *T. discaudata* (RA: 7.70%, MD: 4.24 $\pm$ 11.82 ind. $m^{-3}$). The 3 most frequently occurring species were: *T. turbinata* (occurrence ratio, OR: 52.17%), *T. discaudata* (OR: 35.18%), and *T. stylifera* (OR: 9.88%). A noteworthy discovery in our samples was *E. pacifica*, a marine and brackish water species, collected from 3 stations located near the Kuroshio Branch Current in April 2003. This is the only month when this species was recorded during the investigation period. Variations of dominant species of Temoridae show a clear pattern of seasonal succession. The total abundance was significantly higher in the third quarter (September-November, 116.98 $\pm$ 314.49 ind. $m^{-3}$) than in the fourth quarter (December-February, 24.26 $\pm$ 47.72 ind. $m^{-3}$) ($p = 0.017$, one-way ANOVA). In general, the present results demonstrate that the assemblages of the Temoridae are very much structured by the water masses of the Kuroshio Current and the South China Sea.

Key words. — Temoridae, copepod, phenology, seasonal succession, South China Sea, interplay waters, Kuroshio Current

[4]) Corresponding author; Fax: +886-2-24629464, e-mail: jshwang@mail.ntou.edu.tw

 | DOI:10.1163/9789004465565_018

## RÉSUMÉ

La mer de Chine du Sud est la mer bordière la plus vaste du monde. La mer de Chine du Sud septentrionale possède un système hydrographique complexe affecté par l'interaction entre les eaux du courant de Kuroshio et celles de la mer de Chine du Sud. Pour comprendre les changements de la structure des communautés de la famille de copépodes planctoniques calanoïdes Temoridae par rapport aux zones de rencontre des courants, nous avons collecté des échantillons de plancton dans la mer de Chine du Sud septentrionale, dans les eaux de surface (0 à 2 m) de novembre 2001 à janvier 2007 au cours de 24 croisières de recherches en utilisant un filet à plancton standard « Nord-Pacifique » (vide de maille de 200 $\mu$m, longueur 1,8 m, ouverture de 0,45 m). Sur un total de 253 échantillons, 5 espèces appartenant à 3 genres de Temoridae ont été identifiées: *Eurytemora pacifica* Sato, 1913, *Temora discaudata* Giesbrecht, 1889, *Temora stylifera* (Dana, 1849), *Temora turbinata* (Dana, 1849), et *Temoropia mayumbaensis* Scott T., 1894. Les 3 espèces les plus abondantes étaient: *T. turbinata* (abondance relative, RA: 53,28%; densité moyenne $\pm$ déviation standard, MD: 29,33 $\pm$ 89,53 inds. m$^{-3}$), *T. stylifera* (RA: 36,02%, MD: 19,83 $\pm$ 145,22 inds. m$^{-3}$), et *T. discaudata* (RA: 7,70%, MD: 4,24 $\pm$ 11,82 inds. m$^{-3}$). Les 3 espèces les plus fréquentes étaient: *T. turbinata* (fréquence d'occurrence, OR: 52,17%), *T. discaudata* (OR: 35,18%), et *T. stylifera* (OR: 9,88%). Une découverte remarquable dans les échantillons a été la présence de *E. pacifica*, une espèce d'eau marine et saumâtre, collectée dans 3 stations situées près du courant de Kuroshio en avril 2003. C'est le seul mois au cours duquel cette espèce a été trouvée pendant toute la période de recherche. Les variations des espèces dominantes de Temoridae montrent un modèle clair de succession saisonnière. L'abondance totale était significativement plus élevée dans le troisième quart (septembre-novembre, 116,98 $\pm$ 314,49 inds. m$^{-3}$) que dans le dernier quart (décembre-février, 24,26 $\pm$ 47,72 inds. m$^{-3}$) ($p = 0{,}017$, ANOVA simple). En général, les résultats obtenus démontrent que les assemblages de Temoridae sont fortement structurés par les masses d'eaux du courant de Kuroshio et de la mer de Chine du Sud.


Mots clés. — Temoridae, copépode, phénologie, succession saisonnière, mer de Chine du Sud, eaux d'interaction, courant de Kuroshio

## INTRODUCTION

The South China Sea (SCS) is the largest marginal sea in the world, and covers an area of about 3.4 $\times$ 10$^6$ km$^2$ (Huang et al., 2002; Tseng & Shen, 2003). The northeastern area of the SCS is connected to the Pacific Ocean via the Luzon Channel. This Luzon Channel is an important pathway for the exchange of marine fauna in the Kuroshio Branch Current (KBC, hereafter KC) and South China Sea (Chang et al., 2010; Tseng et al., 2015). The hydrological environment in the northeastern SCS area is affected by the interaction of the KC and SCS water masses (Hwang et al., 2007a). The KC water is warm and saline, and the SCS water is higher saline and warm as well. The KC is a northward flowing current in the western Pacific Ocean. It begins off the east coast of the Philippines and passes along the east coast of Taiwan. Then it flows northeastwards to the Japanese archipelago (Barkley, 1970). In addition, the SCS area lies under the Western Pacific Warm Pool (Huang et al., 2002) which is one of the world's most famous monsoonal climate zones (Tseng & Shen, 2003; Li et al., 2010).

The prevailing East Asian monsoonal winds affect the oceanic currents (Jan et al., 2002). In winter, the North East monsoon prevails, whereas in summer the South West monsoon prevails. Such strong seasonal monsoons play an important role in affecting the structure of zooplankton communities in the waters around Taiwan (Hwang et al., 2004, 2006, 2009; Chang et al., 2010; Chou et al., 2012).

Zooplankton plays an important role in the transfer of energy and nutrients from primary producers to higher consumers in the marine food web (Li et al., 2003; Jiang et al., 2009; Tseng et al., 2009). Among diverse zooplankton taxa, copepods are the most abundant metazoans in the waters around Taiwan, such as in the southern East China Sea (Hwang et al., 2004, 2006; Dur et al., 2007; Chou et al., 2012; Tseng et al., 2012), in the Kuroshio Current (Hsiao et al., 2011c), in the Taiwan Strait (Lo et al., 2004b; Hsiao et al., 2011a), in the northeastern SCS (Hwang et al., 2007a; Tseng et al., 2008a), and the world's oceans in general (Verity & Smetacek, 1996; Schminke, 2007; Kiørboe 2011). Copepods have become suitable bio-indicators of water mass movements. For example, *Calanus sinicus* Brodsky, 1962 is an important indicator species of the China Coastal Current in the western Pacific Ocean (Hwang & Wong, 2005; Yin et al., 2011), and the East China Sea (Molinero et al., 2016, 2018). *Calanoides philippinensis* Kitou & Tanaka (1969) is an indicator species of the SCS water movement in the East China Sea (Tseng et al., 2013b, 2015). *Temora turbinata* (Dana, 1849) is an indicator of warm-water masses and becomes most abundant in coastal areas of northern Taiwan (Hwang et al., 2006, 2009; Dur et al., 2007; Tseng et al., 2013b), as well as dominant in the waters next to nuclear power plants, and abundant during the South East monsoon prevailing period (Hwang et al., 2004; Tseng et al., 2011b). In the above account, its congener *Temora discaudata* Giesbrecht, 1889 was recorded in the same area but not as an important species. There are several reports about the family Temoridae Giesbrecht, 1982 in waters around Taiwan (e.g., Shih & Young, 1995; Liu, 2008; Lian et al., 2018). Most reports focus on the distribution of these copepods, whereas little information is available about the seasonal succession of population densities, phenology, and ecology of this family.

The SCS is as yet poorly understood in terms of marine biota and their ecology (Morton & Blackmore, 2001). This also holds for the role of copepods in this oceanic region (Hwang et al., 2010; Lian et al., 2017). This holds especially for long-term ecological data collected with a consistent method for the same purpose (Chen et al., 2015). In this report, we provide a multi-year investigation of planktonic copepods in the area of Nan-Wan Bay in the northern SCS. Our objectives were to reveal the spatio-temporal changes of the Temoridae, in particular: (1) members of this family, (2) abundance and succession of their assemblage structure, (3) phenology of each species of Temoridae present, and (4) characterize populations related to the interplay waters of the Kuroshio and SCS water masses. These results will provide important information on the ecological role that the Temoridae may play in the northeastern SCS.

## MATERIAL AND METHODS

### Field collections

A total of twelve stations were situated in the area between 21°52.870′-21°56.952′N and 120°43.720′-120°50.850′E; four stations were located in Nan-Wan Bay and eight stations were located at the northern South China Sea (fig. 1). The south side of Nan-Wan Bay facing the Luzon Strait connects to the Philippines and the SCS. Latitude and longitude of the stations and cruises are shown in table I. Field sampling was carried out from November 2001 to January 2007, and comprises 24 research cruises. The year was split into seasonal quarters: spring (March-May), summer (June-August), autumn (September-November), and winter (December-February). The zooplankton samples were collected with a standard North Pacific zooplankton net (mouth diameter 45 cm, length 180 cm, and mesh size 200 $\mu$m), being towed horizontally within the surface zone (0-2 m depth). The water volume filtered was calculated from records of a Hydrobios flow meter that was mounted in the centre of the net mouth. The net was towed at a speed of about 1 knot $h^{-1}$ (about 0.5 m $s^{-1}$) for 10 min. at each station. Samples were preserved in seawater, diluted with 5-10% buffered formaldehyde immediately after they were taken on board. Prior to the plankton tow, conductivity-temperature-depth (CTD) data were obtained.

### Copepod identification and abundance estimates

In the laboratory, zooplankton samples were separated by a Folsom splitter until the subsample contained about 350-500 specimens for taxonomic identification and counting. Subsamples were counted and copepods were identified to species, and the abundance (number of individuals $m^{-3}$) was estimated. The following keys were used for the identification of copepod species: Chen & Zhang (1965), Chen et al. (1974), and Chihara & Murano (1997).

### Statistical analyses

The relationship of copepod abundance with temperature and salinity were tested with the Pearson product moment correlation analyses. The quarterly abundance data of each species collected from all 24 cruises were treated as replicates for temporal analyses. To identify differences in copepod abundance of each species among seasons, one-way ANOVA with a post-hoc Tukey's honestly significant difference test was applied.

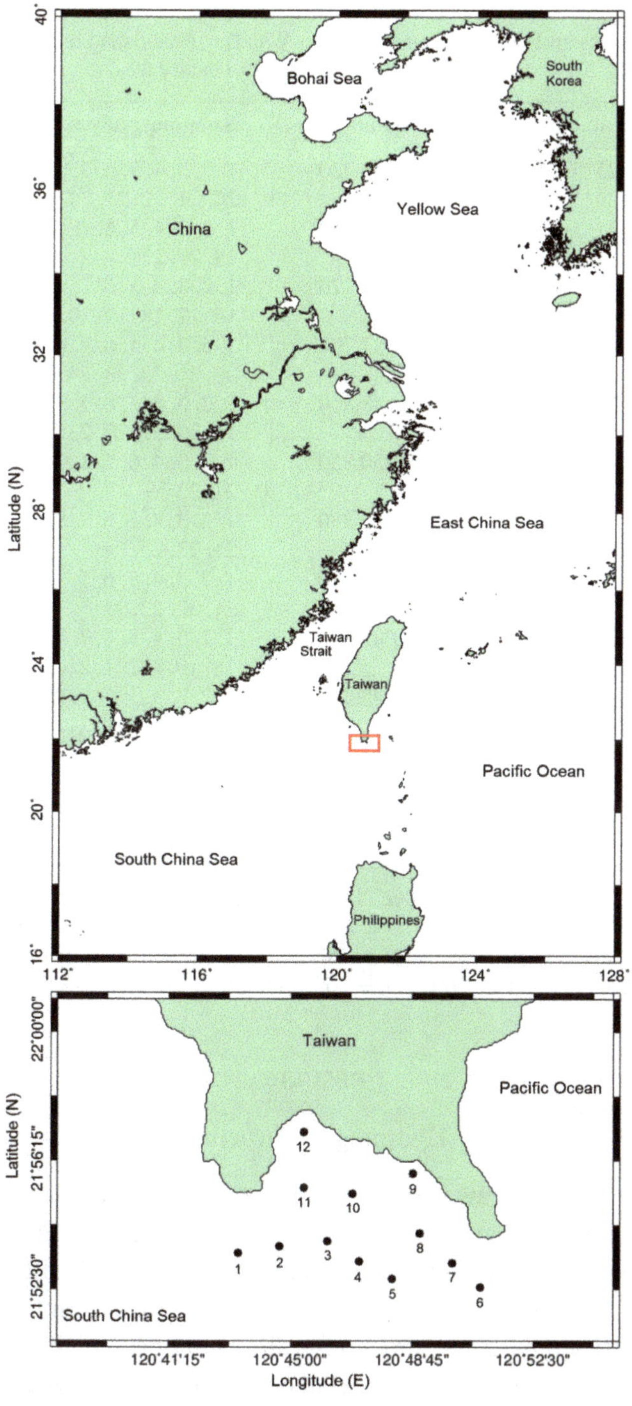


Fig. 1. Sampling stations from November 2001 through January 2007 in Nan-Wan Bay, southern Taiwan.

TABLE I
Coordinates of stations and sampling cruises in Nan-Wan Bay established for zooplankton sampling from November 2001 through January 2007

| Station No. | Latitude (N) | Longitude (E) | Sampling cruises |
|---|---|---|---|
| 1 | 21°53.900′ | 120°43.720′ | 1, 2, 3, 4, 5, 6, 7, 8, 9, 10, 16, 17, 18, 19, 20, 21, 22, 23, 24 |
| 2 | 21°54.904′ | 120°44.740′ | 1, 2, 3, 4, 5, 6, 7, 8, 9, 10, 12, 16, 17, 18, 19, 21, 22 |
| 3 | 21°54.904′ | 120°46.770′ | 1, 2, 3, 4, 5, 6, 7, 8, 9, 10, 11, 12, 13, 15, 16, 17, 18, 19, 20, 21, 22, 23, 24 |
| 4 | 21°53.890′ | 120°47.790′ | 1, 2, 3, 4, 5, 6, 7, 8, 9, 10, 16, 17, 18, 19, 20, 21, 22, 23, 24 |
| 5 | 21°52.870′ | 120°48.810′ | 1, 2, 3, 4, 5, 6, 7, 8, 9, 10, 15, 16, 17, 18, 19, 20, 21, 22, 23, 24 |
| 6 | 21°52.870′ | 120°50.850′ | 1, 2, 3, 4, 6, 7, 8, 9, 10, 12, 17, 18, 19, 20, 21, 22 |
| 7 | 21°53.890′ | 120°50.830′ | 1, 2, 3, 4, 5, 6, 7, 8, 9, 15, 16, 17, 18, 19, 21, 22, 23, 24 |
| 8 | 21°53.900′ | 120°49.820′ | 1, 2, 3, 4, 5, 6, 7, 8, 9, 12, 16, 17, 18, 19, 21, 22, 23, 24 |
| 9 | 21°54.930′ | 120°49.810′ | 1, 2, 3, 4, 5, 6, 7, 8, 9, 11, 12, 13, 16, 17, 18, 19, 20, 21, 22, 23, 24 |
| 10 | 21°55.920′ | 120°47.784′ | 1, 2, 3, 4, 5, 6, 7, 8, 9, 11, 12, 13, 15, 16, 17, 18, 19, 20, 21, 23, 24 |
| 11 | 21°55.930′ | 120°45.760′ | 1, 2, 3, 4, 5, 6, 7, 8, 9, 12, 15, 16, 17, 18, 19, 20, 21, 22, 23, 24 |
| 12 | 21°56.952′ | 120°45.756′ | 1, 2, 3, 4, 5, 6, 7, 8, 9, 12, 15, 16, 17, 18, 20, 21, 22, 23, 24 |

Sampling cruises: 1 = Nov 2001, 2 = Dec 2001, 3 = Mar 2002, 4 = Apr 2002, 5 = Jul 2002, 6 = Sep 2002, 7 = Jan 2003, 8 = Apr 2003, 9 = Aug 2003, 10 = Nov 2003, 11 = Jan 2004, 12 = Apr 2004, 13 = Aug 2004, 14 = Nov 2004, 15 = Jan 2005, 16 = May 2005, 17 = Jun 2005, 18 = Jul 2005, 19 = Oct 2005, 20 = Dec 2005, 21 = Apr 2006, 22 = Jul 2006, 23 = Oct 2006, 24 = Jan 2007.

## RESULTS

### Hydrological structure

The average surface seawater temperature (SST) and salinity recorded during each sampling cruise from November 2001 through January 2007 are shown in fig. 2. The seasonal temperature variations were highly significant with lower values recorded in the period Dec-Feb and peak values during the period Jun-Aug. The lowest temperature throughout the sampling months was recorded in Jan 2005 (22.47 ± 0.12°C; mean ± standard deviation), whereas the highest temperature was recorded in Jul 2005 (29.33 ± 0.42°C) (fig. 2A). The average salinity ranged between 33.26 ± 0.08 (Sep 2002) and 34.62 ± 0.21 (Jan 2005)

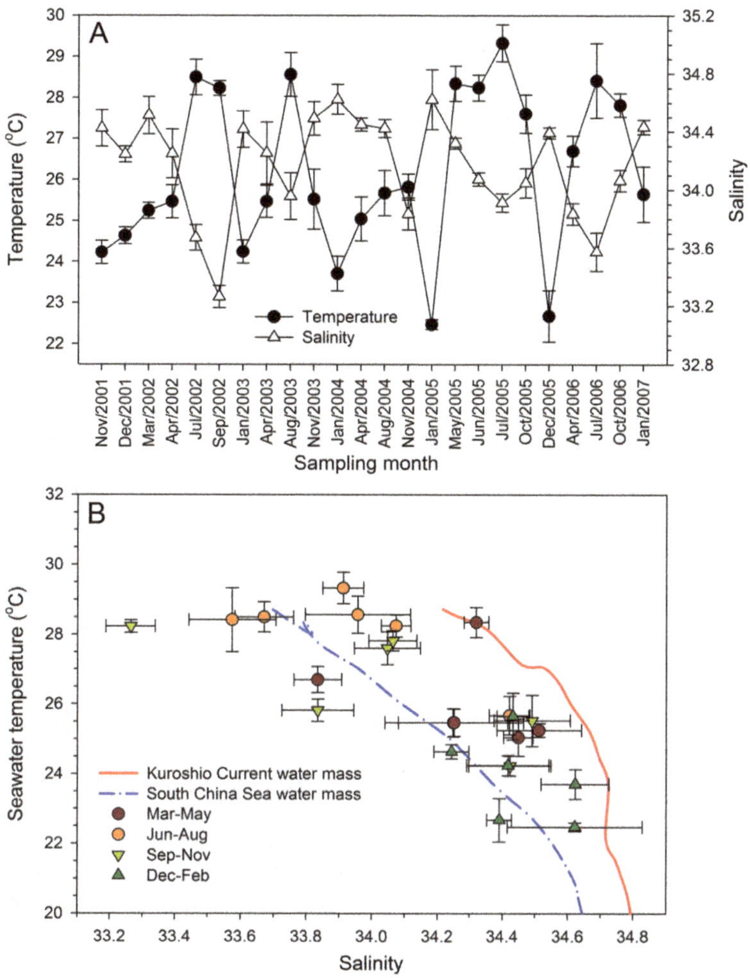


Fig. 2. A, Average temperature and salinity; and B, temperature-salinity diagram comparison with two reference temperature-salinity curves of the Kuroshio Current and South China Sea water masses: of surface seawater on each sampling cruise in Nan-Wan Bay, from November 2001 through January 2007.

(fig. 2A). The correlation between seawater temperature and salinity during the study period showed marked seasonal variations (fig. 2B). The present study applied two reference temperature-salinity (TS) curves of the SCS and the KC, which were obtained from the Department of Ocean Data Bank, Taiwan (http://www.odb.ntu.edu.tw/en/). Seawater temperatures were lower in the period Dec-Feb than during other sampling periods. A comparison with TS reference curves revealed that the water mass properties in the study area were affected by the interplay of both the SCS water and the KC (fig. 2B).

## Identification and characteristics of the Temoridae copepod assemblage

From a total of 253 samples collected from 24 research cruises from Nan-Wan Bay during November 2001 to January 2007, we identified 5 Temoridae species belonging to 3 genera: *Eurytemora pacifica* Sato, 1913, *Temora discaudata*, *Temora stylifera* (Dana, 1849), *Temora turbinata*, and *Temoropia mayumbaensis* Scott T., 1894. *Eurytemora pacifica* is a marine and brackish water species (Walter & Boxshall, 2019) and is a noteworthy discovery in our samples; it was collected from stations 6-8 located near the Kuroshio area in April 2003. The taxonomic key is outlined below and plates of morphological structures of the female copepods for identification are provided in fig. 3.

### KEY FOR THE IDENTIFICATION OF SPECIES OF TEMORIDAE IN NAN-WAN BAY

1a– Female, last thoracic somite posterior end wing-like; female P5 first segment inner proximal margin produced medially into spinous process, distal segment armed with 2 setae; male P5 without inner process; P1 endopods with 1 segment, P2 to P4 endopods with 2 segments .. .......... *Eurytemora pacifica* (fig. 3A)

b– Female, last thoracic somite posterior end not wing-like .......... 2

2a– Posterior corner of prosome rounded .......... 3

b– Posterior corner of prosome spiniform elongated .......... 4

3a– Female urosome 4-segmented and male urosome 5-segmented; female genital somite large and protruding ventrally by about the same amount as the depth of the urosome itself; endopods of P1 2-segmented, P2 to P4 endopods 3-segmented; female P5 3-segmented, with a seta-like endopod, exopod with inner and distal tooth-like processes; caudal rami small, with 3 terminal setae .......... *Temoropia mayumbaensis* (fig. 3E)

b– Caudal rami very long and slender, equal in length to entire abdomen; caudal rami bilaterally symmetrical; female P5 with inner spine on terminal segment considerably shorter and thinner than the 2 terminal spines .......... *Temora turbinata* (fig. 3D)

4a– Caudal rami very long in both female and male; female P5 with inner spine on terminal segment considerably longer and stronger than the 2 terminal spines; caudal rami not bilaterally symmetrical in female .......... *Temora discaudata* (fig. 3B)

b– Caudal rami very long in both female and male; female P5 with inner spine on terminal segment considerably longer and stronger than the 2 terminal spines; caudal rami bilaterally symmetrical in female .......... *Temora stylifera* (fig. 3C)

## Copepod community structure

The quarterly values of the average (mean ± SD) abundance of each species recorded during the study period are shown in table II. The highest proportion of Temoridae species among the whole Calanoida assemblage was recorded in the period Sep-Nov (35.5%), followed by the period Mar-May (22.72%), whereas the lowest proportion was in the period Dec-Feb (6.62%), and comprised 18.79% of total calanoids in all samples collected from November 2001 to January 2007 (table II). When comparing the composition of all samples, the ranking by abundance of dominant species was as follows: *T. turbinata* (relative abundance,

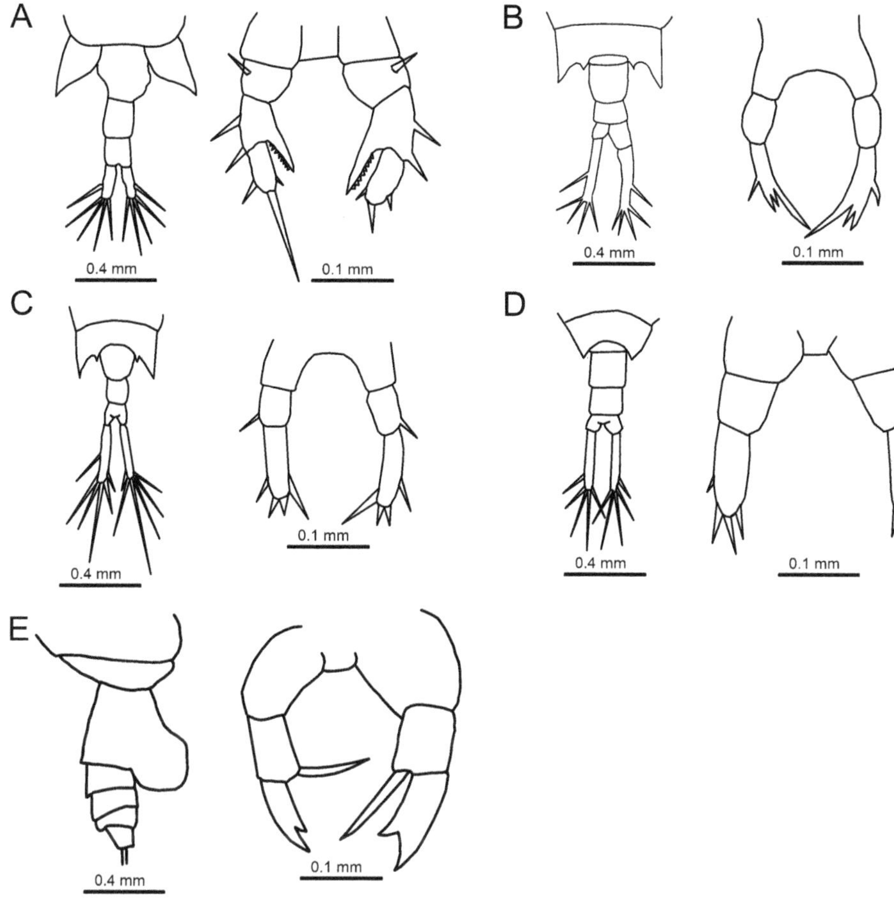


Fig. 3. Morphological characteristics of female urosome: A-D, dorsal view (left) and the fifth swimming leg, posterior view (right) of identified copepod species of the family Temoridae: A, *Eurytemora pacifica* Sato, 1913; B, *Temora discaudata* Giesbrecht, 1889; C, *Temora stylifera* (Dana, 1849); D, *Temora turbinata* (Dana, 1849); and E, lateral view of urosome (left) and posterior view of the fifth swimming leg (right) of *Temoropia mayumbaensis* Scott T., 1894.

RA: 53.28%; mean density ± standard deviation, MD: 29.33 ± 89.53 ind. $m^{-3}$), *T. stylifera* (RA: 36.02%, MD: 19.83 ± 145.22 ind. $m^{-3}$), and *T. discaudata* (RA: 7.70%, MD: 4.24 ± 11.82 ind. $m^{-3}$). The top 3 dominant species together comprised 97.0% of the overall copepod counts. *Temora turbinata* was recorded with the highest occurrence ratio (OR) 52.17%, followed by *T. discaudata* (OR: 35.18%) and *T. stylifera* (OR: 9.88%) (table II).

The variation in abundance and proportion of each species throughout the cruises is shown in fig. 4. *Temora turbinata* was the most dominant species and was identified during most of the sampling cruises, with abundances ranging from 0, in Aug 2004, to 299.0 ± 245.09 (mean ± standard deviation, ind. $m^{-3}$) in Jan 2005.

TABLE II

Number of samples, total abundance of Calanoida copepods (mean ± SD) (ind. $m^{-3}$), number of Calanoida copepods (number $season^{-1}$), identified species of Temoridae copepods (number $season^{-1}$), abundance, relative abundance (RA, %), occurrence frequency (OR, %) of copepods belonging to the Temoridae, and proportion of Temoridae copepods among whole Calanoida assemblage (%) recorded for each season from November 2001 to January 2007

| Sample | Mar-May | Jun-Aug | Sep-Nov | Dec-Feb | All samples | RA | OR |
|---|---|---|---|---|---|---|---|
| Number of samples | 74 | 62 | 50 | 67 | 253 | | |
| Total abundance of Calanoida copepods | 200.72 ± 218.99 | 293.68 ± 259.08 | 329.54 ± 423.86 | 366.73 ± 283.39 | 292.92 ± 300.53 | | |
| Number of Calanoida copepods | 79 | 85 | 63 | 75 | 107 | | |
| Temoridae copepods | | | | | | | |
| Number of species identified | 4 | 4 | 4 | 3 | 5 | | |
| *Eurytemora pacifica* Sato, 1913 | 0.24 ± 1.35 | N.F. | N.F. | N.F. | 0.07 ± 0.74 | 0.13 | 1.19 |
| *Temora discaudata* Giesbrecht, 1889 | 6.16 ± 17.06 | 4.34 ± 12.46 | 2.12 ± 3.93 | 3.61 ± 6.97 | 4.24 ± 11.82 | 7.70 | 35.18 |
| *Temora stylifera* (Dana, 1849) | 0.21 ± 1.28 | 1.34 ± 3.52 | 98.37 ± 317.18 | N.F. | 19.83 ± 145.22 | 36.02 | 9.88 |
| *Temora turbinata* (Dana, 1849) | 39 ± 120.86 | 37.71 ± 106.39 | 16.43 ± 43.27 | 20.54 ± 47.13 | 29.33 ± 89.53 | 53.28 | 52.17 |
| *Temoropia mayumbaensis* Scott T., 1894 | N.F. | 6.28 ± 23.26 | 0.07 ± 0.46 | 0.12 ± 0.69 | 1.58 ± 11.76 | 2.87 | 5.53 |
| Total abundance (mean ± SD) | 45.61 ± 125.83 | 49.66 ± 116.46 | 116.98 ± 314.49 | 24.26 ± 47.72 | 55.05 ± 169.54 | 100.0 | – |
| Proportion of Temoridae copepods among the whole Calanoida assemblage | 22.72 | 16.91 | 35.50 | 6.62 | 18.79 | | |

N.F. = not found.

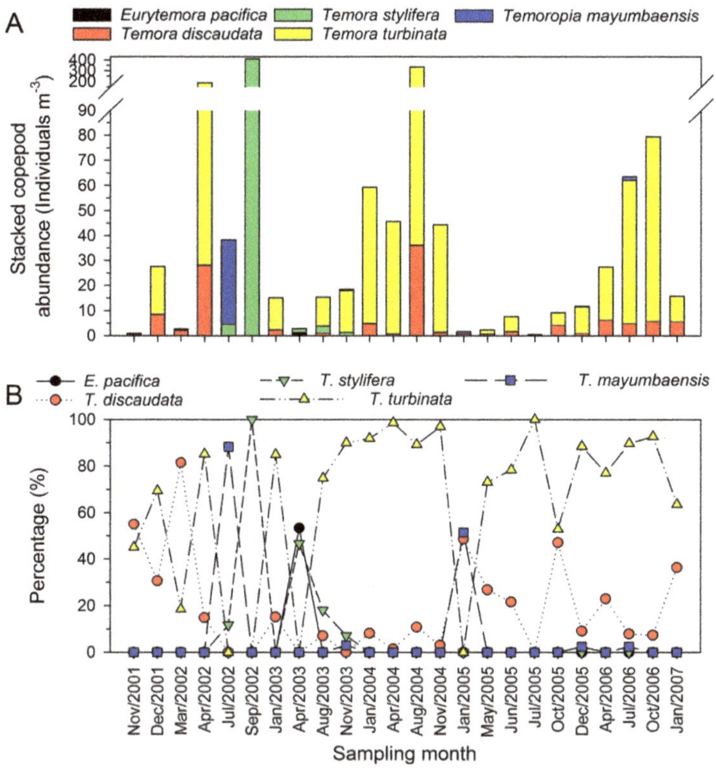


Fig. 4. A, Abundance (individuals $m^{-3}$); and B, proportion; of five Temoridae species on each sampling cruise in Nan-Wan Bay, from November 2001 through January 2007.

The copepod *T. stylifera* was dominant in Sep 2002 with average abundances of $409.21 \pm 556.27$ ind. $m^{-3}$. *Temora discaudata* had high abundances in Apr 2002 and Aug 2004, with average abundance values of $28.17 \pm 34.82$ and $36.17 \pm 27.74$ ind. $m^{-3}$, respectively. The copepod *T. mayumbaensis* was dominant in Jul 2002 with average abundances of $33.85 \pm 47.65$ ind. $m^{-3}$. Total copepod abundance was lower in Jul 2005 with $0.49 \pm 1.21$ ind. $m^{-3}$, whereas the highest abundance was found in Sep 2002, i.e., $409.21 \pm 556.27$ ind. $m^{-3}$ (fig. 4A). Only *T. stylifera* and *T. turbinata* were recorded in the samples of Sep 2002 and Jul 2005 accounting for 100%, respectively. In general, the proportion of each species shows no significant pattern of change (fig. 4B). The abundance and dominance patterns of each species were different at interannual and seasonal scales.

## Phenological analysis

The relationship of abundance with seawater temperature and salinity (fig. 5), and the ranges of seawater temperatures and salinity values of the five species of Temoridae recorded during the research period indicated, that for each species the

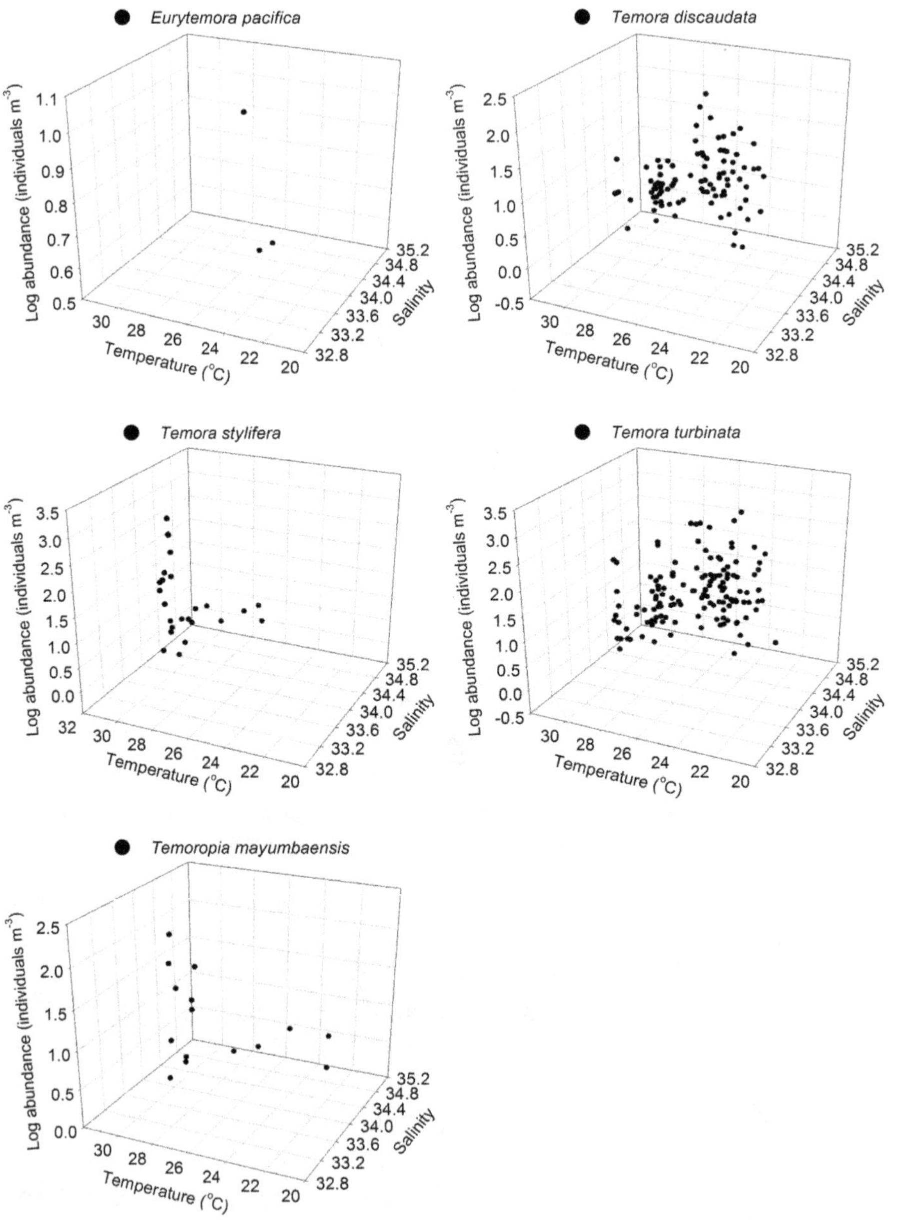


Fig. 5. Relationship of the abundance of five species of Temoridae with seawater temperature and salinity: *Eurytemora pacifica* Sato, 1913; *Temora discaudata* Giesbrecht, 1889; *Temora stylifera* (Dana, 1849); *Temora turbinata* (Dana, 1849); and *Temoropia mayumbaensis* Scott T., 1894.

suitable seawater temperature and salinity were different (table III). *Eurytemora pacifica* was only found at seawater temperatures between 25.06 and 25.40°C. The three species *Temora discaudata*, *T. turbinata*, and *Temoropia mayumbaensis* were

TABLE III
Range (mean ± standard deviation) of seawater temperature (°C) and salinity of the five Temoridae species recorded during November 2001 to January 2007

| Copepod species | Temperature | Salinity |
|---|---|---|
| *Eurytemora pacifica* Sato, 1913 | 25.06-25.40 (25.18 ± 0.19) | 33.83-34.40 (34.10 ± 0.29) |
| *Temora discaudata* Giesbrecht, 1889 | 22.31-29.18 (26.16 ± 1.65) | 33.49-34.66 (34.21 ± 0.28) |
| *Temora stylifera* (Dana, 1849) | 25.19-29.18 (28.08 ± 0.90) | 33.09-34.52 (33.62 ± 0.41) |
| *Temora turbinata* (Dana, 1849) | 22.19-29.66 (26.21 ± 1.77) | 33.48-34.67 (34.20 ± 0.29) |
| *Temoropia mayumbaensis* Scott T., 1894 | 22.07-28.99 (27.12 ± 2.33) | 33.49-34.72 (33.90 ± 0.41) |

found at seawater temperatures above 22.07°C; whereas *Temora stylifera* preferred higher seawater temperatures, i.e., above 25.19°C.

Furthermore, the results of Pearson's correlation analysis found the abundances of *T. stylifera* ($r = 0.133$; $p = 0.035$) and *T. mayumbaensis* ($r = 0.154$; $p = 0.014$) were significantly positively correlated with seawater temperature in the ranges of the present records. Therefore, one-way ANOVA revealed that the abundance of *T. stylifera* in the period Sep-Nov was significantly higher than that from Mar-May ($p = 0.001$), Jun-Aug ($p = 0.002$), and Dec-Feb ($p = 0.001$). The abundance of *T. mayumbaensis* in the period Jun-Aug was significantly higher than that in Mar-May ($p = 0.009$), Sep-Nov ($p = 0.025$), and Dec-Feb ($p = 0.014$). On the other hand, none of the abundances of the Temoridae species showed a significant correlation with salinity. Consequently, a dynamic compositional change in the course of the different months explains the substantial succession in Temoridae assemblages in the northern SCS.

## DISCUSSION

### Copepods of the family Temoridae

The present study recorded five Temoridae species in the waters of the northern SCS. In the western Pacific Ocean, according to a literature review by Shih & Young (1995), 6 species of Temoridae were found in the China seas [plus *Temorites brevis* Sars G.O., 1900, which was misplaced taxonomically in that report, since this species belongs to the family Bathypontiidae]. Liu (2008) recorded 6 species of Temoridae from the China seas [plus *Eurytemora affinis* (Poppe, 1880)]. Recently, Lian et al. (2018) reported 7 species of Temoridae inhabiting the China seas [plus *E. affinis*, *Eurytemora herdmani* Thompson I.C. & Scott A., 1897, *Temora longicornis* (Müller O.F., 1785); minus *T. stylifera*]. Furthermore, Chihara & Murano (1997) reported 6 species of Temoridae from the Japan seas [plus *E.*

*affinis*, *E. herdmani*; minus *T. stylifera*]. Both Chihara & Murano (1997) and Lian et al. (2018) suggested that *T. discaudata* and *T. stylifera* are actually the same species. *Temora discaudata*, therefore, was kept as a valid species. The study of Chang et al. (2010) on copepod communities is much overlapping with the scope of the present study, with 3 species records of the Temoridae. Dur et al. (2007) described the copepod communities in waters around Taiwan (including the SCS); they found 4 species of Temoridae. Summarizing previous reports, eight species of Temoridae were recorded in the western Pacific Ocean (if *T. stylifera* is a valid species). Thus, in the present survey, 62.5% of the Temoridae species of the western Pacific Ocean were recorded.

The distribution of each temorid copepod is the result of complex interactions of biotic and abiotic factors of the environment (McGlynn, 2010). The phenological results from this study showed that the suitable water temperature and salinity for the five copepod species were different (table III and fig. 5). The ability to adapt to the environment also affects the understanding of their distribution. This report reviews the literature on the distribution of five species of Temoridae in the western Pacific (table IV). The historical data indicated that *T. discaudata* and *T. turbinata* had higher reporting rates than the other 3 species; both species were distributed widely in the waters of the western Pacific Ocean (table IV). Rezai et al. (2004) reported *T. discaudata*, *T. stylifera*, and *T. turbinata* in the Strait of Malacca, and Wang et al. (2018) recorded *T. discaudata* and *T. turbinata* in the Lembeh Strait of North Sulawesi, Indonesia, in the tropical seas near the equator. Both *E. pacifica* and *T. mayumbaensis* were only once reported in the SCS (Lian et al., 2018). They are recorded from that area for the second time in the present study, and their distribution extends further to the northeastern SCS.

## Succession of Temoridae copepods

The change of abundance and composition of temorid copepods indicates a seasonal succession in the northern SCS. The data of hydrological parameters revealed that the study area was affected by interplay waters of the SCS and the KC. The area of the KC is known as one of the largest marine ecosystems in the world (Sherman et al. 2004). Previous studies reported that the KC affects various elements of marine life in its range (Yatsu et al., 2013), such as macroalgae (Lin et al., 2018), coral ecosystems (Chen & Shashank, 2009; Yamazaki et al., 2016), planktonic copepods (Hsiao et al., 2004, 2011b; Hwang et al., 2007a; Lee et al., 2009; Chou et al., 2012; Tseng et al., 2013b), jellyfish (Lo et al., 2014c), the squid *Todarodes pacificus* (Steenstrup, 1880) (cf. Bower et al., 1999; Rosa et al., 2001), larval fish assemblages (Lo et al., 2010), and fisheries in general (Chiu & Huang, 1994; Yatsu et al., 2013). The monsoon is another important

TABLE IV

List of the five copepod species of Temoridae found in the present study and recorded earlier in the seawaters around Taiwan

| Species | Yellow Sea | East China Sea | Taiwan Strait | South China Sea | Western Pacific Ocean |
|---|---|---|---|---|---|
| *Eurytemora pacifica* Sato, 1913 | 4, 7, 21, 23, 30, 41, 43 | 30, 23, 42 | 42 | | 5, 23, 28 |
| *Temora discaudata* Giesbrecht, 1889 | 4, 18, 23, 44 | 4, 9, 12, 13, 17, 18, 20, 22, 23, 24, 31, 38, 42, 43, 44 | 1, 6, 22, 23, 25, 26, 27, 36, 37, 40, 42, 43, 44 | 1, 2, 3, 23, 32, 33, 34, 35, 43, 44 | 5, 14, 19, 23, 25, 39, 43, 44 |
| *Temora stylifera* (Dana, 1849) | 30, 44 | 24, 30, 44 | 1, 15, 44 | 1, 2, 3, 16, 30, 43, 44 | 14, 43, 44 |
| *Temora turbinata* (Dana, 1849) | 4, 22, 23, 30, 44 | 4, 8, 9, 10, 11, 12, 13, 17, 18, 20, 22, 23, 24, 30, 31, 38, 42, 43, 44 | 6, 15, 22, 23, 25, 26, 27, 36, 37, 40, 42, 43, 44 | 2, 3, 23, 29, 30, 32, 33, 34, 35, 43, 44 | 5, 14, 23, 39, 43, 44 |
| *Temoropia mayumbaensis* Scott T., 1894 | 22 | 4, 22, 23, 24, 30, 45 | 23 | 23 | 5, 14, 23 |

References: 1 = Chang & Fang (2004); 2 = Chang et al. (2010); 3 = Chen (1982); 4 = Chen & Zhang (1965); 5 = Chihara & Murano (1997); 6 = Chou et al. (2012); 7 = Dai et al. (2018); 8 = Fang et al. (2006); 9 = Fang et al. (2014); 10 = Hsiao & Fang (2013); 11 = Hsiao et al. (2010); 12 = Hsiao et al. (2011c); 13 = Hsiao et al. (2011b); 14 = Hsiao et al. (2011a); 15 = Hsu et al. (2008); 16 = Hwang et al. (2007); 17 = Hwang et al. (2006); 18 = Hwang et al. (2004); 19 = Kotori & Hamaoka (1979); 20 = Lan et al. (1978); 21 = Li & Fang (1990); 22 = Lian & Lin (1978); 23 = Lian et al. (2018); 24 = Liao et al. (2006); 25 = Lin & Lian (1988); 26 = Lo et al. (2004a); 27 = Lo et al. (2004b); 28 = Mori (1937); 29 = Shen & Li (1963); 30 = Shih & Young (1995); 31 = Shih & Chiu (1998); 32 = Tseng et al. (2011a); 33 = Tseng et al. (2008a); 34 = Tseng et al. (2012); 35 = Tseng et al. (2013a); 36 = Tseng et al. (2008b); 37 = Tseng et al. (2011b); 38 = Tseng et al. (2008c); 39 = Tseng et al. (2016); 40 = Wang et al. (2012); 41 = Xiao et al. (1992); 42 = Yu (2005); 43 = Zhang et al. (1978); 44 = Zhang et al. (1982); 45 = Zhu et al. (1991).

environmental factor that affects the direction of water flow and the mixing of water masses. A study in Nan-Wan Bay revealed copepod assemblages with significant differences between the northeasterly monsoon and the southwesterly monsoon periods (Chang et al., 2010). The present study found that abundances of *T. stylifera* and *T. mayumbaensis* were significantly higher in the period Sep-Nov and Jun-Aug than during other sampling periods, respectively. These results, linked with phenology analyses, revealed the proper hydrographic condition of these two species for the upper 2 m of the surface water. Further, this study found that the Temoridae assemblage has a clear seasonal succession, and their abundance could reach about 25 to 30% among the whole Calanoida assemblage in the periods Sep-Nov and Mar-May, respectively. The results showed the importance of those species in the survey area, similar to previous studies in northern Taiwan (Hwang et al., 2004; Tseng et al., 2011b).

Sampling strategies provide an important factor affecting the recorded composition of zooplankton assemblages (Makabe et al., 2012). Tseng et al. (2011a) found that nets of small mesh size significantly affected the abundance of planktonic copepods in the SCS. In their paper, the abundances of *T. discaudata* and *T. turbinata* were significantly higher in samples of a 100 $\mu$m net than with two other larger mesh sizes, 200 $\mu$m and 333 $\mu$m. Wu et al. (2011) found that mesh size affected the abundance and structure of the copepod community in the East China Sea. Similar results were later confirmed for the waters of the South Adriatic Sea (Miloslavić et al., 2014), and the Bohai Sea, China (Chen et al., 2016). We used a 200 $\mu$m mesh size net to collect zooplankton samples, counting mature copepods. Our results suggest using smaller mesh size nets in future surveys in these waters. This helps to collect smaller individuals and larval stages, which would allow also understanding the life history of Temoridae in the waters of the northeastern SCS.

## Remarks on *Eurytemora pacifica*

*Eurytemora pacifica* was recorded in this study in 3 samples collected during April 2003. There are no historical records of *E. pacifica* from the southernmost part of Taiwan. The present study, therefore, signifies the first record of *E. pacifica* in this area, which is affected by waters of the KC and the SCS. *Eurytemora pacifica* is a species inhabiting marine and brackish environments (Razouls et al., 2005-2019). Published plankton records in the western Pacific Ocean report the presence of *E. pacifica* from the Bohai Sea to the SCS (here the precise sampling location was not reported) (Liu, 2008; Lian et al., 2018), the northwestern Pacific Ocean (Lian et al., 2018), and from brackish water areas at high latitudes of the northern Pacific (Japan) (Chihara & Murano, 1997). In addition, *E. pacifica* was found in the estuaries of two rivers in northern and southern Taiwan (Yu, 2005). In the western Pacific, the China Coastal Current plays an important role for the spatial distribution of marine biota. The China Coastal Current has been suggested as a main vector to transport marine organisms to the coastal waters of Taiwan, such as macroalgae (Lin et al., 2018), zooplankton (Tseng et al., 2012), jellyfish (López-López et al., 2013; Tseng et al., 2015), and copepods (Chou et al., 2012). We hypothesize that waters of the China Coastal Current might convey *E. pacifica* from the Bohai Sea and Yellow Sea southward to the coasts of Fujian and Guangdong, China. Then it mixes with the SCS water in the northern SCS through the southern Taiwan Strait, then eastward to the northeast SCS. Based on previous records, *E. pacifica* is distributed in the Bohai Sea, Yellow Sea, the waters of Korea, the archipelago of Japan (Razouls et al., 2005-2019), and southernmost of Taiwan (present study). Phylogeographic studies on *E. pacifica* are needed to better understand the role of geography in shaping differences at population level.

*Eurytemora pacifica* prefers a brackish environment, and most reports of the western Pacific area indicate that the main distribution of this species is restricted to temperate zones: Bohai Sea, Yellow Sea, Korea and Japan (table III). In the present investigation, it occurs in high temperature and high salinity environments. The emergence of this species implies that *E. pacifica* enters the study area from nearby rivers as Yu (2005) recorded it in estuarine waters of the Danshuei River (southwestern East China Sea) and the Kaoping River (southeastern Taiwan Strait). Another possibility is that inland rivers adjacent to the study area are the particular habitat of *E. pacifica* in the northern SCS. However, the distribution of *E. pacifica* in the waters around Taiwan is still inadequately studied.

In conclusion, the present study reveals a succession of the copepod assemblages of species of the family Temoridae. It also confirms the hypothesis of Chang et al. (2010) that copepod assemblages change with water conditions at the northeastern margin of the SCS, where water masses of the KC and those of the SCS are mixing.

## ACKNOWLEDGEMENTS

The first two authors, LI-CHUN TSENG and YAN-GUO WANG, contributed equally to this study. We are grateful for financial support from the Ministry of Science and Technology (MOST) of Taiwan through grants nos. MOST 107-2811-M-019-004 and MOST 108-2811-M-019-504 to L.-C. Tseng, as well as grants nos. MOST 106-2621-M-019-001, MOST 107-2621-M-019-001 and MOST 108-2621-M-019-003 to J.-S. Hwang. Finally, we would like to acknowledge the assistance provided by the captain and crew of the "Ocean Research Vessel III" for their assistance during the research cruises.

Conflict of interest.— No potential conflict of interest was reported by the authors.

## ETHICAL APPROVAL

All applicable international, national, and/or institutional guidelines for the care and use of animals were followed.

## REFERENCES

BARKLEY, R. A., 1970. The Kuroshio Current. Science Journal, **6**: 54-60.

BOWER, J. R., Y. NAKAMURA, K. MORI, J. YAMAMOTO, Y. ISODA & Y. SAKURAI, 1999. Distribution of *Todarodes pacificus* (Cephalopoda: Ommastrephidae) paralarvae near the Kuroshio off southern Kyushu, Japan. Marine Biology, **135**: 99-106.

CHANG, W.-B. & L.-S. FANG, 2004. Temporal and spatial variations in the species composition, distribution, and abundance of copepods in Kaohsiung Harbor, Taiwan. Zoological Studies, **43**: 454-463.

CHANG, W.-B., L.-C. TSENG & H.-U. DAHMS, 2010. Abundance, distribution and community structure of planktonic copepods in surface waters of a semi-enclosed embayment of Taiwan during monsoon transition. Zoological Studies, **49**: 735-748.

CHEN, B., 1982. The species composition and distribution of the planktonic copepods in the Xisha and Zhongsha Island China. Acta Scientiarum Naturalium Universitatis Amoiensis, **21**: 209-217. [In Chinese.]

CHEN, C. A. & K. SHASHANK, 2009. Taiwan as a connective stepping-stone in the Kuroshio Triangle and the conservation of coral ecosystems under the impacts of climate change. Kuroshio Science, **3**: 15-22.

CHEN, H., Y.-C. LIAO, C.-Y. CHEN, J.-I. TSAI, L.-S. CHEN & K.-T. SHAO, 2015. Long-term monitoring dataset of fish assemblages impinged at nuclear power plants in northern Taiwan. Scientific Data, **2**: 150071.

CHEN, H., H. YU & G. LIU, 2016. Comparison of copepod collection efficiencies by three commonly used plankton nets: a case study in Bohai Sea, China. Journal of Ocean University of China, **15**: 1007-1013.

CHEN, Q.-C. & S.-Z. ZHANG, 1965. The planktonic copepods of the Yellow Sea and the East China Sea. I. Calanoida. Studia Marina Sinica, **7**: 20-131. [In Chinese, with English summary.]

CHEN, Q.-C., S.-Z. ZHANG & C.-S. ZHU, 1974. On planktonic copepods of the Yellow Sea and the East China Sea. II. Cyclopoida and Harpacticoida. Studia Marina Sinica, **9**: 27-76.

CHIHARA, M. & M. MURANO, 1997. An illustrated guide to marine plankton in Japan. (Tokyo University Press, Tokyo).

CHIU, T.-S. & J.-B. HUANG, 1994. Kuroshio and its related fisheries of Taiwan: review and prospective inferred from ichthyoplankton. Journal of the Fisheries Society of Taiwan, **21**: 157-168.

CHOU, C., L.-C. TSENG, C.-H. OU, Q.-C. CHEN & J.-S. HWANG, 2012. Seasonal succession of planktonic copepods in bight environments of northeastern Taiwan. Zoological Studies, **51**: 1380-1396.

DAI, Y.-Y., C.-Q. HOU, S. YANG, B.-L. ZHANG, Y. YU, S. WANG, W. CHEN, E.-J. FANG, Y. GAO, B. GUO, W.-Y. YANG & J.-W. ZHANG, 2018. Community structure of zooplankton in artificial reef area of Tianjin seashore and its relationship with environment factors. Transactions of Oceanology and Limnology, **164**: 165-172. [In Chinese.]

DUR, G., J.-S. HWANG, S. SOUISSI, L.-C. TSENG, C.-H. WU, S.-H. HSIAO & Q.-C. CHEN, 2007. An overview of the influence of hydrodynamics on the spatial and temporal patterns of calanoid copepod communities around Taiwan. Journal of Plankton Research, **29**(Supplement 1): i97-i116.

FANG, T.-H., S.-H. HSIAO & F.-H. NAN, 2014. Nineteen trace elements in marine copepods collected from the coastal waters off northeastern Taiwan. Continental Shelf Research, **91**: 70-81.

FANG, T.-H., J.-S. HWANG, S.-H. HSIAO & H.-Y. CHEN, 2006. Trace metals in seawater and copepods in the ocean outfall area off the northern Taiwan coast. Marine Environmental Research, **61**: 224-243.

HSIAO, S.-H. & T.-H. FANG, 2013. Hg bioaccumulation in marine copepods around hydrothermal vents and the adjacent marine environment in northeastern Taiwan. Marine Pollution Bulletin, **74**: 175-182.

HSIAO, S.-H., T.-H. FANG & J.-S. HWANG, 2011a. Copepod species and their trace metal contents in coastal northern Taiwan. Journal of Marine Systems, **88**: 232-238.

HSIAO, S.-H., T.-H. FANG, C.-T. SHIH & J.-S. HWANG, 2011b. Effects of Kuroshio Current on copepod assemblages in Taiwan. Zoological Studies, **50**: 475-490.

HSIAO, S.-H., J.-S. HWANG & T.-H. FANG, 2010. The heterogeneity of the contents of trace metals in the dominant copepod species in the seawater around northern Taiwan. Crustaceana, **83**: 179-194.

HSIAO, S.-H., S. KÂ, T.-H. FANG & J.-S. HWANG, 2011c. Zooplankton assemblages as indicators of seasonal changes in water masses in the boundary waters between the East China Sea and the Taiwan Strait. Hydrobiologia, **666**: 317-330.

HSIAO, S.-H., C.-Y. LEE, C.-T. SHIH & J.-S. HWANG, 2004. Calanoid copepods of the Kuroshio Current east of Taiwan, with a note on the presence of *Calanus jashnovi* Hulsemann, 1994. Zoological Studies, **43**: 323-331.

HSU, P.-K., W.-T. LO & C.-T. SHIH, 2008. The coupling of copepod assemblages and hydrography in a eutrophic lagoon in Taiwan: seasonal and spatial variations. Zoological Studies, **47**: 172-184.

HUANG, C.-C., M.-T. CHEN, M.-Y. LEE, K.-Y. WEI & C.-Y. HUANG, 2002. Planktic foraminifer faunal sea surface temperature records of the past two glacial terminations in the South China Sea near Wan-An shallow (IMAGES core MD972151). Western Pacific Earth Sciences, **2**: 1-14.

HWANG, J.-S., H.-U. DAHMS, L.-C. TSENG & Q.-C. CHEN, 2007a. Intrusions of the Kuroshio Current in the northern South China Sea affect copepod assemblages of the Luzon Strait. Journal of Experimental Marine Biology, **352**: 12-27.

HWANG, J.-S., R. KUMAR, H.-U. DAHMS, L.-C. TSENG & Q.-C. CHEN, 2007b. Mesh size affects abundance estimates of *Oithona* spp. (Copepoda, Cyclopoida). Crustaceana, **80**: 827-837.

HWANG, J.-S., R. KUMAR, H.-U. DAHMS, L.-C. TSENG & Q.-C. CHEN, 2010. Interannual, seasonal, and diurnal variation in vertical and horizontal distribution patterns of 6 *Oithona* spp. (Copepoda: Cyclopoida) in the South China Sea. Zoological Studies, **49**: 220-229.

HWANG, J.-S., S. SOUISSI, H.-U. DAHMS, L.-C. TSENG, F. G. SCHMITT & Q.-C. CHEN, 2009. Rank-abundance allocations as a tool to analyze planktonic copepod assemblages off the Danshuei estuary (Taiwan). Zoological Studies, **48**: 49-62.

HWANG, J.-S., S. SOUISSI, L.-C. TSENG, L. SEURONG, F. G. SCHMITT, L.-S. FANG, S.-H. PENG, C.-W. WU, S.-H. HSIAO, W.-H. TWAN, T.-P. WEI, R. KUMAR, T.-H. FANG, Q.-C. CHEN & C.-K. WONG, 2006. A 5-year study of the influence of the northeast and southwest monsoons on copepod assemblages in the boundary coastal waters between the East China Sea and the Taiwan Strait. Journal of Plankton Research, **28**: 943-958.

HWANG, J.-S., Y.-Y. TU, L.-C. TSENG, L.-S. FANG, S. SOUISSI, T.-H. FANG, W.-T. LO, W.-H. TWAN, S.-H. HSAIO, C.-H. WU, S.-H. PENG, T.-P. WEI & Q.-C. CHEN, 2004. Taxonomic composition and seasonal distribution of copepod assemblages from waters adjacent to nuclear power plant I and II in northern Taiwan. Journal of Marine Science and Technology-Taiwan, **12**: 380-391.

HWANG, J.-S. & C.-K. WONG, 2005. The China Coastal Current as a driving force for transporting *Calanus sinicus* (Copepoda: Calanoida) from its population centers to waters of Taiwan and Hong Kong during the NE monsoon period in winter. Journal of Plankton Research, **27**: 205-210.

JAN, S., J. WANG, C. S. CHERN & S. Y. CHAO, 2002. Seasonal variation of the circulation in the Taiwan Strait. Journal of Marine Systems, **35**: 249-268.

JIANG, Z.-B., J.-N. ZENG, Q.-Z. CHEN, Y.-J. HUANG, Y.-B. LIAO, X.-Q. XU & P. ZHENG, 2009. Potential impact of rising seawater temperature on copepods due to coastal power plants in subtropical areas. Journal of Experimental Marine Biology and Ecology, **368**: 196-201.

KIØRBOE, T., 2011. What makes pelagic copepods so successful? Journal of Plankton Research, **33**: 677-685.

KOTORI, M. & S. HAMAOKA, 1979. Plankton investigations in Ishikari Bay, Hokkaido. III. Faunal and taxonomical notes on Copepoda. The Scientific Reports of Hokkaido Fisheries Experimental Station, Japan, **21**: 1-16.

LAN, Y.-C., M.-A. LEE, W.-Y. CHEN, F.-J. HSIEH, J.-Y. PAN, D.-C. LIU & W.-C. SU, 2008. Seasonal relationships between the copepod community and hydrographic conditions in the southern East China Sea. ICES Journal of Marine Science, **65**: 462-468.

LEE, C.-Y., D.-C. LIU & W.-C. SU, 2009. Seasonal and spatial variations in the planktonic copepod community of Ilan Bay and adjacent Kuroshio waters off northeastern Taiwan. Zoological Studies, **48**: 151-161.

LI, C., R. WANG & S. SUN, 2003. Grazing impact of copepods on phytoplankton in the Bohai Sea. Estuarine Coastal and Shelf Science, **58**: 487-498.

LI, S. & J.-C. FANG, 1990. Larval stage of marine planktonic copepods of China. (China Ocean Press, Beijing). [In Chinese.]

LI, W., C. LUO, D. WANG & T. LEI, 2010. Diurnal variations of precipitation over the South China Sea. Meteorology and Atmospheric Physics, **109**: 33-46.

LIAN, G.-S. & J.-M. LIN, 1978. Studies on the calanoid copepods of the south Yellow Sea and the East China Sea. Marine Science and Technology, **11**: 59-112. [In Chinese, with English summary.]

LIAN, G.-S., Y.-G. WANG, R.-X. SUN & J.-S. HWANG, 2018. Species diversity of marine planktonic copepods in China's seas. (China Ocean Press, Beijing).

LIAN, X., Y. TAN, L. HUANG & L. ZHOU, 2017. Striking taxonomic differences in summer zooplankton in the northern South China Sea: implication of an extreme cold anomaly. Acta Oceanologica Sinica, **36**: 87-96.

LIAO, C.-H., W.-J. CHANG, M.-A. LEE & K.-T. LEE, 2006. Summer distribution and diversity of copepods in the upwelling waters of the southeastern East China Sea. Zoological Studies, **45**: 378-394.

LIN, S.-M., L.-C. TSENG, JR., P.-O. ANG, J. BOLTON & L.-C. LIU, 2018. Long term study on seasonal changes in floristic composition and structure of marine macroalgal communities along the coast of northern Taiwan, southern East China Sea. Marine Biology, **165**: 83.

LIN, Y.-H. & G.-S. LIAN, 1988. Ecology of planktonic copepods in the Taiwan Strait. Journal of Oceanography in Taiwan Strait, **7**: 248-255. [In Chinese.]

LIU, R., 2008. Checklist of marine biota of China seas. (Science Press, Academia Sinica, Beijing). [In Chinese.]

LO, W.-T., C.-L. CHUNG & C.-T. SHIH, 2004a. Seasonal distribution of copepods in the Tapong Bay, southwestern Taiwan. Zoological Studies, **43**: 464-474.

LO, W.-T., H.-Y. HSIEH, L.-J. WU, H.-B. JIAN, D.-C. LIU & W.-C. SU, 2010. Comparison of larval fish assemblages between, during and after northeasterly monsoon in the waters around Taiwan, western North Pacific. Journal of Plankton Research, **32**: 1079-1095.

LO, W.-T., J.-S. HWANG & Q.-C. CHEN, 2004b. Spatial distributions of copepod assemblages relative to hydrological conditions in the southeastern Taiwan Strait. Zoological Studies, **43**: 218-228.

LO, W.-T., S.-F. YU & H.-Y. HSIEH, 2014c. Hydrographic processes driven by seasonal monsoon system affect siphonophore assemblages in tropical-subtropical waters (western north Pacific Ocean). Plos ONE, **9**: e100085.

LÓPEZ-LÓPEZ, L., J. C. MOLINERO, L.-C. TSENG, Q.-C. CHEN & J.-S. HWANG, 2013. Seasonal variability of the gelatinous carnivore zooplankton community in northern Taiwan. Journal of Plankton Research, **35**: 677-683.

MAKABE, R., A. TANIMURA & M. FUKUCHI, 2012. Comparison of mesh size effects on mesozooplankton collection efficiency in the Southern Ocean. Journal of Plankton Research, **34**: 432-436.

MCGLYNN, T., 2010. Effects of biogeography on community diversity. Nature Education Knowledge, **3**: 42.

MILOSLAVIĆ, M., D. LUČIĆ, B. GANGAI & I. ONOFRI, 2014. Mesh size effects on mesozooplankton community structure in a semi-enclosed coastal area and surrounding sea (south Adriatic Sea). Marine Ecology, **35**: 445-455.

MOLINERO, J. C., L.-C. TSENG, C. LÓPEZ-ABBATE, E. RAMIREZ-ROMERO & J.-S. HWANG, 2018. Interannual changes in zooplankton echo subtropical and high latitude climate effects in the southern East China Sea. Plos ONE, **13**: e0197382.

MOLINERO, J. C., L.-C. TSENG, L. LÓPEZ-LÓPEZ, U. SOMMER, S. SOUISSI & J.-S. HWANG, 2016. Climate driven winter variations of *Calanus sinicus* abundance in the East China Sea. Fisheries Oceanography, **25**: 555-564.

MORI, T., 1937. Pelagic Copepoda from the neighbouring waters of Japan: 1-150. (Soyo, Tokyo).

MORTON, B. & G. BLACKMORE, 2001. South China Sea. Marine Pollution Bulletin, **42**: 1236-1263.

RAZOULS, C., F. DE BOVÉE, J. KOUWENBERG & N. DESREUMAUX, 2005-2019. Diversity and geographic distribution of marine planktonic copepods (sensu lato). Sorbonne University, CNRS. http://copepodes.obs-banyuls.fr/en (accessed 6 May 2019).

REZAI, H., F. M. YUSOFF, A. ARSHAD, A. KAWAMURA, S. NISHIDA & B. H. R. OTHMAN, 2004. Spatial and temporal distribution of copepods in the Straits of Malacca. Zoological Studies, **43**: 486-489.

ROSA, A. L., J. YAMAMOTO & Y. SAKURAI, 2011. Effects of environmental variability on the spawning areas, catch, and recruitment of the Japanese common squid, *Todarodes pacificus* (Cephalopoda: Ommastrephidae), from the 1970s to the 2000s. ICES Journal of Marine Science, **68**: 1114-1121.

SCHMINKE, H. K., 2007. Entomology for the copepodologist. Journal of Plankton Research, **29**: 149-162.

SHEN, J.-R. & F.-X. LI, 1963. The estuarine Copepoda of Chiekong and Zaikong rivers, Kwangtung Province, China. Acta Zoologica Sinica, **15**: 571-596. [In Chinese, with English summary.]

SHERMAN, K., P. CELONE & S. ADAMS, 2004. NOAA Fisheries Service's Large Marine Ecosystems Program: status report. (NOAA Technical Memorandum NMFS-NE-183. NOAA, NEFSC).

SHIH, C.-T. & T.-S. CHIU, 1998. Copepod diversity in the water masses of the southern East China Sea north of Taiwan. Journal of Marine Systems, **15**: 533-542.

SHIH, C.-T. & S.-S. YOUNG, 1995. A checklist of free-living copepods, including those associated with invertebrates, reported from the adjacent seas of Taiwan. Acta Zoologica Taiwanica, **6**: 65-81.

TSENG, L.-C., H.-U. DAHMS, Q.-C. CHEN & J.-S. HWANG, 2008a. Copepod assemblages of the northern South China Sea. Crustaceana, **81**: 1-22.

TSENG, L.-C., H.-U. DAHMS, Q.-C. CHEN & J.-S. HWANG, 2009. Copepod feeding study in the upper layer of the tropical South China Sea. Helgoland Marine Research, **63**: 327-337.

TSENG, L.-C., H.-U. DAHMS, Q.-C. CHEN & J.-S. HWANG, 2012. Mesozooplankton and copepod community structures in the southern East China Sea: the status during the monsoonal transition period in September. Helgoland Marine Research, **66**: 621-634.

TSENG, L.-C., H.-U. DAHMS, Q.-C. CHEN & J.-S. HWANG, 2013a. Geospatial variability of autumn community structure on epipelagic zooplankton in the upper layer of the northern South China Sea. Zoological Studies, **52**: 2.

TSENG, L.-C., H.-U. DAHMS, J.-J. HUNG, Q.-C. CHEN & J.-S. HWANG, 2011a. Can different mesh sizes affect the results of copepod community studies? Journal of Experimental Marine Biology and Ecology, **398**: 47-55.

TSENG, L.-C., S.-H. HSIAO, S. K. SARKAR, B. D. BHATTACHARYA, Q.-C. CHEN & J.-S. HWANG, 2016. Influence of Kuroshio water on the annual copepod community structure in an estuary in the northwest Pacific Ocean. Continental Shelf Research, **118**: 165-176.

TSENG, L.-C., J.-J. HUNG, Q.-C. CHEN & J.-S. HWANG, 2013b. Seasonality of the copepod assemblages associated with interplay waters off northeastern Taiwan. Helgoland Marine Research, **67**: 507-520.

TSENG, L.-C., J.-J. HUNG, J. C. MOLINERO, Q.-C. CHEN & J.-S. HWANG, 2015. Indicator species and seasonal succession of planktonic copepod assemblages driven by the interplay of subtropical and temperate waters in the southern East China Sea. Crustaceana, **88**: 96-112.

TSENG, L.-C., R. KUMAR, Q.-C. CHEN & J.-S. HWANG, 2011b. Faunal shift between two copepod congeners (*Temora discaudata* and *T. turbinata*) in the vicinity of two nuclear power plants in southern East China Sea: spatiotemporal patterns of population trajectories over a decade. Hydrobiologia, **666**: 301-315.

TSENG, L.-C., R. KUMAR, H.-U. DAHMS, C. T. CHEN, S. SOUISSI, Q.-C. CHEN & J.-S. HWANG, 2008b. Copepod community structure over a marine outfall area in the north-eastern South China Sea. Journal of the Marine Biological Association of the United Kingdom, **88**: 955-966.

TSENG, L.-C., S. SOUISSI, H.-U. DAHMS, Q.-C. CHEN & J.-S. HWANG, 2008c. Copepod communities related to water masses in the southwest East China Sea. Helgoland Marine Research, **62**: 153-165.

TSENG, R. S. & Y. T. SHEN, 2003. Lagrangian observations of surface flow patterns in the vicinity of Taiwan. Deep-Sea Research II, **50**: 1107-1116.

VERITY, P. G. & V. SMETACEK, 1996. Organism life cycles, predation, and the structure of marine pelagic ecosystems. Marine Ecology Progress Series, **130**: 277-293.

WALTER, T. C. & G. BOXSHALL, 2019. World of Copepods database. *Eurytemora pacifica* Sato, 1913. Available online at: http://www.marinespecies.org/aphia.php?p=taxdetails&id=232028 (accessed 9 May 2019).

WANG, Y.-G., X.-Y. CHEN, B.-P. XING, R.-X. SUN, N. FITRIA, P. XIANG, C.-G. WANG & M. LIN, 2018. Zooplankton composition and distribution in the Lembeh Strait of North Sulawesi, Indonesia. Acta Oceanologica Sinica, **37**: 35-44.

WANG, Y.-G., J.-H. LIN, C.-G. WANG & M. LIN, 2012. Species composition and distribution characteristics of pelagic copepods in the northern sea of Fujian during withdraw of Zhe-Min coastal current. Huanjing Kexue, **33**: 1839-1845. [In Chinese.]

WU, C.-J., C.-M. SHIN & K.-P. CHIANG, 2011. Does the mesh size of the plankton net affect the result of statistical analyses of the relationship between the copepod community and water masses? Crustaceana, **84**: 1069-1083.

XIAO, Y.-C., S.-W. GAO & H.-Q. ZHANG, 1992. Zooplankton. In: R. Y. LIU (ed.), Living resources of Jiaozhou Bay: 170-203. (Science Press, Beijing). [In Chinese.]

YAMAZAKI, A., T. WATANABE, U. TSUNOGAI, F. IWASE & H. YAMANO, 2016. A 150-year variation of the Kuroshio transport inferred from coral nitrogen isotope signature. Paleoceanography and Paleoclimatology, **1**: 838-846.

YATSU, A., S. CHIBA, Y. YAMANAKA, S.-I. ITO, Y. SHIMIZU, M. KAERIYAMA & Y. WATANABE, 2013. Climate forcing and the Kuroshio/Oyashio ecosystem. ICES Journal of Marine Science, **70**: 922-933.

YIN, J., L. HUANG, K. LI, S. LIAN, C. LI & Q. LIN, 2011. Abundance distribution and seasonal variations of *Calanus sinicus* (Copepoda: Calanoida) in the northwest continental shelf of South China Sea. Continental Shelf Research, **31**: 1447-1456.

YU, R.-Y., 2005. Seasonal distribution of copepods in relation to environmental factors in Tanshui and Kaoping estuaries. (M.Sc. Thesis, National Sun Yat-sen University, Kaohsiung, Taiwan).

ZHENG, Z., S. LI & S. LI, 1978. Species composition and geographical distribution of marine planktonic copepods in China. Journal of Xiamen University, (Natural Science) **2**: 51-63.

ZHENG, Z., S.-J. LI, S. LI & B.-Y. CHEN, 1982. On the distribution of planktonic copepods in Taiwan Strait. Journal of Oceanography in Taiwan Strait, **1**: 69-79. [In Chinese.]

ZHU, C.-S., J.-Q. HUANG & S.-J. LI, 1991. Studies on the ecology of copepods in Minnan-Taiwan Bank Fishing Ground. In: Miunan-Taiwan bank fishing ground upwelling ecosystem study: 440-455. (Science Press, Beijing). [In Chinese.]

First received 12 June 2019.
Final version accepted 13 December 2019.

[When citing this volume, please refer to *Crustaceana* 93 (2020), 241-547]

*Proceedings of the* Eurytemora *conference, St. Petersburg, 2019*

# *PSEUDODIAPTOMUS JAPONICUS* KIKUCHI, 1928 (COPEPODA, PSEUDODIAPTOMIDAE), A BRACKISH-WATER COPEPOD FORMERLY KNOWN AS *P. INOPINUS* BURCKHARDT, 1913, ON SAKHALIN ISLAND (RUSSIAN FAR EAST)

BY

DENIS S. ZAVARZIN[1])
Sakhalin branch of the Federal State Budget Scientific Institution «Russian Federal Research Institute of Fisheries and Oceanography» («SakhNIRO»), Yuzhno-Sakhalinsk, Russia

## ABSTRACT

We determined the taxonomic status of *Pseudodiaptomus inopinus* on Sakhalin using specimens from different localities, because it is currently known as a complex of species. The specimens examined appeared to belong to *P. japonicus*, based on the structure of the female genital operculum. However, a few females lacked spinules on the medial surface of the first exopodal segment of the female P5, which has been described as a character distinguishing *P. japonicus* from its two sibling species. The species is distributed in brackish waters from the very south to the very north of Sakhalin Island. The northernmost locality on Sakhalin is also the northernmost record of the species.

## RÉSUMÉ

Nous avons déterminé le statut taxonomique de *Pseudodiaptomus inopinus* de Sakhaline à partir de spécimens de localités différentes, l'espèce étant actuellement connue comme un complexe d'espèces. Les spécimens examinés semblent appartenir à *P. japonicus*, sur la base de la structure de l'opercule génital femelle. Cependant, quelques femelles ne présentaient pas de sétules sur la surface médiane du premier segment de l'exopodite de la P5 femelle, qui a été décrit comme un caractère distinguant *P. japonicus* de ses deux espèces apparentées. L'espèce est présente dans les eaux saumâtres de l'extrême sud à l'extrême nord de l'île de Sakhaline. La localité la plus septentrionale de Sakhaline est aussi la mention la plus septentrionale de l'espèce.

## INTRODUCTION

*Pseudodiaptomus inopinus* Burckhardt, 1913, has been recorded from the waters of Sakhalin. Some authors (e.g., Borutskii et al., 1991) described the species

[1]) e-mail: zdsmail@inbox.ru

 | DOI:10.1163/9789004465565_019

on Sakhalin as *Schmackeria inopina* (Burckhardt, 1913). However, "*P. inopinus*" is known as a complex of species (Sakaguchi & Ueda, 2010, 2011; Soh et al., 2012). *P. inopinus* from Japan and Korea were recently re-examined and reallocated into some different species (Sakaguchi & Ueda, 2018; Ueda & Sakaguchi, 2019), whereas the taxonomic status of this species in Sakhalin waters is still unclear and therefore the distribution of the species on Sakhalin is still questioned. In this paper, we tried to eliminate this gap. Moreover, new features have been found in the morphological structure of individuals from Sakhalin.

## MATERIAL AND METHODS

The material for this study originated from zooplankton samples collected by the author and colleagues during the hydrobiological and combined expeditions of SakhNIRO in 2000-2013 in eight water bodies of Sakhalin: Lake Tunaicha (46°44′41″N 143°13′54″E), Lake Ptich'e (46°24′5″N 143°33′33″E), Lyutoga River estuary (46°41′52″N 142°31′3″E), Busse Lagoon (46°31′43″N 143°19′55″E), Pil'tun Bay (53°6′21″N 143°10′10″E), Chaivo Bay (52°28′20″N 143°12′19″E), Nabil'skii Bay (51°35′39″N 143°17′19″E), and Baikal Bay (53°28′48″N 142°24′22″E) (fig. 1). Sampling was carried out by vertical hauls from bottom to surface with a Juday plankton net (mouth diameter 18 cm, mesh size 82 $\mu$m). The samples were fixed in 5% buffered formalin solution. Microscopic examination and photographs were performed using an Olympus BX51 and an Olympus SZX10 microscope, equipped with an Olympus DP25 camera and Olympus CellSens Standard 1.5 software. Final photos were made by combining of 10-60 images at different focus depths using "CombineZP" (https://combinezp.software.informer.com). Drawings were made from photographs and/or with a drawing tube. Measurements were performed with an ocular micrometer and Olympus CellSens Standard 1.5 software. For the terminology of the copepod body and appendages, we followed Huys & Boxshall (1991).

## RESULTS AND DISCUSSION

"*Pseudodiaptomus inopinus*" from Sakhalin reservoirs was identifiable to *P. japonicus* Kikuchi, 1928, based on the structure of the genital operculum, i.e., with short posterior processes (distance $a < b$ in fig. 2A) (Ueda & Sakaguchi, 2019). This structure is characteristic for *P. japonicus* and differentiates the species from its two sibling species, *P. inopinus* s. str. and *P. yamato* Ueda & Sakaguchi, 2019, which have the genital operculum with long posterior processes (Burckhardt, 1913; Ueda & Sakaguchi, 2019).

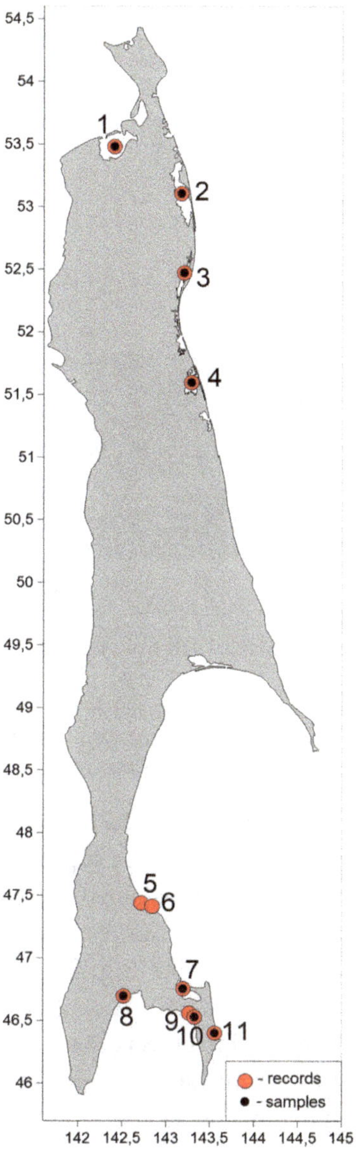


Fig. 1. Records of *Pseudodiaptomus japonicus* Kikuchi, 1928 on Sakhalin, and sampling sites of the present study. 1, Baikal Bay (Nemchinova, 2011, as *Schmackeria inopina*); 2, Pil'tun Bay; 3, Chaivo Bay; 4, Nabil'skii Bay (Kafanov et al., 2003, as *S. inopina*); 5, Lake Lebyazh'e; 6, the sea coastal area in the southeast of the island (Kos, 1985, as *P. inopinus*); 7, Lake Tunaicha (Zavarzin, 2003, 2004, 2005, as *S. inopina*); 8, Lyutoga River estuary (Labay et al., 2015, as *P. inopinus*); 9, Lake Vyselkovoe (Borutskii & Bogoslovskii, 1964, as *P. inopinus*); 10, Busse Lagoon (Kos, 1985, as *P. inopinus*); 11, Lake Ptich'e (Zavarzin & Atamanova, 2014, as *P. inopinus*).

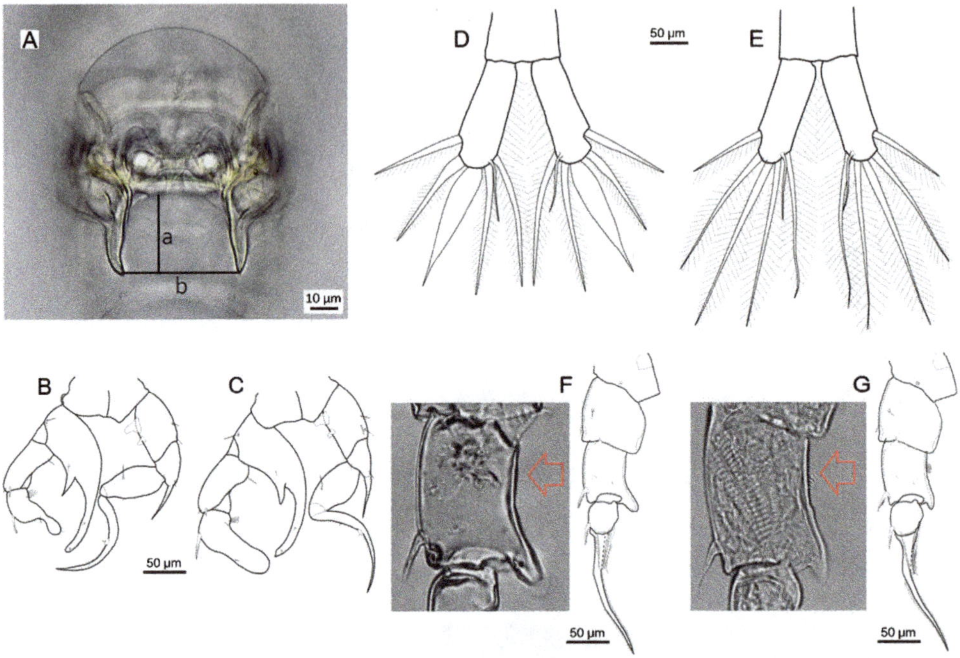


Fig. 2. *Pseudodiaptomus japonicus* Kikuchi, 1928 from Lake Tunaicha: A, female, genital operculum, ventral; B, male, p5 "thumb-shaped" form; C, male, p5 "paddle-shaped" form; D, female, anal somite and caudal rami with thick setae form; E, female, anal somite and caudal rami with thin setae form; F, female, left p5 without setae on inner surface form; G, female, left p5 with setae on inner surface form.

Two morphs of male P5, i.e., thumb-shaped P5 (fig. 2B) and paddle-shaped P5 (fig. 2C) (Burckhardt, 1913), which have been suggested as intraspecific variation (Soh et al., 2012), were found. The thumb-shaped morph was much more common.

Two morphotypes also were found in the female caudal setae. One is typical, that is the outer terminal seta is spike-like (fig. 2D), and the other, atypical one, is devoid of such a spine-like seta (fig. 2E). A similar atypical specimen was also reported from the Yukiura River in Japan (Sakaguchi & Ueda, 2010).

Among 18 female specimens from Lake Tunaicha examined, two females were devoid of medial spinules on the medial surface of both P5 Exp1 (fig. 2F, arrowed). The presence of these spinules (fig. 2G, arrowed) was described as a critical character of *P. japonicus* by Ueda & Sakaguchi (2019), by which the species is distinguishable from *P. yamato*. The number of these spinules in the other 16 specimens from the lake varied between two and seven for each leg, regardless of the morphotype. These facts indicate that presence of the spinules on the medial surface of the female P5 Exp1 is not a critical character of *P. japonicus*.

The body length of *P. japonicus* on Sakhalin varied among reservoirs. To show the differences in size, we compared the length of specimens from northern and

TABLE I

*Pseudodiaptomus japonicus* Kikuchi, 1928; body length in different water bodies

| Water body, date | Sex | Body length (mm) | *n* |
|---|---|---|---|
| Chaivo Bay (northern Sakhalin, Russia), 22 September 2002 | ♀ | 1.48-1.5 | 4 |
| | ♂ | 1.28-1.33 | 10 |
| Lake Tunaicha (southern Sakhalin, Russia), 4 September 2003 | ♀ | 1.18-1.35 | 14 |
| | ♂ | 1.08-1.23 | 5 |
| Lake Suigetsu (Honshu, Japan), 18 September 2010* | ♀ | 1.19-1.26 | 4 |
| | ♂ | 0.96-1.19 | 7 |

*Data from Ueda & Sakaguchi (2019).

southern inland water bodies, and also with published data from Japan (table I). To exclude seasonal fluctuations in size, we used individuals from samples taken in September. The results show a trend of size increment with latitude.

*Pseudodiaptomus japonicus* is a brackish-water species, occurring in estuaries, lagoons, coastal marshes, and tidal embankments. Although the species is recorded not only in brackish-water but in fresh waters in Korea as well (Chang, 2014), it has not been recorded from fresh waters on Sakahalin. In waters of Sakhalin, the species often co-occurs with *Sinocalanus tenellus* (Kikuchi, 1928) and *Eurytemora* cf. *affinis* (Poppe, 1880).

This species is widespread in northern East Asia: along the coast of Japan except for the warm Kuroshio Current coasts of the middle and western mainland (Ueda & Sakaguchi, 2019), Korea (Chang, 2014), Sakhalin and the Japan Sea coast of mainland Russia (Smirnov, 1929; present study). The species has been recorded from 11 seashore reservoirs on Sakhalin (see fig. 1). Among these localities, Baikal Bay is the new northernmost limit of *P. japonicus*; the former known northern limit was the Akkeshi River, Hokkaido, Japan (Ueda & Sakaguchi, 2019).

## ACKNOWLEDGEMENTS

The author is deeply grateful to Dr. Vyacheslav Labay ("SakhNIRO", Russia) for organizing the expeditions, to Dr. Hiroshi Ueda (Usa Marine Biological Institute, Kochi University, Japan) for valuable advice and revision of the text, to anonymous reviewers, as well as to all colleagues who participated in collecting material.

## REFERENCES

BORUTSKII, E. V. & A. S. BOGOSLOVSKII, 1964. Zooplankton ozer yuzhnogo Sakhalina (Zooplankton of southern Sakhalin lakes). In: Ozera yuzhnogo Sakhalina i ikh ichtyofauna (Lakes of southern Sakhalin and their ichthyofauna): 97-140. (Moscow State University, Moscow). [In Russian.]

BORUTSKII, E. V., L. A. STEPANOVA & M. S. KOS, 1991. Opredelitel' Calanoida presnykh vod SSSR (Identification guide to Calanoida from fresh waters of the USSR): 1-504. (Nauka, Leningrad [now St.-Petersburg]). [In Russian.]

BURCKHARDT, G., 1913. Zooplankton aus ost- und süd-asiatischen Binnengewässern. Zoologische Jahrbücher, (Abteilung für Systematik, Ökologie und Geographie der Tiere) **34**: 341-471.

CHANG, C. Y., 2014. Continental Calanoida. Invertebrate Fauna of Korea, **21**(35): 1-60. (National Institute of Biological Resourses, Incheon).

HUYS, R. & G. A. BOXSHALL, 1991. Copepod evolution: 1-468. (The Ray Society, London; nr. **159** of the series).

KAFANOV, A. I., V. S. LABAY & N. A. PECHENEVA, 2003. Biota and macrobenthic communities of the northeast Sakhalin lagoons: 1-176. (Sakhalin Research Institute of Fisheries and Oceanography, Yuzhno-Sakhalinsk). [In Russian.]

KOS, M. S., 1985. K faune Calanoida (Copepoda) pribrezhnykh raionov Yuzhnogo Sakhalina (To fauna Calanoida (Copepoda) of coastal southern Sakhalin regions). Researches Fauna of Seas, **30**(38): 225-258. [In Russian.]

LABAY, V. S., L. A. ZHIVOGLJADOVA, A. V. POLTEVA, I. V. MOTYLKOVA, N. V. KONOVALOVA, D. S. ZAVARZIN, L. N. BARANCHUK-CHERVONNYI, A. V. KORDJUKOV, D. S. DAIROVA, V. D. NIKITIN, A. A. ZHIVOGLJADOV, N. K. ZAVARZINA & S. N. SAFRONOV, 2015. Watercourses of Sakhalin Island: life in the running water: 1-236. (Sakhalin Regional Museum, Yuzhno-Sakhalinsk). [In Russian.]

NEMCHINOVA, I. A., 2011. Structure and quantitative description of zooplankton in Baikal Bay off north western Sakhalin in July 2009. In: Proceedings of the 26$^{th}$ international symposium on Okhotsk Sea and sea ice (Mombetsu, Hokkaido, Japan, 20-25 February 2011): 49-54. (Mombetsu).

SAKAGUCHI, S. O. & H. UEDA, 2010. A new species of *Pseudodiaptomus* (Copepoda: Calanoida) from Japan, with notes on the closely related *P. inopinus* Burckhardt, 1913 from Kyushu Island. Zootaxa, **2623**: 52-68.

SAKAGUCHI, S. O. & H. UEDA, 2011. Morphological divergence of *Pseudodiaptomus inopinus* Burckhardt, 1913 (Copepoda: Calanoida) between the Japan Sea and Pacific coasts of western Japan. Plankton Benthos Research, **6**: 124-128.

SAKAGUCHI, S. O. & H. UEDA, 2018. Genetic analysis on *Pseudodiaptomus inopinus* (Copepoda, Calanoida) species complex in Japan: revival of the species name of *P. japonicus* Kikuchi, 1928. Plankton Benthos Research, **13**(4): 173-179.

SMIRNOV, S. S., 1929. Beiträge zur Copepodenfauna Ostasiens. Zool. Anz., **81**: 317-329.

SOH, H. Y., S. W. KWON, W. LEE & Y. H. YOON, 2012. A new *Pseudodiaptomus* (Copepoda, Calanoida) from Korea supported by molecular data. Zootaxa, **3368**: 229-244.

UEDA, H. & S. O. SAKAGUCHI, 2019. *Pseudodiaptomus yamato* n. sp. (Copepoda, Calanoida) endemic to Japan, with redescriptions of the two closely related species *P. inopinus* Burckhardt and *P. japonicus* Kikuchi. Plankton Benthos Research, **14**: 29-38.

ZAVARZIN, D. S., 2003. Seasonal dynamics of zooplankton from the Tunaycha Lake (southern Sakhalin). Transactions of Sakhalin Scientific Research Institute of Fisheries and Oceanography, **5**: 106-112. [In Russian.]

ZAVARZIN, D. S., 2004. Composition and spatial distribution of zooplankton communities from the Tunaicha Lake (southern Sakhalin) by the 2001 summer survey data. Transactions of Sakhalin Scientific Research Institute of Fisheries and Oceanography, **6**: 331-338. [In Russian.]

ZAVARZIN, D. S., 2005. Some aspects of seasonal dynamic of zooplankton from Tunaycha Lake (southern Sakhalin) at the contemporary stage. Vladimir Ya. Levanidov's Biennial Memorial Meetings, **3**: 95-105. [In Russian.]

ZAVARZIN, D. S. & I. A. ATAMANOVA, 2014. Zooplankton seasonal dynamics in Ptichye Lake and adjoining sea coastal waters of southern Sakhalin. Vladimir Ya. Levanidov's Biennial Memorial Meetings, **6**: 239-249. [In Russian.]

First received 27 August 2019.
Final version accepted 23 October 2019.

# TAXONOMIC INDEX

www.ingramcontent.com/pod-product-compliance
Lightning Source LLC
Chambersburg PA
CBHW071229230926
51849CB00001B/60